DAYLIGHT

and its spectrum

DAYLIGHT

and its spectrum

SECOND EDITION

S. T. HENDERSON
M.A., Ph.D. (Cantab.), B.Sc. (Lond.)

Adam Hilger Ltd
Bristol

1st edition, 1970
2nd edition, 1977

British Library Cataloguing in Publication Data

Henderson, Stanley Thomas
Daylight and its spectrum: 2nd ed.
1. Daylight
I. Title
535   QC455

ISBN 0–85274–343–2

Published by Adam Hilger, Ltd,
Techno House, Redcliffe Way, Bristol BS1 6NX
A company now owned by The Institute of Physics

Printed by W & J Mackay Limited, Chatham, Kent ME5 8TD

PREFACE

The material in this book began to accumulate when I was asked to be chairman of a British Standards Institution technical committee in 1960. The requirement was to revise an old and little-used specification for artificial daylight. The committee soon found that the plentiful earlier work on the colour or spectrum of daylight included very little conducted in this country, and that unsuitable for the purpose in hand. We thought it desirable to rectify this omission. This led, among other projects, to my own work on the spectral distribution, carried out in the laboratories of Thorn Electrical Industries Ltd with the help of an enthusiastic colleague, Mr Donald Hodgkiss. The developments which followed are described in this book. In this study of daylight there was a large element of curiosity about a complex and variable natural pheno-menon which the observer could not influence, and the collection of information opened many vistas into fields far removed from those encountered in my earlier work on phosphors and light sources.

A proposal from Adam Hilger Ltd in 1968 to write a book on daylight and its spectrum provided a new opportunity to resume the search into a widely scattered literature, and to make an attempt to present the whole as a coherent picture. I believe I have assembled most of the significant work of the last ninety years. It is related in approximately chronological order in Chapters 7 to 13, preceded by chapters on essential background topics.

These introductory chapters are less complete in detail than the descriptions of spectral work. Some of them, for example that on the solar constant, might well be expanded with interesting material not essential to the main argument. However, curtailment was necessary in order to include in the book some account of artificial daylight. The details of apparatus and equipment have not been treated very deeply, partly because a high degree of sophistication in this matter has not proved essential in practice, and indeed sometimes adds to the observer's difficulties by yielding more information than he can conveniently use. Scattering and polarization of light in the atmosphere, important as they are, have also been treated briefly. The interested reader will find many useful books on these and other aspects of daylight, for example those mentioned in references 7, 17, 31, 41, 47 and 449. A wider coverage is given in *Solar Radiation*, edited by the late Professor N. Robinson (Elsevier, 1966), and this also is to be recommended.

Daylight might be considered as restricted to the visible part of the spectrum. The great practical importance of the ultra-violet region justifies its inclusion, and much space is devoted to it in the present book, while the infra-red region is less significant and has not been so fully discussed.

The literature references were intended to extend to the end of 1968, and a few only for 1969 have been included. To assist in placing well-known names in perspective, Appendix 2 has been added giving the dates of the more interesting authors and contributors to the subject (where available, and excluding those still living).

It is a pleasure to thank many friends and colleagues for providing reprints, photo-copies, diagrams, calculations, quotations, library facilities or other assistance. They include Dr I. Ando, Mrs A. F. Andrews, Dr C. J. Bartleson, Dr P. Bener, Mr R. W. Brocklebank, Mrs J. F. Bull, Mrs J. M. Cowlard, Dr R. Dogniaux, M. Frank Elgar, Miss M. B. Halstead, Dr B. Hisdal, Mr D. Hodgkiss, Dr D. Labs, Dr D. L. MacAdam, Dr E. W. Maddison, Dr I. A. Magnus, Mr K. McLaren, Dr T. H. Morton, Dr Y. Nayatani, Dr I. Nimeroff, Dr J. N. Ott, Col. M. C. Perceval-Price, Mr P. Petherbridge, Dr R. Peyturaux, Mr I. T. Pitt, Professor M. H. Rees, Dr V. D. P. Sastri, Dr A. W. S. Tarrant, Dr M. P. Thekaekara, Dr D. L. Thomas, Mr G. T. Winch. Special thanks are due to Mr Neville Goodman and Mr David Tomlinson of Adam Hilger Ltd for their continued help in the preparation of the book. Their encouragement has made the task an agreeable one.

Acknowledgements are made to the following for permission to reproduce copyright material:

The Trustees of the British Museum (Plates 2, 3, 6, 8).

Taylor and Francis Ltd (Fig. 4, ref. 204).

The Institute of Physics and the Physical Society (Figs. 34–36, ref. 330, 1963 paper; Figs. 38, 39, ref. 330, 1964 paper; Fig. 64, ref. 386).

The Illuminating Engineering Society of London (Figs. 84, 86, 87, ref. 629).

The University of Surrey and Dr A. W. S. Tarrant (Figs. 49, 50, ref. 353).

Pergamon Press Ltd (Fig. 78, ref. 434).

Editions du C.N.R.S., France (Fig. 57, ref. 372).

The Administrator of the *Revue d'Optique Theorique et Instrumentale* (Fig. 32, ref. 319).

Springer-Verlag, Berlin, Heidelberg, New York (Fig. 77, ref. 173; Fig. 60, ref. 373).

Akademische Verlagsgesellschaft Leipzig (Fig. 10, ref. 248; Fig. 11, ref. 249).

Ronan Picture Library and The Royal Astronomical Society (Plates 1, 4, 5).

Dr P. Bener (Fig. 25, ref. 306; Fig. 26, ref. 305; Figs. 27, 28, ref. 307; Fig. 29, ref. 309).

Helvetica Physica Acta (Fig. 13, ref. 263).

Svenska Geofysiska Föreningen (Figs. 22, 23, ref. 299).

Dr M. P. Thekaekara (Figs. 55, 56, ref. 88).

Naval Research Laboratory, Washington and Dr J. A. Curcio (Figs. 53, 54, ref. 357).

National Bureau of Standards, Washington (Fig. 14, ref. 275; Figs. 15, 16, ref. 279).

The Illuminating Engineering Society of New York (Fig. 5, ref. 212; Fig. 18, ref. 598).

Journal of the Optical Society of America (Fig. 12, ref. 256; Figs. 17–19, ref. 28; Fig. 21, ref. 291; Fig. 33, ref. 321; Fig. 40, ref. 333; Fig. 41, ref. 336; Fig. 43, ref. 343; Fig. 63, ref. 334; Figs. 68, 73, ref. 335).

Applied Optics (Fig. 61, ref. 383).

Science (Figs. 65, 66, ref. 391, copyright 1966 by the American Association for the Advancement of Science).

University of Chicago Press (Fig. 8, ref. 233; Fig. 9, ref. 234; Fig. 76, ref. 412).

The Franklin Institute (Plate 7 and Fig. 80, ref. 591).

Kollmorgen Corporation ('Super skylight' curve, Fig. 83).

Some of the diagrams used have had extra curves added, and some scales and units have been altered to improve consistency.

The following have made textual quotations available.

Smithsonian Institution Press (ref. 224).

University of Toronto Press (ref. 497).

Harvard University Press and Heinemann (London), publishers of the Loeb
   Classical Library (refs. 1, 580).

Jonathan Cape; Harcourt, Brace and World, Inc. (N.Y.); and the estate of Edward
   McCurdy (refs. 8, 495 and the quotation beginning Chapter 2).

The Clarendon Press, Oxford (ref. 2).

The Royal Society (refs. 14, 38 and the quotation beginning Chapter 16).

Editions Somogy, Paris; Thames and Hudson, London (quotation beginning
   Chapter 12).

Taylor and Francis Ltd (ref. 204).

The figure and reference numbers above have been altered to conform with renumbering in the second edition.

PREFACE TO THE SECOND EDITION

This edition includes much new material, partly from early work and partly from the many publications of the last seven years. Chapters 5, 10 to 15, and 18 (formerly 17) have been altered most; a new chapter (16) on whiteness has been added since this subject is of much interest at present and is related to daylight. Many new literature references and several tables and figures are included, and some necessary corrections and revisions have been made.

I am grateful to Dr V. Hisdal and Dr V. D. P. Sastri for allowing me to use unpublished information on their work; to Dr P. Bener, Miss M. B. Halstead, Miss P. H. M. Schocktee and Dr M. P. Thekaekara for valuable help in the form of information, discussions, reprints and computations used in the book. Others who have provided useful material are Dr A. Berger, Dr P. D. Carman, Dr F. J. J. Clarke, Dr E. Ganz, Dr F. Grum, Professor I. A. Magnus, Dr D. F. Robertson and Dr R. Sève. Acknowledgments are also due for permission to use copyright material in some of the new figures:

Professor M. Bossolasco and Birkhäuser Verlag, Basel (Fig. 2, ref. 183 (*b*)).

Dr P. Bener and Pergamon Press, Oxford (Fig. 27, ref. 414).

Dr P. Bener (Fig. 30, ref. 311): research sponsored in part by AFCRL through the
   U.S. Air Force, contract F 61052-67-C-0029.

Dr P. Bener (Fig. 31, ref. 84): research sponsored by the U.S. Army, contract
   DAJA 37-68-C 1017.

Dr V. D. P. Sastri and The Institute of Physics (Figs. 44, 45, ref. 344).

Dr Y. Kurioka, Electrotechnical Laboratory, Osaka (Fig. 74 (part), ref. 348).

British Standards Institution for extracts from BS 1376:1974 (complete copies from BSI, 2 Park St., London). Commission Internationale de l'Eclairage, Paris, for extracts from CIE Publication no. 2.2 (1975). Mr W. N. Sproson and the British Broadcasting Corporation (Fig. 79).

This edition has been prepared with the valuable assistance of Mr K. J. Hall of Adam Hilger Ltd.

CONTENTS

1

The Rise of Spectroscopy

We all *know* what light is; but it is not easy to *tell* what it is.
SAMUEL JOHNSON in 1776, according to BOSWELL.

It is obvious, however, that whatever side we take concerning the nature of light, many, indeed almost all the circumstances concerning it, are incomprehensible, and beyond the reach of human understanding.
Encyclopaedia Britannica, article on 'Light', 1792.

The purpose of this book is to follow the progress of the study of daylight, selecting representative contributions from the great amount of published work. Necessarily much has to be omitted, and some of the related topics are treated more briefly than they deserve. The practical applications of the accumulated knowledge are discussed towards the end of the book.

Origins of the subject can be traced far back in time, though the early ideas seemed to have little to do with daylight and its spectrum. They were in fact concerned with the rainbow and how it was produced, and the history of this question, extending over many centuries before the relation of the rainbow to sunlight was established, contains a host of speculations on light, optics and colour, on astronomy and meteorology. Some of these ideas are still acceptable, some were irrational or even perverse. They are worth a brief survey, for they suggest the great difficulties our predecessors had in understanding what now seems simple. On the other hand, these ideas sometimes show that our modern solutions owe much to the past, and not all the old ways of thought on physical subjects should be dismissed as unscientific or obsolete or uninteresting.

The sun is of course central to our subject. In this chapter we concentrate on light and the spectrum, for this was essential knowledge before any progress could be made in analysing the sun's contribution to the earth, and the vital optical role of the atmosphere in our existence.

ANCIENT AND MEDIAEVAL IDEAS

In the times of the classical Greek civilization there was a good deal of speculation on light and colour by those who wrote or left a tradition of their teaching about the appearances of the surrounding world. There was little reliance on deliberate experiment instead of passive observation, and conflicting views had to be resolved, if at all, by persuasive argument. Many of the famous men among the ancients had opinions which have some bearing on our present subject. For example, concerning the behaviour of light, Plato thought that reflection occurred when rays from the object and from the eye met at a mirror. Euclid imagined light to be a real but invisible thread proceeding from the eye to the object seen. Aristotle and many others held the view that light travelled, probably with infinite speed, from the sun to an object and thence to the eye. As for the sun itself, its divine status was long defended, for instance against Anaxagoras who described it as an incandescent stone and was persecuted for this irreligious view. Socrates supported him and paid the penalty for this and many other heresies. The sun illuminated the moon, according to Thales of Miletus. Some centuries later Aristarchus called the moon a mirror: he was apparently the earliest astronomer to place the sun at the centre of our planetary system (and to escape punishment for this impious opinion). On the other hand, some philosophers like Anaximenes considered the moon to be self-luminous. This confusion of irreconcilable ideas in astronomy persisted for at least two thousand years. Faster progress in geometrical optics was due to the Arab scientists, who became the leaders in this field when Greek and Roman intellectual efforts on natural phenomena had declined. In the more abstruse subject of colour we find, not surprisingly, that a satisfactory explanation leading to further progress took as long to find as the fundamentals of, say, the mechanics of the solar system.

In early Greek times the idea arose, and persisted for a very long time, that colours were the result of contamination of white light by earthly or atmospheric influences. The number of colours was a matter of serious concern, partly because they could be distinguished in the rainbow and must therefore be fundamental in some way, partly because of the magical properties of certain numbers. The authority of Aristotle, persisting for many centuries, supported a three-colour system. 'These colours are almost the only ones that painters cannot manufacture; for they produce some colours by a mixture of others, but red, green and blue cannot be produced in this way, and these are the colours of the rainbow—though between the red and green band often appears a yellow one. . . . When the sight is fairly strong the colour [of white light] changes to red, when it is less strong to green, and when it is weaker still to blue. There is no further change of colour, the complete process consisting, like most others, of three stages. . . . This is why the rainbow is three-coloured. . . . The yellow colour that appears in the rainbow is due to the contrast of two others; for red in contrast to green appears light.'[1] Others preferred a four- or five-colour theory, but there was a contrary belief that the rainbow included a very large number of colours. Most famous among those holding this opinion was Seneca, the tutor to the emperor Nero. About the same time the elder Pliny, better known as the author of a *Natural History*, pronounced on the diverse colouring in the rainbow: it was due to the mixture of clouds, fire and air.

Whatever the means of producing colour may be, there has been a recurring idea that it is subjective. About 100 B.C. Posidonius the Syrian, distinguished by making the nearest estimate in antiquity of the sun's distance (though too small by nearly half), believed that colours were an illusion. Galileo was more explicit: 'I hold that tastes, colours, smells and the like exist only in the being which feels, which being removed, these qualities themselves do vanish.'[2] Newton's views were still more precise: 'The rays to speak properly are not coloured. In them there is nothing else than a certain power and disposition to stir up a sensation of this or that Colour . . . so Colours in the Object are nothing but a disposition to reflect this or that sort of rays more copiously than the rest.'[3] He also referred to light producing the 'Phantasms of Colours' in the mind.[4] Eddington was shorter and downright: 'The wave is a reality . . . the colour is mere mind-spinning.'[5]

There have been many other attempts to explain the origin of colour, and some of the more objective ones are discussed later. Here we may emphasize that Newton's words show him to be aware of the subjective aspect of colour, while his analytical, unemotional approach was well justified by the practical consequences of his theories. This approach, however, was intensely disliked by Goethe, whose experiments on prismatic boundary colours drew attention to an undeveloped area in the study of spectra, but whose views on the nature of white light and colours might have been more carefully considered in his time if he had not so violently assailed the eighteenth century's scientific hero.

Colour was of course not the main issue or even a very important one in the development of spectroscopy. The advance of optics from mediaeval times was more significant, particularly the stimulus given by the Arabic work *The Treasury of Optics*, written about the year 1000 by Alhazen, a practitioner in mathematics and optics in Cairo. This is one of the famous scientific works which affected European work for centuries. It became more accessible when a complete Latin edition was published in Basel in 1572. Among its many achievements the book helped to dispel the long-standing confusion between reflection and refraction.

One other disputed phenomenon was solved in the seventeenth century, at the beginning of the modern period of experimental science. This was the velocity of light, to become more important at a later time. Common observation and such experiments as had been made suggested an infinite velocity. The leading philosophers from Aristotle to Kepler, Galileo and Descartes could hardly be expected to think otherwise. But here again there had been dissentients, like Empedocles (of earth, air, fire and water fame) and Alhazen of Cairo. Descartes believed, inconsistently, in variable velocities to account for the law of refraction, then correctly formulated for the first time. Roemer's work on the observation of Jupiter's satellites settled the matter in 1676: at least a finite velocity was established. Whether it is constant is another matter.

PROGRESS FROM GALILEO TO NEWTON

About 1600 astronomy was advancing beyond the study of the movements of nearby celestial objects, considered as perfect spherical bodies, towards some knowledge of their structure and surface features, particularly in the easily observed sun and moon. Sunspots were seen, and reported in print about 1610. This was a nearly simultaneous 'discovery' by Galileo in Florence, Scheiner in Ingolstadt, and Johann Fabricius in East Friesland, though there must have been many unreported sightings by the unaided eye as well as those on record from much earlier times.[6] The moon's surface had been plain to see for centuries without the development of any widespread belief in a rough surface, so powerful were the prevailing ideas of perfection in heavenly bodies and so heavily weighted was the philosophical idea against the evidence of the eyes. Even Leonardo da Vinci (1500) wrote of the sphere of water largely covering the moon, and of how reflections from the ripples on this water caused the rough appearance. Galileo increased his offence in describing sunspots by the statement that the moon's surface was rough and mountainous. He was not the first to observe the moon by telescope, but his observations received much publicity and notoriety.

The accumulation of telescopic evidence brought the end to many strange beliefs. This alone would not have led to the developments of the eighteenth and nineteenth centuries. One more basic idea was needed, founded on experiment, one in which Newton played the major role but not the earliest or the only significant role. It was the separation of light into a spectrum. Once this technique had been established and made capable of measurement, the way was open to the modern age of discovery in astronomy; in fact astrophysics had arrived.

THE SPECTRUM OBSERVED

It can hardly be said that the spectrum was ever discovered. The rainbow, although an imperfect form of the physicist's nearly pure spectrum, had been seen, admired and speculated upon, probably long before the time of written records. The Greeks had a great deal to say about it and, to cut short a long and fascinating history,[7] it was in 1637 that Descartes, mathematician, physicist and philosopher, produced a theory of its origin which remained serviceable for 150 years and indeed is not yet entirely obsolete though lacking later refinements based on the wave theory of light. This is in spite of his now untenable theory of the origin of colours by the rotation of spherical particles which transmit light through matter. His successor, Newton, also believed in a theory of light which we do not now accept, but by conclusive experiments and penetrating thought about them he produced a satisfactory theory of colour and the spectrum; this with his greater achievements in mathematics brought him contemporary fame which obscured many others, including Descartes, who had made substantial contributions to knowledge of the spectrum. Some of them came close to anticipating Newton's more abstract conclusions, and their published works are still worth attention to show the mixture of occasional accurate observation and frequent metaphysical speculation which was characteristic of the age.

Man-made spectra have been known at least since the thirteenth century. Experiments on globes of water to imitate raindrops were often made by those trying to explain the rainbow. For example, Leonardo da Vinci wrote: 'If you place this glass full of water on the level of the window, so that the sun's rays strike it on the opposite side, you will then see the aforesaid colours [of the rainbow] producing themselves, in the impression made by the solar rays which have penetrated through this glass of water, and terminated upon the floor in a dark place at the foot of the window; and since here the eye is not employed we can say with certainty that these colours do not derive in any way from the eye.'[8]

Descartes is specially important. In an appendix to his *Discours de la Méthode* (1637) he formulated the hitherto elusive law of refraction, possibly independently of Snell. He also gave a diagram (Plate 1) showing rays from the sun impinging on a glass 'prism or triangle', with the refracted rays forming a rainbow-hued patch on a cloth or white paper placed on the vertical wall.[9] The green, blue and violet colours were seen towards H, the red, orange and yellow towards F, and widening the aperture DE destroyed the colours near G.

Marcus Marci of Kronland in Bohemia, a court physician in Prague, was another of Newton's immediate predecessors. He wrote on medical subjects, also on the rainbow (*Thaumantias*, 1648), described refraction through a triangular prism and the persistence of colours, once separated, in subsequent refractions. He believed the different colours to be the result of slightly differing angles of incidence at the first refraction, which caused different pathlengths in the refracting medium. Plate 2 reproduces one of his (perspective?) drawings, where a small difference in the angles DKB and HFK ($\frac{1}{2}°$ due to the size of the sun's disc) gives a shorter path FG for the purple light appearing at E, and a longer one KI for the red light appearing at L. Both refractions were said to be essential to produce colour.[10]

Another court physician interested in the production of colours by refraction was Marin de la Chambre (1650). Like Marci he thought pathlength important, but that interaction between refracted and internally reflected rays was necessary, as in Plate 8.[11] 'The reflected rays mixing with the luminous mass traversing the triangle, infect it with their colours.' He tried to explain how the red rays emerged nearest the prism apex where the opacity was least, red being the brightest colour ('la couleur la plus claire et la plus haute de toutes'), while green and purple rays emerged in order lower down the prism face. This is not at all convincing [Plate 8(*b*)]. La Chambre's explanation of the rainbow contributed nothing to progress, but in France he was considered not less eminent than Descartes, as Boyer relates.[7]

Francesco Grimaldi of Bologna was one of the many Jesuit experimental physicists, as they could rightly be called. He used globes of water to investigate refraction and its bearing on the rainbow. He knew that rays of coloured light did not change by one refraction, but still believed colours to be illusory. He demonstrated coloured fringes on the base of a tank of water when a narrow beam of white light fell obliquely on the surface, and with the next diagram of his book,[12] shown in Plate 3, explained at length how red light appeared along KL, blue or violet along HN, yellow or green in between; likewise red at VZ, blue at ÆY. Grimaldi's book was published in 1665, just as Newton was entering his creative years. His most important discovery was of diffraction, destined to produce refinements in Descartes' rainbow theory long after its first revision by Young, who used the principles of optical interference in his contribution to the subject.

Newton's experiments with prisms and his interpretations of them to explain the physical origin of colours, are now part of basic optical theory. Plate 4 shows one of his diagrams[13] which like the explanation in the text had admirable simplicity compared with the complex and sometimes inconsistent arguments of Marci and the others. In this example Newton showed how the beam of sunlight Fϕ was dispersed by the first prism into the spectrum *pqrst* (violet to red), converged by the lens, recombined by the second prism to a beam of white light EY, redispersed at the third prism to the spectrum PQRST; then if green light at *r* was intercepted, the green at R would vanish, and similarly for other colours.

Contemporaries and successors of Newton favoured a wave theory of light rather than his corpuscular theory. With Newton's great prestige, his views on this subject may have retarded the pace of development in physical optics, which was very slow until the nineteenth century. Newton had established the essential fact that light rays of a given refrangibility were unalterable in that respect, whether separated in a spectrum or still undispersed from other rays. At that time the discovery of colours produced from white light seemed to be the most important, as it was the 'oddest' discovery, in Newton's own description of it. In practical value we can see now that the colour coding of the rays, their identification by observation, was more useful, for it made the experiments convincing in a way that mere measurements of angles of refraction could not have done.

WOLLASTON, FRAUNHOFER AND THEIR SUCCESSORS

Newton's prism had insufficient dispersion. The use of a circular entrance aperture made his spectra far from pure in the first experiments, and even with a slit aperture he failed to see the dark lines in the spectrum of sunlight. This crucial discovery came 150 years later. In the meantime the spectra of flames had been observed by a number of experimenters. Thomas Young, the first to measure wavelengths of light, had seen what we now call the D lines in the spectrum of a candle flame and, with his own interests concentrated on interference phenomena, had thought the lines to be due to the same cause. This scarcely delayed the slow progress along the main line of advance, for in 1802 his contemporary Wollaston performed Newton's single prism experiment, probably with a more parallel beam of light, and saw the spectrum divided into red, yellow–green, blue and violet sections: 'I cannot conclude these observations on dispersion, without remarking that the colours into which a beam of white light is separable by refraction, appears to me to be neither 7, as they usually are seen in the rainbow, nor reducible by any means (that I can find) to 3, as some persons have conceived; but that, by employing a very narrow pencil of light, 4 primary divisions of the prismatic spectrum may be seen, with a degree of distinctness that, I believe, has not been described nor observed before.'[14] Besides these divisions, Wollaston saw 'other distinct dark lines, *f* and *g*, either of which in an imperfect experiment might be mistaken for the boundaries of these colours'. In Plate 5, taken from Wollaston's paper, A was somewhat confused, B was the distinct boundary between red and yellow–green, D and E were distinct limits to the violet. C was not so distinct, separating green and blue. It is probable that B was Fraunhofer D, and A, f, C, g, D and E can possibly be identified with Fraunhofer B, E, b, F, G and H respectively. From the size of the colour patches which he measured it appears that Wollaston saw far into the violet but only a much restricted red band; also that he thought f and g were in some way different from the colour divisions. These simple observations began spectroscopy as we know it, the source of most of our knowledge of the sun and stars, sunlight and skylight. Not until 1946, when it became possible to place observing instruments beyond the earth's atmosphere, was there another advance of comparable importance.

Wollaston's publication might have been forgotten if Fraunhofer had not made an independent discovery about 1815. Like Wollaston, he was measuring dispersive powers of materials, in this case glasses for possible combination in an achromatic telescope. To view the spectrum he added a telescope to the prism and slit, and with an oil lamp saw a bright streak in the yellow region of the spectrum and identified it as R. Observing sunlight to find a similar bright streak, he saw instead 'almost innumerable strong and weak vertical lines which however are darker than the other part of the coloured image; some seem to be almost completely black'.[15] His picture of the spectrum (Plate 6) showed the main lines named from A to H, ending at I in the ultra-violet. The K line was strong but described as one of a pair at H and not separately named at this stage. The D lines were resolved and shown to be identical in position with the bright line R. Fraunhofer counted 574 lines between B and H. Later he measured wavelengths with his own primitive diffraction gratings—another notable invention; and he was able to identify the D lines in the light from clouds, the moon, and the planet Venus.

In 1859 Kirchhoff's law appeared. According to this, all temperature radiators at the same temperature have the same quotient of radiant emittance to absorption factor for a given frequency of radiation. In present day terms, this means that at each temperature and for each wavelength, the directional spectral emissivity is equal to the spectral absorptance for radiation incident in the same direction. Soon after this the implications of the law were studied by Kirchhoff and Bunsen in collaboration; they established the relation between dark and bright lines of the same wavelength, and the unique correspondence of wavelength and the element concerned. The spectrum became a fixed frame of reference with possibilities of detecting the physical state of distant light sources, their chemical composition and, by Doppler's principle (1842), their motion in the line of sight. At this time the unity of the universe was accepted in the sense that gravitation operated equally on all celestial bodies as well as the earth, but of material evidence about the other bodies there was none, except a few meteorites. Thus by a few simple experiments, as they now seem, the spectroscopic pioneers ended this almost total isolation of the earth from the rest of the universe.

To some degree Kirchhoff had been anticipated in these achievements by Foucault, better known for his pendulum and velocity of light experiments. Foucault had in 1848 observed that the bright spectral lines of a flame containing sodium were reversed and appeared dark when a brighter arc was placed behind the flame. He did not pursue the subject. Stokes also deserves an honourable mention for his discovery of the Kirchhoff principles without ever claiming any credit. Kirchhoff himself superimposed the spectra of sunlight and the light from a gas flame containing salt, and again confirmed the identity of the position of the dark and bright D lines. He deduced that the sun's atmosphere provided the necessary absorption for reversal. There were several others who made observations on coloured flames and spectral lines, but their contributions were overshadowed by Kirchhoff's.

SPECTROSCOPY BECOMES USEFUL

Line identification by wavelength was firmly established by A. J. Ångström's map of a thousand lines in the solar spectrum (1868), and culminated for a time in Rowland's map of 21 000 lines, measured with great accuracy (1897). Many more lines have been added since, and successive atlases produced. The Utrecht atlas covering the range 361·2 to 877·2 nm is an important one,[16] and other sectional publications are mentioned in later chapters. Some years ago progress was reported on the revision of the solar spectrum atlas,[17] at which time 22 000 lines were listed, and only 71 per cent of these had been identified even incompletely. They proved that sixty-two terrestrial elements were certainly present in the sun, with four others still doubtful. Ten years later only two more elements had been positively identified. Many lines are due to simple molecules like CN, CH, C_2, OH, MgH and NH. In an atlas of the 300 to 365 nm region, Mitchell and Mohler included nearly 2000 lines, some new and some revised from earlier lists.[18]

William Herschel, a great maker of telescopes, was one of the first to view the spectra of the brighter stars. He also discovered the infra-red part of the solar spectrum in 1800, and his son John Herschel found it to be a source of dark lines in 1840. The ultra-violet, discovered by Ritter in 1801, similarly provided its own multitude of absorption lines. Till recently atmospheric absorption limited the exploration of this region: for example, the very strong magnesium lines at wavelengths just shorter than 290 nm could not be observed (Chapter 3).

Two particular identifications of spectral lines made advances in knowledge of the sun's constitution. The bright emission line at 587·6 nm, not identical with the sodium lines, was observed by a visual spectroscope in a solar prominence, revealed during an eclipse in 1868. The observer, Janssen, reported this to Lockyer, who gave the name helium to the element responsible; much later it was found in the earth's atmosphere. The second case was an emission line at 530·3 nm seen in the spectrum of the solar corona in an eclipse in 1869. Charles A. Young and Harkness observed this independently; the element was named coronium, but until 1898 the line was confused with a line from the chromosphere and a coincident one in the photosphere, 1474 K as it was then known on Fraunhofer's empirical scale of the dark lines. With better photographs Lockyer was able to distinguish the known (iron) line from the coronium line.[19] Whereas the former arises from Fe II, the latter was not identified for more than seventy years when Edlén showed its origin to be a highly ionized state of iron, Fe XIV. The eclipse of 1870 provided an elegant demonstration of the theory of emission and absorption. At the moment of totality Young saw the dark Fraunhofer lines on a bright background change to bright lines on a darker background, so observing the 'flash spectrum' of the reversing layer in the sun's atmosphere (Chapter 2).

A further demonstration of the possibilities of spectra occurred in 1868 when Huggins observed the Doppler effect in the star Sirius. This principle is now basic to the 'red shift' of distant objects and the cosmogonies developed from the spectral measurements. Finally we may remember, in this most fruitful epoch, the work of Respighi and Secchi who began in 1869 to classify stars by their spectral similarities and differences.

Another kind of red shift is the change of photons in a gravitational field. According to general relativity theory the wavelengths of light emanating from the sun should suffer a red shift when observed on earth, in amount equivalent to a Doppler shift of 0·64 km/s compared with a terrestrial source. This means an increase at 500 nm by about 10^{-3} nm, and though difficult to detect on account of other solar red shifts[20] the change is not negligible in view of the precision of atlases of the solar spectrum. These usually quote wavelengths to 0·001 Å or 10^{-4} nm. Larger shifts of the same origin occur for stars of higher density than the sun, for example about thirty times as much for the white dwarf companion of Sirius.

AIDS TO SPECTROSCOPY

One other indispensable technique arrived somewhat later than might have been wished, but still with great effect on the speed of progress in that pre-balloon, pre-rocket, pre-satellite age of solar investigation. This was photography, without which the retarded state of spectroscopy, especially in the ultra-violet, is almost impossible to imagine. The first applications in this field were made by Draper, who photographed the moon by a half-hour exposure in 1840, and by Edmond Becquerel who photographed the solar spectrum as far into the ultraviolet as 340 nm, both using daguerrotype plates. In 1839 the largely forgotten experiments of John Herschel produced a fugitive print of the solar spectrum on a silver chloride paper, which almost unbelievably showed the spectrum in 'sombre, but unequivocal tints, imitating those of the spectrum itself'. Herschel mentioned the photographic effect of rays beyond the violet part of the spectrum, a 'new prismatic colour' preferably called 'lavender . . . rather for the purpose of abbreviating the uncouth appellation of ultra-violet'.[21] Also about 1840 Fox Talbot made the first fundamental discoveries in negative–positive reproduction with his Calotype process. In 1851 the wet collodion plate on glass appeared, giving spectroscopists a comparatively sensitive tool to record their observations and extend them in the ultra-violet direction. In the seventies operations were simplified and accelerated by the gelatin dry plate; about 1884 sensitization to green and yellow light was achieved, then further steps to the infra-red. The limit was 700 nm about 1905, 800 nm about 1919, 900 nm about 1925, and 1·3 μm by 1934. To supplement this versatile chemical detector of radiation, several physical detectors have in turn improved the power of non-photographic spectroscopy. They are the thermopile (1830), the bolometer (1880) which first made spectroradiometry possible as distinct from spectroscopy, the vacuum emission photocell (1889), the photo-voltaic cell (Se 1876, Cu_2O 1904), photoconductive detectors of many kinds (Se 1873, sulphides from 1920, Si 1946), and the photo-multiplier (1936). The Golay pneumatic cell of 1947 used a new principle, coinciding with the new age of space research.

Phosphors have had a limited use in spectroscopy. In the earlier days Stokes used a fluorescent plate to detect ultra-violet lines in the spectrum of the aluminium spark (1862). Draper in 1881 described the effects of the solar spectrum on a layer of luminous paint of unspecified nature. This was apparently excited by the blue and ultra-violet region of the spectrum, and showed acceleration of the phosphorescence (*Ausleuchten*) in the bands between 800 and 1000 nm.[22] Soon after, in 1883, Henri Becquerel of radioactivity fame found that phosphors would detect even more of the solar spectrum, down to 1·4 μm. In the rocket flights of 1948 the phosphor $CaSO_4$:Mn was used to record short ultra-violet radiation from the sun, probably the Lyman α line.[23] X-radiation from the corona was measured similarly during a total eclipse in 1961.[24] The photon counter (1950) is an alternative to the use of phosphors in measuring power at the shorter wavelengths. The use of organic scintillators has been applied to the detection of ultra-violet radiation at less than 390 nm, particularly in simulated solar radiation work on spacecraft.[25]

The various forms of telescope and other devices mentioned brought solar spectroscopy up to the space age, when fresh revelations began and old topics were revived. Some of the conclusions from this new approach are included in the description of the sun which follows, but the main foundation of our knowledge is still the work of Newton and Fraunhofer. From their simple prisms and gratings came a long line of spectroscopes and monochromators, changing little in principle though much in detail and accuracy, which will not be described in detail. One more recent method of considerable potential, the application of interferometry to spectroscopy, has been used in some work on the daylight spectrum (Chapter 11), and also slightly outside the terrestrial spectrum to the exploration by rocket of the magnesium doublet near 280 nm. For most purposes the older methods of measurement are adequate.

2

The Sun

The sun has substance, shape, movement, radiance, heat and genera-
tive power; and these qualities all emanate from itself without its
diminution . . . But I marvel greatly that Socrates should have
spoken with disparagement of that body, and that he should have
said that it resembled a burning stone . . . I do not perceive in the
whole universe a body greater and more powerful than this, and its
light illumines all the celestial bodies which are distributed throughout the universe.

LEONARDO DA VINCI, *The Notebooks*. (trs. E. MCCURDY)

To men living on the earth the sun has meant many things through
the centuries. Unlike those imponderables, light and colour, it was
evidently real. It became an object of wonder and fear, often of wor-
ship; it generated superstition and strange beliefs, and surmise about
our earth's place in the universe. For long its connection with daylight
was not appreciated. Light was one phenomenon but the sun was
another for it was not always present in the sky when daylight still
prevailed. Many religions gave the sun a separate origin, for example
in the book of Genesis, where light was created on the first day but the
sun not till the fourth day. This idea is no longer a mere curiosity from
the past: theories of the 'Big Bang' origin of our present universe
require a primaeval high temperature concentration of matter emitting
vast amounts of radiation. Only later after expansion would this form
stars like the sun. In the last few centuries the sun has been examined
by instruments to help our limited sight, and finally by the methods of
analytical thought which have so changed our ideas of the world by
concentrating on how things happen, rather than why they happen. In
the context of physics no doubt now remains of the sun's place in our
local planetary system, its provision of all but a minute fraction of our
light and heat, and of its effects on our atmosphere, climate and radio
environment. Its own structure is understood in more and more detail
as observations multiply on earth, and are gathered above the

atmosphere, by refined and complex equipment not thought of even forty years ago.

THE SUN'S PHYSICAL CHARACTERISTICS

We turn now to a matter of fact, an account of the sun and the ways in which it determines our light and heat and living, not forgetting that its place in the universe still remains mysterious, in spite of the influx of so much coded radiation from the sun itself and the universe beyond.

The sun is very nearly spherical, of diameter $1 \cdot 39 \times 10^6$ km and mean distance from the earth of $1 \cdot 50 \times 10^8$ km. It has a density of only $1 \cdot 4$ g/cm^3 because it is almost entirely composed of hydrogen and helium. The former predominates to about 75 per cent of the total mass, and by its conversion to helium in the central core, at $1 \cdot 5 \times 10^7$ K and about 10^{11} atmospheres pressure, provides the energy radiated from the outer layers. The main reaction is probably

$$^1H \rightarrow {}^2H \rightarrow {}^3He \rightarrow {}^4He$$

with other possible routes such as

$$^1H \rightarrow {}^2H \rightarrow {}^3He \rightarrow {}^7Be \rightarrow {}^7Li \rightarrow {}^4He.$$

The energy is largely produced as gamma-radiation which is degraded into a spectrum spread out to the longer wavelengths by absorption and re-emission processes throughout the sun's bulk. The outer layers are in a constant state of violent turbulence due to gravitational, magnetic and, predominantly, thermal forces, and the detailed picture of the structure and radiation has been developed primarily by spectroscopy, with some help from visual and photographic methods applied to the sun as a whole. The sun's total power is $3 \cdot 8 \times 10^{20}$ MW, and this results in a loss of mass of $4 \cdot 2 \times 10^6$ tonne/s.

At the earth's distance of about 107 sun diameters, and outside the atmosphere, the sun provides a surface density of radiant flux equal to about $1 \cdot 35$ kW/m^2, the much measured but somewhat imprecise solar constant (Chapter 5). The illuminance at the same distance, outside the earth's atmosphere, is of some interest but has been given many differing values due to the difficulties of measurement within the atmosphere. A determination of the luminance of the sun's disc, made in Ottawa, led to a calculated illuminance of $12 \cdot 2$ klm/ft^2 or 131 klx;[26]

extrapolation from ground level illuminance in South Africa gave 145 klx;[27] extrapolation of a spectral distribution of daylight at an elevated site in Arizona gave 137 klx;[28] and in a year's observations at Madrid an appreciable number of the measurements of vertical and horizontal illuminance were higher than 130 klx.[29] These values were found before 1960. From a more recent NASA determination of the solar constant the calculated illuminance is 125 klx which is equivalent to a luminous efficacy of 88 lm/W.[30] For a full radiator the maximum possible efficacy is 95 lm/W at 6600 K, and 91·5 lm/W for a full radiator at 5755 K, the nominal solar temperature adopted in this chapter (see below). No artificial source giving white light of this quality has yet approached this maximum, though there are more efficient methods of light production than purely thermal radiation. Efficacies of about 90 lm/W can be achieved but only with a reduction in the quality by more or less serious alterations in the emitted spectrum. For light units and definitions see Appendix 1.

In the remainder of this chapter the numerical values become less certain, and there are differences of opinion among writers. The values quoted may be described as reasonable for descriptive purposes but not necessarily authoritative. For more exact details and references to many variable factors not discussed here, see Allen.[31]

Our knowledge of these solar quantities is comparatively recent. Less than 200 years ago William Herschel, discoverer of the planet Uranus and of the infra-red, thought like other contemporaries that sunspots revealed the darker lower levels of the sun. He claimed that the surface might be cool and dark, protected by clouds from the fiery outer layer consisting of 'lucid clouds swimming in the transparent atmosphere of the sun; or rather, of luminous decompositions taking place within that atmosphere'. Further, 'we have great reason to look upon the sun as a most magnificent habitable globe'.[32] Seventy years later similar opinions were current on the luminous clouds and sunspots, but the temperature of the solid surface was now admitted to be very high and possible inhabitants were no longer mentioned.[33]

THE PHOTOSPHERE

The visible 'surface' of the sun, the photosphere, is the level where temperature fall from the outside reaches a minimum and a rapid increase with depth begins. Here also the transparency of the gaseous material increases sharply outwards and decreases sharply inwards. An apparently well defined disc results from these changing optical absorptions. The pressure also changes and is roughly 5 torr at the apparent surface.

The rapidly increasing absorption has the further effect of causing the disc to appear less bright from the centre outwards, and this 'limb darkening' effect is more marked as the wavelength of observation decreases through the visible spectrum. The high luminance makes all outer parts of the sun's incandescent atmosphere invisible without special optical devices, or at the rare occasion of a total eclipse. Outside our atmosphere the mean luminance of the disc is about 1.8×10^5 stilb, a value not beyond the reach of our light production techniques. Certain arcs in xenon at high pressure do achieve this luminance at their brightest parts. Thouret *et al.*, and Lienhard, claimed for 10 to 30 kW arcs a peak luminance as high as 10^6 stilb over very small areas.[34, 35]

The photosphere is the source of almost all the radiation received by the earth at sea level. The other parts of the sun either modify the light by absorption, or make additions to it which do not arrive at the earth's surface, or are negligibly small in terms of energy.

The first 1500 km above the photosphere, in the region of minimum temperature, is the reversing layer where cooler gases of many terrestrial elements partly absorb, at their own characteristic frequencies, the radiations from the hotter gases below. So arise the Fraunhofer lines, appearing to be dark by contrast with the brighter continuous backgrounds due largely to unquantized transitions in the hydrogen atoms and the ion H^- at visible wavelengths, with contributions from metal emissions in the ultra-violet.[36] Hydrogen provides strong absorptions in the ultra-violet where the Balmer series commencing at Fraunhofer C (656·3 nm), continuing with F, G' and h (486·1, 434·0, 410·2 nm), builds up to a band, the 'Balmer discontinuity' occurring at 364·7 nm. Line profiles provide evidence on the temperatures and abundances of elements in the sun. For a concise and informative account, see Unsöld.[37] There has been much investigation into the lowering of the general level of the extraterrestrial spectrum, especially in the ultra-violet, by the assumed presence of many faint and unidentified absorption lines (Chapter 11).

THE CHROMOSPHERE

The reversing layer is generally considered to be part of the chromo-sphere, a gaseous atmosphere recognizable to 10 000 or 15 000 km above the photosphere, or 1 to 2 per cent of the sun's radius. Its density falls with increasing height but the temperature rises until it merges into the corona at 10^5 K or 10^6 K. It has been described as 'an assemblage of innumerable small jets of flame' ejected from the photo-sphere. At times some of these attain heights of several times the normal, when they are known as prominences. The spectrum of the chromosphere is dominated by the emission lines of hydrogen, helium and Ca II, especially by hydrogen Balmer α (656·3 nm) which gives the chromosphere its red colour if seen with the photosphere masked. It was named by Lockyer for this reason. The visible emission lines make too small a contribution to the whole sun spectrum to be noticeable in our daylight, and therefore hydrogen α, for example, remains an absorption line. Most of the chromosphere emission is in the short wavelength ultra-violet region including the hydrogen resonance line, α of the Lyman series at 121·6 nm which assumed a predominant role in spectral measurements by rocket. It was first detected on earth, in a gas discharge, by Lyman in 1904, but not in solar radiation till a rocket flight in 1952.

THE CORONA

The last distinct component of the sun is the corona, an unsymmetrical extension of the sun's atmosphere only visible with high quality optical screening of the disc, first achieved by Lyot in 1931 at the Pic du Midi observatory, or in total eclipses which have been thoroughly exploited for this purpose in recent years. Strangely enough the impressive appearance of the corona has been very little described in earlier writings though Halley mentioned it in 1715, and thought it due to the moon's atmosphere,[38] as did Kepler before him. The first eclipse used for extensive study was in 1842. The corona is essentially a hydrogen plasma at 1 to 2×10^6 K. In spite of its visible extension to at least a sun diameter from the surface, and to several times this distance spectrographically, its total visible light emission is only a millionth of the whole sun's, or half that of the full moon.

Radiation from the corona includes a small proportion of very short wavelength emission lines, ultra-violet and X-ray; also a number of lines from highly ionized states of iron, calcium and nickel, the strongest being Fe XIV at 530·3 nm (coronium) discussed in Chapter 1. The next in intensity are Fe XIII lines at 1074·7 and 1079·8 nm. Coronal light mainly consists, however, of photosphere emission scattered by electrons (the K component) or by dust particles (the F component). In total the light closely resembles normal daylight in colour and consists of a continuous spectrum with both emission and absorption lines superimposed. The more interesting parts of this are indistinguishable at the earth's surface or are previously absorbed, not unlike the chromospheric emission. They are not of merely academic interest in the production of terrestial daylight because of their far-reaching effects on the earth's atmosphere.

THE SUN'S TEMPERATURE AND EMISSION SPECTRUM

Temperatures estimated for different parts of the sun are based either on absorption or emission line profiles, or more often by fitting full radiation curves to observed sections of continuous spectral emission curves, and assuming that they do in fact arise from full radiators. This is clearly not valid for visible radiation from a transparent body like the corona, or from the chromosphere. Emission curves in the ultra-violet, even when corrected for emission and absorption lines and bands, yield different temperatures for the equivalent full radiator according to the spectral range chosen, and the same applies to a lesser extent in the visible spectrum. The literature on this subject is sometimes confusing because of the different types of temperature quoted; these include brightness, colour, distribution, radiation, real and electron temperatures.

The range 5500 to 6200 K covers most estimates of the sun's equivalent temperature made in the last fifty years or so, though about 1880 Pickering[39] suggested 22 000 °C and Rosetti[40] the range of 10 000 to 20 000 °C. Goody[41] showed the moderate agreement of the spectral power distribution of a full radiator at 5785 K with that of the sun between 100 nm and 10 nm, while Allen[42] proposed 5800 K, derived by a process of equalizing the areas of the full radiator curve and the best available curve for solar radiation extrapolated to outside the atmosphere. A similar calculation treats the sun as a full radiator with a known radiant flux (the solar constant S). By the Stefan–Boltzmann law, the sun at temperature T, of radius r and distance D from the earth, produces an irradiance at the earth's distance of

$$S = \sigma T^4 \frac{4\pi r^2}{4\pi D^2}$$

where

$$\sigma = 5{\cdot}67 \times 10^{-8} \text{ W/m}^2\text{K}^4.$$

If

$$S = 1{\cdot}35 \text{ kW/m}^2$$

then

$$T = 5755 \text{ K}.$$

The luminance of a full radiator at 5755 K is $1{\cdot}82 \times 10^5$ stilb, while the mean for the sun's disc is $1{\cdot}84 \times 10^5$ stilb, calculated from the solar constant.[30] The sun's main radiation pattern arises from a zone of very variable temperature, from the reversing layer at about 4500 K to a depth of a few hundred km below the apparent surface, where optical absorption is complete and the temperature near 10 000 K. The value of 5755 K can be no more than a useful equivalent when the radiating body varies so much in temperature, and has a spectrum differing appreciably from that of a full radiator. The near agreements in luminance and luminous efficacy values, though striking, are perhaps fortuitous as there is still uncertainty in the measurements. From Table 9 and Fig. 73 in a later chapter it is apparent that the measured proportion of solar radiation in the visible region is somewhat higher than for the equivalent full radiator, and this should make the luminous efficacy even higher than the value quoted above.

A full radiator at 5755 K has its peak emission close to 500 nm on a wavelength scale. In the ultra-violet region the flux decreases steadily. Up to 1946 the constitution of the sun's radiation spectrum was known only as far as about 230 nm.[43] With later rocket experiments it was confirmed that outside the atmosphere the solar flux continued to decrease down to about 160 nm. Here the character of the spectrum changes with the appearance of emission lines superimposed on the continuum. The latter, now derived from emitters of higher temperature than the photosphere, is too large in comparison with the rest of the spectrum to fit the Planckian curve for 5755 K. The principal line is the Lyman α line of hydrogen at 121·6 nm, which is found to be a disturbing background to all interstellar observations. Other emission lines in the sun's spectrum are the Lyman α of He II at 30·4 nm, and the He I line at 58·4 nm. Later discoveries in this region have been described by Goldberg.[44] Extending from the short wavelength ultra-violet is a band of X-radiation with its maximum between 1 and 10 nm. The whole emission of the sun below 200 nm is roughly 0·01 per cent of the total flux, but its importance in astrophysics is far greater than this figure suggests.

Below 100 nm the solar flux is liable to large increases during periods when the sun shows more activity at sunspot maxima, or at times of sudden brief outbursts of energy when the corona may have an effective temperature of 5×10^6 K. The increases are still a very small fraction of the total radiant flux and cannot affect the value of the solar constant within the uncertainty of its measurement.

The infra-red end of the solar spectrum has not attracted so much attention as the visible and ultra-violet regions. With the bolometer the exploration of the infra-red region at high resolution became possible, but the complications due to atmospheric absorption and re-radiation (Chapter 3) have limited direct observation to small spectral regions. It has generally been accepted that the shape of the full radiator for 6000 K fits the solar spectrum well beyond 1·25 μm. Minor corrections to this figure have been made since Moon adopted it (Chapter 8). This part of the solar flux is important in making corrections to the solar constant at wavelengths beyond 2 μm. A full radiator at 5755 K emits about 40 per cent of its radiation at wavelengths greater than 830 nm and 6 per cent beyond 2 μm. Further details are given in Table 9. Infrared emission is not discussed in detail since it has little to do with daylight.

Last in this survey of the components of the sun's radiation is the radio emission from the corona, one component of which was first detected in this country in 1942 as interference on radar installations and attributed to sunspot emission, while at about the same time the normal thermal emission from the quiet sun was detected in the U.S.A. A similar discovery had been recorded by Jansky in publications commencing in 1932; he had found a source of radio emission located near the centre of the galaxy. In 1943, following his earlier discoveries of 'cosmic static' from different parts of the sky, Reber identified the sun as a source at a wavelength of about 1·87 m (160 MHz).[45] Details of this beginning of radio astronomy are related by Hey.[46]

The sun's emission extends over centimetre and metre wavelengths, in quantity highly variable as the corona itself is known to be, and amounting to only 10^{-14} of the total solar flux. Taking the emission per nm at 500 nm from a full radiator at 5755 K as unity, the emission from a full radiator of the same size at 10^6 K is 2×10^{-6} for a 10 mm band at a wavelength of 10 mm, or 2×10^{-14} for a 1 m band at a wavelength of 1 m. These quantities of radiation are in fact exceeded by 10^3 to 10^4 times;[47] even so they emphasize the sensitivity of detection at the radio wavelengths. Quantitative measurements, if attributed to a full radiator as large as the sun's visible disc, suggest a temperature between 10^4 and 10^6 K under normal 'quiet sun' conditions. At metre wavelengths the emission increases again by thousands of times in short-period storms due to solar flares or sunspots, in a much more spectacular way than is observed at short ultra-violet wavelengths. At such times the calculated temperatures exceed that of the sun's centre. The radio emission cannot in fact have its origin in purely thermal radiation, but arises from the plasma surrounding the sun, with synchrotron radiation in the magnetic field of sunspots largely responsible for the high output. In some distant celestial objects the relatively enormous radio emission was difficult to explain even by this means, until the idea of the neutron star showed other possibilities.

Details of these invisible radiations, and of information derived from the sun's surface structure and its changes, are of no great significance to daylight reaching the earth and are not discussed more than superficially in the following chapters. The present theory of solar structure and radiation is much more complicated than the preceding treatment and has been reviewed at length by E. G. Gibson.[48] Though not yet formulated into an accepted theory, there is evidence that the sun's interior may be cooler and more turbulent than now assumed, with a much longer presumptive life and, perhaps, a more variable output of radiation. Though not important in daylight, the X-ray, short ultra-violet, infra-red and radio regions of the spectrum have become of predominant importance in astronomy and astrophysics.

In the radiation from the stars there is so little measurable power that their spectra have not been examined in as great detail as the solar spectrum. The sun's main spectral type (G) is found in many other stars, among them one component of the binary star Capella, one of the brightest in the sky. Characteristic of stars in this part of the main sequence are the Ca II absorptions (Fraunhofer H and K) which persist as two of the most intense lines in the terrestrial daylight spectrum. Other stars which were originally thought to be similar by Fraunhofer, with his simple apparatus, are now differently classified, though his experiments were sensitive enough to find a difference between the sun's spectrum and the spectra of Sirius and Castor (both type A). An example of what can now be achieved by scanning stellar spectra is provided by the absolute irradiance measurements of Willstrop[49] using a 1P21 photomultiplier. His data for the star HD20766 were recalculated by Labs and Neckel,[50] who showed them to be in extraordinarily close agreement with their own curve for the sun, namely within 5 per cent between 400 and 650 nm.

3

The Earth's Atmosphere

This most excellent canopy, the air, look you, this brave o'erhanging
firmament, this majestical roof fretted with golden fire, why, it
appears no other thing to me but a foul and pestilent congregation
of vapours.

WILLIAM SHAKESPEARE, *Hamlet*, Act 2, scene 2.

We now consider the effect of our atmosphere on the solar radiation
described in Chapter 2. The spectral extremes, important as small
energy contributions when the total solar radiation is to be calculated,
have a very different significance in their reactions with the atmosphere.
So powerful are the effects that the daylight we enjoy is radically
altered from the incident flux in the extent, shape and power content
of the spectrum, transmitted with consequent changes in the composition and temperature structure of the atmosphere, in the temperature
of the earth's surface, and indirectly in climatic conditions on earth.
Without these interactions between radiation and atmosphere, vegetable and animal life on earth could not assume their present physical
forms.

From the direction of the sun, but outside the atmosphere, the earth
is seen to reflect diffusely a fraction of the incident radiation. This
fraction, the 'albedo', is roughly 40 per cent. The emission is reflected
from clouds, from the earth's surface and the sea, and in part scattered
in the atmosphere. Of the remaining inward flux about 45 per cent
reaches the earth's surface to be absorbed, and about 15 per cent is
absorbed in the atmosphere. This 60 per cent of the incident power is
all re-radiated into space at infra-red wavelengths differing in spectral
distribution from the original solar spectrum. The earth remains very
accurately in balance with respect to energy input and output, and its
average conditions and temperature must have remained thus for
centuries. It is estimated that an excess of only 0·1 per cent of incident
radiation above that radiated would raise the temperature of the whole

atmosphere by 6 K in a year. In the following we consider first the interaction between radiation and the highest levels of the atmosphere, then the effects in the lower atmosphere and at the earth's surface. In the next chapter the scattering effects by which radiation is merely redirected are described.

As in Chapter 2, the values of heights, temperatures and some wavelengths in the following surveys are intended to be illustrative rather than the latest definitive values. They omit changes with earth latitude, between day and night conditions, and a large variability in recorded data due to the fact that experiments cannot cover the whole vast field available. Terminology also is not entirely agreed between the physicists, meteorologists and astrophysicists whose interests meet here.

THE UPPER ATMOSPHERE

The earth's atmosphere scarcely varies in composition up to a height of 100 km, where the pressure is 2×10^{-4} torr, except for water vapour which is almost entirely contained in the lowest 10 km and is far from uniformly distributed. At 250 km the pressure is reduced by another factor of 10^3. Here diffusion under gravity is beginning to separate the constituents, whose molecules now have mean free paths of hundreds of metres and do not collide often enough to establish a Maxwellian distribution of velocities, or to reach equilibrium in the chemical reactions which are possible. Below this level occur most of the reactions between the gases and solar radiation. Above it the temperature is at a fairly uniform level of 1000 to 1500 K (according to the phase of solar activity at the time), persisting for many hundred km upwards. This temperature is reached by a gradual rise from a minimum of about 170 K at the 'mesopause' (85 km) through the 'thermosphere' to the high value mentioned. The rise is due to absorption in this zone of the X-ray and very short wavelength ultra-violet components of the incident flux. These photons are sufficiently energetic to ionize the gases, chiefly to O_2^+, N_2^+ and O^+. The latter is the predominant component by reason of extensive prior dissociation of oxygen molecules. Nitrogen atoms are much less numerous because of the higher energy required for dissociation, 9·8 eV compared with 5·1 eV for oxygen, or wavelength limits of 127 and 243 nm respectively.

Since the power in the solar spectrum increases steadily with increasing wavelength over this range, oxygen molecules suffer far more dissociation than do nitrogen molecules.

The by-product of these ionizations is a concentration of free electrons, of no importance to the daylight reaching the earth, but of very great significance in making radio communication possible in normal times and vulnerable during excess solar activity. Several named levels between 70 and 300 km in this 'ionosphere' have been attributed to ionization by photons from different parts of the solar spectrum. Exploration of this atmospheric zone continues by rocket and satellite, and by radio pulses from the ground ('ionosondes').[51]

Some extra production of ions results from the 'solar wind', the constant outward streaming of the sun's higher layers, in effect a stream of protons and electrons entering the atmosphere at about 300 km/s.[52] This makes another contribution to the high temperature gradient in the thermosphere.

OZONE IN THE ATMOSPHERE

Before solar radiation at the short ultra-violet wavelengths had been directly observed its presence was suspected, and this was closely involved with the detection of ozone in the atmosphere. Laboratory investigations on ozone by Chappuis (1880) had shown its weak visible absorption bands, and more significantly Hartley (1881) measured the complex of strong absorption bands in the ultra-violet spectrum. Hartley came to the conclusion that ozone was a normal constituent of the atmosphere, and thought that the sharp ultra-violet limit to the solar spectrum was due to absorption by this ozone. In 1913 Fabry and Buisson, using their own measurements on ozone absorption, showed that the observed spectrum limit for direct sunlight, if due to ozone, could be expressed in terms of the zenith angle of the sun z, which determines the atmospheric pathlength traversed by the light.[53] Cornu had previously found a formula for the minimum observable wavelength of the spectrum λ_{min} from measurements on his photographs. If solar altitude h is used instead of zenith angle z, the formula is equivalent to

$$\sin h = \cos z = M \exp\left[-m(\lambda_{min}-\lambda_0]\right.$$

where M, m and λ_0 are constants. With $M = 0.49$, $\lambda_0 = 300$ nm, m was altered according to the progress of Cornu's experiments to the eventual value of 0.11256, found from measurements on high altitude photographs.[54] This reduces his formula, in common logarithms and ångström units, to

$$\lambda_{min} = \text{Constant} - 205 \log \cos z.$$

Fabry and Buisson, considering the reduction of a photographic image to the limit of detection, found

$$\lambda_{min} = \text{Constant} - 177 \log \cos z$$

and claimed even closer agreement between the numerical factors if variable atmospheric attenuation with wavelength was also taken into account. This near agreement with Cornu's work supported the hypothesis that absorption by ozone limited the spectrum. By 1921 Fabry and Buisson had made extensive measurements of the ozone content of the atmosphere, using photography to provide absorption values at the wavelengths affected.[55] Since then knowledge of the quantity and distribution of ozone has become indispensable in meteorology, with a large literature.[56]

An explanation of the production of ozone from atomic oxygen was given in detail by Chapman in 1930.[57] The exact mechanism is still discussed because of the large number of possible reactions, and a lack of decisive evidence for alternative paths. Incidentally, the prediction made at this time of the predominant dissociation of molecular to atomic oxygen at about 100 km, was not verified directly until post-war rocket work began. Undoubtedly several processes of formation and decomposition interact to maintain the permanent ozone concentration. The photochemical processes have been the most closely examined and, where possible, subjected to experiment. Non-equilibrium conditions, disturbances by atmospheric turbulence and temperature gradients, and gravitational diffusion in the more rarified layers complicate this question.

The primary photochemical reaction is the dissociation of O_2 to $2O$ which requires a minimum of 5.1 eV, or a photon of maximum wavelength 243 nm. Absorption occurs in the Herzberg system which shows bands from 259.5 nm with a continuum below 242.9 nm. The Schumann–Runge system of strong absorption bands from 190.2 nm down to the continuum below 175·9 nm also takes a prominent part in the dissociation.

The different photon energies result in different configurations of oxygen atoms, two at the ^3P ground state by the smaller Herzberg quanta, one each of ^3P and the metastable ^1D by the Schumann–Runge system. Different reactions may then follow. These absorptions occur mostly above 50 km, in the lowest zone of what is conventionally known as the upper atmosphere. The amount of dissociation is extensive, having regard to the low molecular densities at this height. Above 100 km the number of oxygen atoms equals that of oxygen molecules, and exceeds it more and more with increasing height. The molecular oxygen is replenished from below by turbulent mixing, while part of the atomic oxygen moves upwards to suffer some ionization, and part downwards to react with oxygen molecules. This produces ozone, and the reaction must occur in the presence of a third body to remove the heat liberated. The typical reaction is

$$O + O_2 + M = O_3 + M$$

and occurs largely in a zone at 10 to 40 km above the earth (the 'ozonosphere'). Some of the oxygen atoms are lost by recombination in the presence of a third body. Ozone formation also occurs by the reaction of excited oxygen molecules:

$$O_2^* + O_2 = O_3 + O.$$

The ozone is continually decomposed by the absorption of photons between 200 and about 350 nm, most favourably at the peak absorption of the Hartley bands, 255 nm, and to a smaller extent in their weak extension to longer wavelengths, the Huggins band, which has a peak near 310 nm and can be recognized to 350 nm. There is also some absorption in the Chappuis bands, with a broad maximum near 602 nm, and in an infra-red band at 9·6 μm. The energy degraded to heat in these interactions reverses the normal fall of atmospheric temperature with height through the 'troposphere' (6·5 K/km). From the 'tropopause' at 10 to 15 km there ensues a rise of temperature through the 'stratosphere' to a maximum of 270 K at the 'stratopause' (50 km). Most of the residual ozone is in this zone of reversed temperature gradient, its removal proceeding by reactions such as:

$$O_3 = O_2 + O$$
$$O + O_3 = 2O_2$$
$$2O_3 = 3O_2*$$
$$O_2* + O_3 = 2O_2 + O.$$

Reaction with reducing agents polluting the lower atmosphere, and then by vegetation at ground level, reduces the concentration to a mere trace. This was detected by Houzeau in 1858,[58] long before these photochemical reactions were thought of. The concentration near the ground is tending to rise in cities due to the contents of motor car exhaust gases.

Ozone formation is likely to follow ionization in which solar protons, cosmic rays, X-rays and the constant electrostatic discharges in the atmosphere take part. Other possible reactions such as

$$O_2*^+ + O_2 = O_3 + O$$

have been proposed.[59] It has been suggested that a large fraction of the ozone is formed by lightning discharges. A recent account of the chemistry considered the effects of trace compounds of H, N and Cl.[709]

In the permanent band of ozone-bearing atmosphere between 10 and 50 km height the maximum concentration is 10 parts per million, reached at about 25 km. The total amount is variable in the northern hemisphere, equal to a layer from about 2·5 to 4·0 mm thick at NTP. The Dobson unit, named after a famous worker in this field, is 10^{-2} mm of ozone at NTP; in the older literature the centimetre unit is customary. The distribution of ozone with height has been measured by rocket, and compares reasonably with the calculated yields of photochemical equations if reactions of water vapour are included.[60]

This quantity is readily measured in total and in distribution, serving as a tracer for air circulation and movements in the upper atmosphere, apart from its relation to solar activity. Quantitative determination is now a standard meteorological practice, made by measurements of atmospheric transmission in the near ultra-violet spectrum, comparing solar flux at non-absorbing and weakly absorbing wavelengths, for example, 310 and 325 nm. Dobson has recounted the history of this development, with the more complicated system of ultra-violet wavelengths now used to eliminate the effects of atmospheric scattering.[61] New units and details of measurement from the work of the International Geophysical Year were described by Dütsch.[62] Wardle, Walshaw and Wormell made a spectrograph for ozone measurement, and this could be used in moonlight, or starlight with photographic recording.[63]

Variations in the total ozone content of the atmosphere are continuous, and regular seasonal and latitude patterns are known.[64, 65] Though present only as a tenuous layer, it amounts to 3 to 4×10^9 tonnes in a total atmosphere mass of about 5×10^{15} tonnes. The most obvious effect of its presence is the deep blue colour of the zenith sky at twilight (Chapter 15).

BELOW THE OZONE LAYER

The persisting effect of the atmosphere is to set up a permanent barrier to the passage of radiation at wavelengths below about 290 nm by the ozone formed within it. Any photons of wavelengths shorter than 200 nm which may have penetrated from higher levels will be stopped by the very steeply rising absorption in the Hartley bands. By this filtering action the atmosphere has determined the types of life evolving on the earth. Its other functions are equally vital: to provide oxygen and carbon dioxide for animal and vegetable life; and to prevent temperature extremes between day and night by the absorption of radiation from the earth, with water vapour playing a major role. The present source of the oxygen is photosynthesis. Its earliest appearance was discussed by Berkner and Marshall,[66] the main problem being to account for enough oxygen, and consequently for enough ozone, to screen plant life on land from the ultra-violet radiation. Photosynthesis must have begun under the protection of a sufficient depth of water when the atmosphere had 1% or less of its present oxygen content. The subject has been treated by several other authors.[67, 68]

Short wavelength ultra-violet radiation causes other reactions in the upper atmosphere. They include the dissociation and ionization of nitrogen, the dissociation of carbon dioxide to the monoxide and oxygen atoms, and of water to H and OH, the reaction of methane with oxygen, and the ionization of nitrogen dioxide which is another rare constituent at high levels. These reactions have little importance as far as daylight is concerned. However, ultra-violet radiation which penetrates to the earth's surface causes further photochemical reactions when the atmosphere contains local concentrations of nitrogen dioxide, as in motor car exhaust gases. These gases also contain hydrocarbons and give rise to irritant aldehydes and other organic compounds, and by reactions involving nitric oxide produce ozone at concentrations far above normal. Together these constitute 'photochemical smog' as in Los Angeles. Several mechanisms have been suggested among the many possible reactions of the pollutants, oxygen and minor constituents of the atmosphere including SO_2, NH_3, NO, N_2O, CH_4, CO.[69-71] Unusually high concentrations of ozone (over 100 parts per billion or several times the normal value) observed in London during sunny weather in 1972[72] suggest that similar smog conditions are possible in other cities with the growth of pollution by car exhaust gases.

THE LIMIT OF ATMOSPHERIC TRANSMISSION

The exact wavelength limit has been investigated for a long time, even though it has no fundamental importance and depends on the elevation of the observing site, and on the ozone content and turbidity of the atmosphere. Cornu and Götz have been the main interested parties, each for a number of years. Cornu's formula relating λ_{min} to h has been discussed, with the ingenious but somewhat circumstantial proof arising from it and attributing the absorption to ozone. Cornu elaborated the formula by a term introducing the elevation, based on a decrease of 1 nm in λ_{min} for a rise of 868 m. This value was confirmed by the study of other spectrograph plates exposed at high levels by Simony.[73] Cornu believed the atmosphere to be responsible for the cut-off, though not by its water vapour content, nor of course by ozone which had not been established as the cause at that time. Abbot, Fowle and Aldrich suggested the cause to be absorption in the solar envelope,[74] being misled by the near constancy of λ_{min} reported by Miethe and Lehmann for very different elevations and geographical sites.[75]

Table 1 includes most of the records for λ_{min}. They were all for direct sunlight except where stated otherwise. Nearly all were measured by photography, with Cornu's experiments unique in using wet collodion plates.

Different instrumental and photographic sensitivities, and the presence of scattered light in spectrometers, have no doubt affected some of these measurements. Götz discussed methods and results in several papers.[81–83] More recent measurements in the U.S.S.R. found λ_{min} near 290 nm, and the variation with solar altitude $\lambda_{min} = 290 - 25 \cdot 8 \log \sin h$.[84] Though the height of the absorbing ozone layer should cause little change of λ_{min} with altitude of the observing site, elevation

TABLE 1: Minimum wavelengths observed in the solar spectrum.

Author	Year	Place	Altitude (m)	λ_{min} (nm)	Reference
Hoelper	1928	Norderney	~ 0	294·7	76
Götz	1929	Arosa	1860	298·9[a]	82
		Spitzbergen	23	294·5	130
Hoelper	1926	Riezlern	1150	294·3	227
Kechlibarov and Andreychin	1962	Sofia	590	294	316
Dorno	1909	Davos	1590	293·9	243
Cornu	1879	Riffelberg	2570	293·2	54
	1878	Courtenay	170	293; 294·8[b]	80, 73
Hirt *et al.*	1957	Stamford, Conn.	~ 0	293[c]	320
Simony	1888	Tenerife	3700	292·2	73
Pettit	1924–1931	Mount Wilson	1727	290; 295[d]	233
	1934–1939	Mount Wilson	1727	292	234
Miethe and Lehmann	1908	Berlin	50	291·3	75
		Aswan	116	291·6	
		Zermatt	1620	291·3	
		Gornergrat	3136	291·1	
		Monte Rosa	4559	291·2	
Abbot *et al.*	1909	Mount Whitney	4420	290[e]	74
Wigand	1912	Balloon	9000	289·6; 289·4[d]	77, 81
Fabry and Buisson	1920	Marseille	~ 0	288·5	78
Götz and Casparis	1941	Jungfraujoch	3600	287·7	83
Regener and Regener	1934	Balloon	31000	287·5	79
Götz	1930	Arosa	1860	286·3	82

(a) For moonlight. (b) Mean of many observations. (c) By photomultiplier. (d) For skylight.
(e) By bolometer: an estimated limit.

above sea level has a large effect on shorter wavelength radiation when h is low; for example, sky radiance at 297·5 nm is doubled from zero to 6 km elevation if $h \leqslant 15°$. The effect of ozone concentration is also important. Calculations on direct solar radiation show that a reduction of ozone content to half the normal level will extend the wavelength limit by about 5 nm.[85] The minimum found in this way was 281·5 nm for $h = 60°$ at the equator, assuming a minimum detectable irradiance for spectrophotometry of 10^{-7} W/m² nm, and 50 per cent of the normal ozone. Other calculated values for normal ozone were a few nm greater than the minimum shown in Table 1.

Large variations of flux occur near the limiting wavelength. Being changes in a small absolute proportion of the radiation they are of therapeutic rather than physical interest (Chapter 17). One effect of the curtailment of the terrestrial daylight spectrum is the exclusion of the strong Mg II absorptions at 279·6 and 280·2 nm. These are even more intense than the prominent Ca II absorptions at 393·4 and 396·8 nm. This freedom from radiation below 290 nm does not mean that this radiation will not penetrate the lower atmosphere if generated where there is no ozone. Nazarova reported that the maximum optical range for horizontal propagation near the ground occurs in the region of 242·4 to 290 nm.[86]

ABSORPTION OF INFRA-RED RADIATION IN THE ATMOSPHERE

At the other end of the visible spectrum the atmosphere has potent effects on the incident radiation. The first stage of the process is absorption by ozone, carbon dioxide and water, which are minor constituents of the atmosphere only in quantity. The absorptions are calculable to some accuracy from the known bands and concentrations of the three components.

Ozone absorbs in a moderately narrow band centred at 9·6 μm, with minor bands at 4·7 and 14·2 μm. Its absorption is far exceeded by those of carbon dioxide and water.

Carbon dioxide absorbs powerfully in narrow bands at 2·7 and 4·3 μm, and again from 13 to 17 μm. At the peaks of the first two of these bands, a pathlength of 1 km in air at sea level containing a normal concentration (0·033 per cent CO_2 by volume) results in absorptions of 58 and 100 per cent respectively.

Water vapour, the chief barrier to atmospheric transparency in the infra-red, is characterized by strong bands centred on 1·4, 1·85, 2·7 and 6·3 μm and a wide rotational band from about 14 μm to 1 mm.

Considering again a 1 km path at sea level, if the air is saturated with water vapour at 0 °C the absorptions in the first four bands up to 6·3 μm are respectively 99 per cent at the peak, 28 per cent at the peak, 100 per cent between 2·6 and 2·9 μm, 100 per cent between 5·5 and 7 μm. The humidity corresponds to nearly 5 mm of precipitable water in the 1 km path: this quantity is normally measured for a vertical column through the whole atmosphere, when 5 mm is about half the value common at sea level, and 30 mm would cause very humid conditions.

The vertical temperature gradient in the atmosphere reduces the water vapour content to near zero at about 15 km elevation. All the powerful absorptive effects occur below this level. The temperature rise in the stratosphere allows an appreciable vapour pressure of water to remain at that level, and absorption bands have been revealed in solar spectra recorded on high level flights. Stauffer and Strong estimated the precipitable water above 11 km to be 0·023 mm[87] and independently Thekaekara found a value of 0·028 mm at 11·6 km.[88, 89]

Beyond the wide rotational band the atmosphere becomes transparent again, but not until millimetre wavelengths are reached. By a Golay detector Sinton was able to explore this region of the sun's emission and also that from the moon.[90]

Herzing adapted the measurement procedures for ozone to the infrared bands of water vapour (Chapter 7). Other absorption bands for the rarer constituents, nitrous oxide, methane and carbon monoxide, have proved useful in work on radiation in the upper atmosphere and the airglow (Chapter 15).

Liquid water in the form of clouds has several different effects on the atmosphere and on incident light. Temperature changes are caused by the heat liberated in condensation, light absorption is increased in the infra-red region, and the planetary albedo is increased by reflection from the upper surfaces.

Recent spectral atlases record the complexity of atmospheric infrared absorption, which includes many hundreds of lines apart from the main bands so far described. Curcio and his colleagues covered the visible spectrum as well, with separate sections for 440 to 550 nm, 540 to 852 nm, and 851 nm to 1·16 μm, measured on 16 and 27 km paths over water and near sea level.[91-93] Bradford, Farmer and Todd collated information on the 3·5 to 5·5 μm region and extrapolated it up to altitudes of 30 km.[94, 95] Delbouille began measurements in 1958 at the Jungfraujoch for a high resolution atlas, including some ultra-violet and visible sections of the spectrum and the infra-red from 750 nm to 1·2 μm.[96-98]

Though the absorptions discussed in this chapter have their largest effects in the ultra-violet and infra-red regions of the spectra, all well-resolved spectra of the visible region show small bands due to water vapour, and some due to molecular oxygen. The approximate positions of the bands are at 594, 652 and 723 nm for water vapour, and 629, 688 and 759 nm for oxygen. The last is a very strong absorption. The photometric consequence of atmospheric absorptions is that the ratio of illuminance to total irradiance under both clear and overcast skies is increased to the level of 100 to 120 lm/W, or even higher if skylight is measured separately.[27, 99] These values are much greater than the extraterrestrial luminous efficacy (Chapter 2), as a comparison of curve A with D and E in Fig. 1 confirms.

A more extensive discussion of absorption by the atmospheric gases has been given by Goody,[100] and useful tables and diagrams by Allen.[101]

INFRA-RED EMISSION IN THE ATMOSPHERE

Absorption of incident infra-red radiation in the atmosphere is followed by its re-radiation. The sources of radiation are now at terrestrial, not solar temperatures and the spectral power distribution is therefore quite different. Clouds, atmosphere and the earth's surface emit as full radiators or, perhaps more accurately, as grey bodies of very nearly unit emissivity, and at temperatures near 290 K. The peak emission of a full radiator at 290 K is 10 μm, and an emission band of the expected shape has been demonstrated by Bell *et al.* from the horizon of a clear night sky at high altitude.[102] Between 8 and 12 μm the atmosphere is fairly transparent except for the small ozone absorption band at 9·6 μm, and consequently some radiation can pass outwards to balance part of that received by the earth. Elsewhere in this band of secondary radiation some absorption by water vapour and carbon dioxide occurs, causing a rise of temperature and increased radiation. The downward part of this warms the earth's surface a little more. This is the 'green-house effect', or the trapping of radiation by conversion to wavelengths not transmissible by the glass of the greenhouse; in this case the water vapour and carbon dioxide in the atmosphere represent the glass.

The warmed atmosphere radiates upwards as well, and eventually all the 15 per cent of solar flux primarily absorbed in the atmosphere, and the 45 per cent absorbed by the earth's surface, are radiated into outer space. By comparison the heat flow from the earth's hot interior is negligible, being about 0·05 W/m² or one ten-thousandth of the solar power absorbed.

The value of 290 K quoted is essentially an observed value, including the greenhouse effect, otherwise the effective mean temperature of the earth's surface would be about 245 K according to a simple calculation. If the earth is regarded as a thermal radiator of temperature T, radius R and albedo A, then the solar input is $\pi R^2 S(1-A)$, where S is the solar constant. This is equated to the output according to the Stefan–Boltzmann law,

$$\pi R^2 S(1-A) = 4\pi R^2 \sigma T^4$$

where

$$\sigma = 5\cdot67 \times 10^{-8} \text{ W/m}^2 \text{ K}^4.$$

If $S = 1\cdot35$ kW/m² and $A = 0\cdot4$, then $T = 244$ K. This calculation was discussed in more detail by Jastrow and Rasool.[103] The radiation balance of the earth is subject to so many disturbing interactions of the surface, clouds, atmosphere and incident radiation that no simple theory can be expected to give an accurate quantitative statement. The temperatures quoted are means for the whole earth, and local variations are large, as in the Antarctic where the albedo is very high and water vapour concentration low. The calculations are also made uncertain by the doubtful value of the albedo, quoted between 0·35 and 0·45 by most authors, and as low as 0·3 by Raschke and Pasternak from the measurements made in a recent satellite experiment.[104]

An analysis of the processes in the earth's radiation balance was given by Möller in a review of the optics of the lower atmosphere.[105] Hunten provided a complementary paper on the upper atmosphere.[106] Radiation effects in the atmosphere are treated at length by Kondratiev.[84] A shorter account with details of atmospheric gases and climatic effects is given by Barry and Chorley.[107]

In the next chapter light scattering processes in the atmosphere are considered.

4

Scattering of Light
in the Atmosphere

The exhalations whizzing in the air
Give so much light that I may read by them.
WILLIAM SHAKESPEARE, *Julius Caesar*, Act 2, scene 1.

RAYLEIGH SCATTERING

Radiation from the sun is affected by scattering without change of wavelength, as well as by reflection and absorption in the atmosphere. The first type of scattering to receive mathematical treatment was that due to very small particles. In 1871 Rayleigh showed that, for spherical particles very small compared with the wavelengths of visible light, the amount of scattering depends on the inverse fourth power of the wavelength, and on the direction of scattering. Polarization also occurs and was closely examined by Rayleigh.[108] The theory is particularly applicable to the ultra-violet and visible regions of the daylight spectrum, where the effects are easily observed in the comparative absence of photochemical and absorption processes.

It should be noted that, though Rayleigh scattering is now understood to apply very well to the gas molecules and atoms in the atmosphere, this was not assumed by Rayleigh. While discounting the current theories of scattering by water particles, he was unsure of how the scattering occurred, as it evidently did in the blue sky. His only suggestion, not a very firm one, was that minute salt particles suspended in the air might be responsible.

Repeated attempts have been made, with reasonable but never perfect success, to fit theory and observation together. A number of circumstances prevented the exact confirmation of Rayleigh's theory. It applied only to a single scattering process, whereas multiple scattering occurs in the atmosphere. In addition, sunlight is scattered by

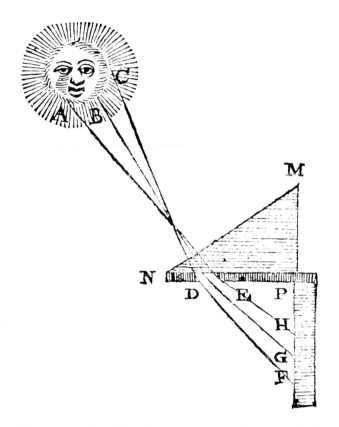

PLATE 1. The spectrum of sunlight (Descartes, *Discours de la Méthode*). (See p. 6.)

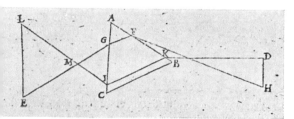

PLATE 2 (above). Production of colours by refraction Marci, *Thaumantias*). (See p. 6.)

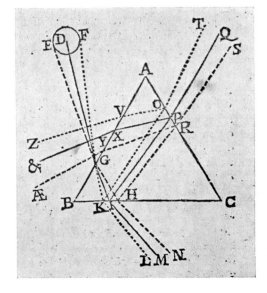

PLATE 3 (right). Refraction by a prism (Grimaldi, *Physico-nathesis de Lumine, Coloribus et Iride*). (See p. 7.)

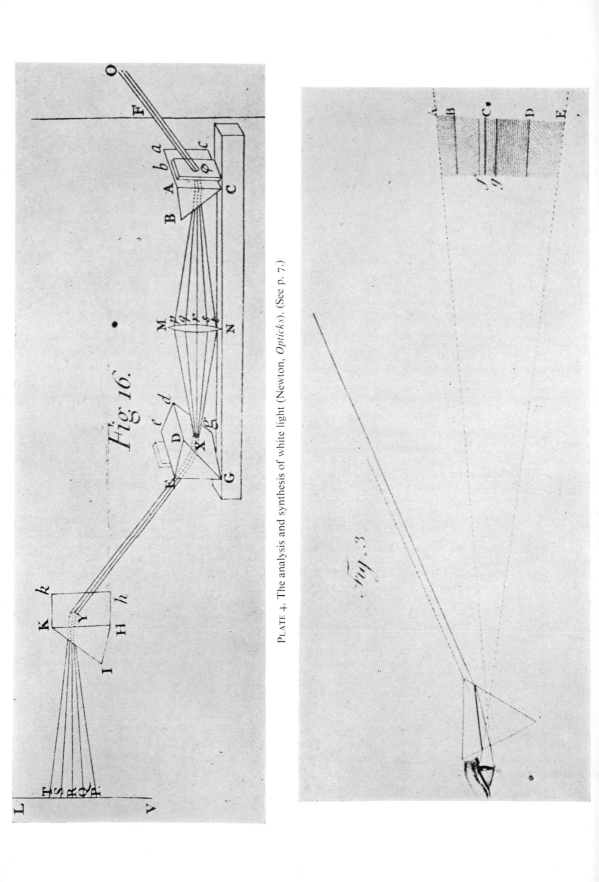

PLATE 4. The analysis and synthesis of white light (Newton, *Opticks*). (See p. 7.)

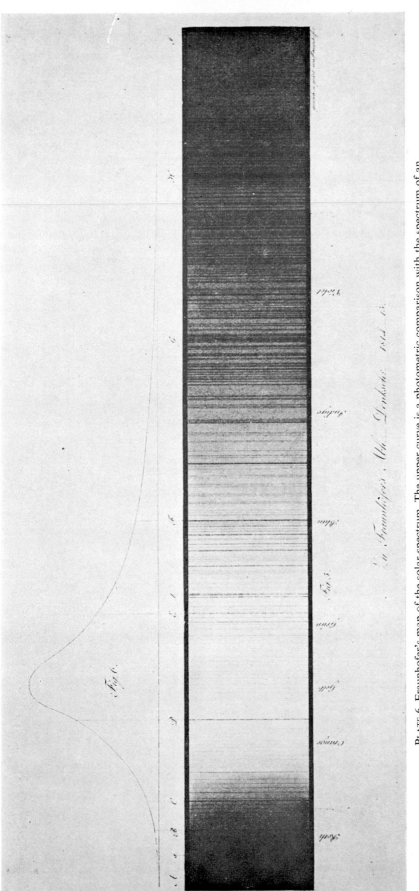

PLATE 6. Fraunhofer's map of the solar spectrum. The upper curve is a photometric comparison with the spectrum of an oil lamp (*Denkschr. Königl. Akad. Wiss. München*). (See p. 9.)

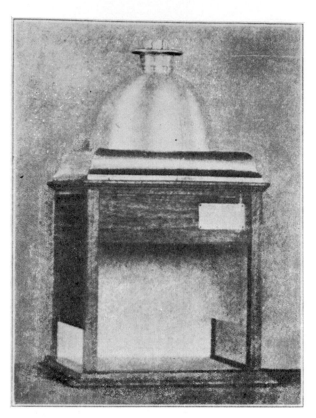

PLATE 7. Colour-matching booth with artificial daylight provided by gas mantle and colour filters (Ives, *J. Franklin Inst.*). (See p. 273.)

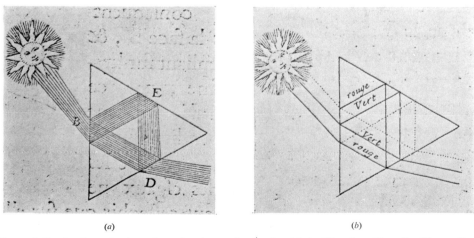

(a)   (b)

PLATE 8. Production of colours by refraction and reflection (de la Chambre, *Nouvelles Observations et Coniectures sur l'Iris*). (See p. 7.)

suspended particles as large as, or larger than, the wavelengths of visible light, where Rayleigh's theory does not apply. In some of the early work it appears that the spectral distribution curve for the blue sky was expected to show an inverse fourth power distribution, though this must depend on the spectrum of the incident sunlight which was not well known at the time. The reports of these investigators are worth reading, but are sometimes hardly worth the effort of trying to unravel their experiments in the hope of discovering useful information in modern terms.

The blue colour of the sky, considered more fully in Chapter 15, is the most obvious result of Rayleigh scattering. Because the scattering occurs in all directions, some of the incident sunlight is redirected outwards from the atmosphere, forming one component of the earth's albedo. It produces the bluish halo seen from outside the atmosphere, making earthlight on average more blue than moonlight where no such scattering occurs.

The quantitative aspect of scattering is conveniently expressed as an extinction or attenuation of incident light according to the principle first established by Bouguer.[109] He expressed it in this form: when light traverses different thicknesses of the same body, the difference of the logarithms of the quantities of light (incident and emergent) is always in the same ratio to the thickness of the medium traversed. Both types of logarithm are used in the mathematical statement, but usually it is expressed as:

$$I/I_0 = \exp\left(-\alpha l\right)$$

for a thickness l and an absorption coefficient α. This law was also discovered by Lambert, expressed by him more mathematically, and is often given his name. Bouguer's discovery was the earlier, and his complete work was published posthumously in the same year as Lambert's statement of the law (1760).

The coefficient α usually varies with wavelength, and for Rayleigh scattering is proportional to $1/\lambda^4$. Necessary modifications to the theory appear later, but its use in a simple form has been most important in calculations of the solar constant (Chapter 5). For this the logarithm of H_λ, the relative spectral power distribution of sunlight, is plotted versus the air mass m, or sec z where z is the sun's zenith angle. Each wavelength gives a nearly straight line, the more so for clearer and more stable atmospheres. By extrapolation to $m=0$, the lines give the extraterrestrial values $H_{0\lambda}$ required for solar constant calculations, and their slopes varying with wavelength provide the atmospheric extinction coefficients. It is desirable to extend the lines as far as possible. Some observations have been made up to $m = 10$ or more, where z is nearly 90° and sec z is no longer nearly proportional to m owing to refraction in the air. Correction factors are used in such cases.[110, 111] The lines are called Bouguer lines, or in the U.S.A. Langley lines. Each shows the total of all extinction processes occurring at a given wavelength, not merely the Rayleigh scattering. The additional processes required the simple Rayleigh theory to be modified.

Later chapters give examples of the use of the linear relation between H_λ and m: Fabry and Buisson, and Plaskett (Chapter 7); Stair, and Dunkelman and Scolnik, and Peyturaux (Chapter 9); Bener (Chapter 10); Stair and Ellis, and Thekaekara, and Guttmann (Chapter 11). Peyturaux suggested recently that the method might be unreliable because of systematic variations in the atmosphere at the observing site (Chapter 11); but, if so, this should not have affected simultaneously all the different sites used by Langley, Abbot and their colleagues at the Smithsonian Astrophysical Observatory (SAO) who used this method more than anyone else (Chapters 5, 7 to 9).

Modifications to the Rayleigh theory began soon after its appearance. From his many observations on the diffused light of the sky Crova thought that there might be a separation of particles in the atmosphere by their size and rate of fall.[112] The expression for the amount of scattered light would then contain a term which was not merely the number of particles N of radius r per unit volume, but $N/\sqrt{r}$, or with certain assumptions $N/\sqrt{\lambda}$. The Rayleigh factor became $\lambda^{-4.5}$ instead of λ^{-4}, and this agreed moderately well with some of Crova's observations made over the limited spectral range of 510 to 635 nm. He quoted measurements by Vogel and by Rayleigh showing remarkably close agreement with the λ^{-4} law.

One of the nearest experimental approximations to Rayleigh scattering, measured under very careful control, occurred in the high altitude measurements of Dunkelman and Scolnik (Chapter 9). By this time the application of Rayleigh theory had been transformed by Chandrasekhar's study of multiple scattering, and simplified by tables for calculation.[113–115]

Another modification was introduced to allow for the refractive dispersion of air.[116] The Rayleigh extinction coefficient, k_R for air mass $m = 1$, is given by

$$k_R = 4{\cdot}3 \times 10^4 (n-1)^2 \lambda^{-4}$$

where the refractive index n varies with λ. This may be simplified to[117]

$$k_R = 3{\cdot}60 \times 10^{-3} \lambda^{-4 \cdot 05}.$$

If in addition the non-spherical shape of atmospheric molecules is included, the approximate formula becomes

$$k_R = 3{\cdot}86 \times 10^{-3} \lambda^{-4 \cdot 05}.$$

In these formulae λ is measured in μm.

Other modifications of the theory for simpler calculations were discussed by Penndorf.[118] The formula most used is that for the volume scattering coefficient β, which for standard atmospheric pressure and the normal temperature range is given approximately by

$$\beta = \frac{32\pi^3 (n-1)^2}{3\lambda^4 N} \frac{6 + 3\rho}{6 - 7\rho}$$

where ρ is a depolarization factor and N the number of gas molecules per cm^3. This appearance of molecular concentration in the formula suggests that this variable could be determined from the scattering coefficient measured directly. Fowle showed that this did in fact give a value of the Loschmidt number very close to the value accepted at that time, $2{\cdot}705 \times 10^{19}$.[119]

NON-RAYLEIGH SCATTERING

Crova's work preceded the development of a theory of scattering by larger particles. Even when this was formulated there remained considerable reliance on the Rayleigh theory alone, though with difficulties. About 1930 Duclaux and Gindre found that they could not reconcile their observations on atmospheric transmission and sky colours with Rayleigh theory because the observed extinction coefficients were too low.[120, 121] Their measurements included some made on the summit of Mont Blanc. They used narrow-band colour filters, and from the calculated luminances of rocks, snow and sky determined contrast values which should have been insufficient to allow the particular features to be seen. In practice there was sufficient contrast for visibility. The conclusion followed that the atmospheric absorption included two processes of the type $1/\lambda^a$. In one $a=0$, an absorption independent of wavelength. In the second $a \geqslant 2.5$ producing a 'Tyndall fog', named after the physicist who made artificial blue skies from chemical aerosols. Duclaux and Gindre found that their observations agreed with exponents between 0·68 and 2·96 on different days, but a single exponent would not fit the whole spectrum. Zettwuch (Chapter 7) had similar difficulties. Clearly the position was unsatisfactory, but a solution was at hand, in the two invaluable papers by A. Ångström.

It is certain that the atmosphere is never free from changing clouds of suspended particles, solid or liquid, even at high altitudes. Their presence accounts for the zodiacal light (Chapter 15), and similar clouds must exist around the sun causing the F-corona light (Chapter 2). As far as terrestrial daylight is concerned, the existence of these aerosols has always been a disturbing factor in measurements which involve atmospheric transmission. It is often difficult to detect them, and their constant variation is a major cause of the variability of daylight. Some idea of their relative importance was shown by Tousey and Hulburt, who found that attenuation of sunlight in the atmosphere above a height of 3 km was 35 per cent above that calculated for Rayleigh scattering.[122] Dunkelman and Scolnik found their best day's observations at 2·4 km elevation were still 30 per cent above the theoretical prediction. Kondratiev *et al.* considered that extinction by aerosols exceeded that by molecular scattering and water vapour absorption above 30 km.[123] Coulson's experiments on smog in Los Angeles showed that forward scattering by aerosol particles, producing a solar aureole, was probably always present to some extent, even in very clear conditions;[124] an example of high level scattering appears in 'noctilucent' clouds whose spectral emission with a maximum at 480 nm has been attributed to an aerosol of ice particles at a height of about 80 km.[125]

The mathematical treatment of scattering by larger spherical particles was first effective in the hands of Mie.[126] The formulae he developed from experiments on gold sols have been used to some extent in application to daylight, particularly for the ultra-violet component, as in Götz and Schönmann's measurements at Arosa (Chapter 7). More commonly an empirical approach has been found sufficient. For example, Fowle expressed extinction by the formula

$$I/I_0 = \exp\left[-(a_1 + a_2)m\right]$$

where a_1 referred to molecular scattering and a_2 to other causes, with air mass m. It was realized that, as particles in a cloud or aerosol increased in size, the scattering became less dependent on wavelength as in a jet of cooling steam which scatters light, first blue, then more and more white. The Rayleigh exponent could be considered to change from 4 to zero. Ångström's valuable contribution was in establishing a reasonable way of using a changing exponent for aerosol scattering.

ÅNGSTRÖM'S THEORY

In his first paper A. Ångström discussed other attempts, including K. J. Ångström's (1906), to explain absorption in the atmosphere, which was known to make much greater changes in the incident solar radiation than could be attributed to variation in the solar constant.[127] If the calculated vertical transmission of the atmosphere at wavelength λ is T instead of the observed value U, then according to Ångström

$$U/T = \exp\left(-\beta/\lambda^a\right).$$

The exponent a was found to vary from 1·0 to 1·5 by recalculation of data from the SAO (Chapter 7). In the second paper[128] the extinction at a given wavelength was given in the form:

$$H_\lambda/H_{0\lambda} = q^m \exp\left[-m\beta_0/\lambda^a \exp\left(-\delta h\right)\right]$$

where $q = \exp\left(-a_1\right)$ for Rayleigh scattering at $m = 1$, β_0 is a new turbidity coefficient with λ measured in μm, a is taken as 1·3 for particles predominantly 1 μm in diameter, δ is a constant (0·7), and h is the height above sea level in km. Ångström published many curves showing the calculated decrease of solar spectral irradiance with changes in h, m and β. Variations of β with latitude, season and height were estimated from measurements at a number of other locations. At Washington, D.C. and Stockholm, values of β were 0·158 and 0·112 respectively for the maxima in May or June, and 0·05 and 0·028 for minima in November or December. Davos had much clearer air, with

β varying only from 0·034 to 0·016. The coefficient β has been much used since, though as Ångström admitted, it is not independent of m if water absorption also occurs. It is not the same as the volume scattering coefficient β used above in the discussion of Rayleigh scattering.

Among uncomplicated applications of the theory was that of Elvegard and Sjöstcdt. They attempted to formulate spectral distributions for normal sunlight, including one which clearly separated molecular and aerosol scattering

$$H_\lambda/H_{0\lambda} = \exp\left[-\, a(b/\lambda^4 + c/\lambda^{1\cdot3})\right].$$

Other formulae are discussed in Chapter 8.

Using the theory of aerosols supplementing Rayleigh scattering, Bullrich, de Bary and Möller calculated the colours to be expected in the sky.[129] Their coefficient c_1 for the aerosol component was derived in a different way from Ångström's β, but was related to it. For a number of positions in the sky relative to the sun at different altitudes, and for different values of c_1, the authors obtained values of dominant wavelength, luminance and saturation, and considered the possible effects of multiple scattering and ground reflection (Chapter 9).

A curious example of the failure of the theory of extinction occurred in the course of work by Götz in Spitzbergen.[130] The intention was to correlate ozone concentration and distribution with changes in the ultra-violet part of the daylight spectrum, including the short wavelength limit. Some aspects of this are considered in Chapter 8. Here the interesting feature is that Götz measured total transmission coefficients q in the region of 302 to 326 nm, with q varying from 0·04 to 0·5 in this range. The value of q was the product of three transmission factors, respectively for ozone at the known concentration, for Rayleigh scattering, and for the aerosol concentration, the first two of these being known while the third could be measured. The method worked reasonably for previous measurements at Arosa, where turbidity or the aerosol factor was 0·9 at 323 nm, which is equivalent to $\beta=0\cdot024$ in Ångström's notation. At Spitzbergen, where solar irradiance at a given solar altitude h was greater than at Arosa, this calculation failed because it showed the total transmission to be greater than that allowed by Rayleigh scattering alone. Their conclusion resembled that of Duclaux and Gindre. The measurements in the Spitzbergen work were based on spectrum photographs taken by crossed quartz spectrographs and were therefore somewhat indirect. There are other atmospheric transmission measurements exceeding those allowed by Rayleigh scattering, as in a few of Bener's observations on ultra-violet at Davos in 1966 and on the Weissfluh in 1967 (Chapter 10).

OTHER SCATTERING FORMULAE

Ångström's developments in the theory and calculation of scattering were supplemented in 1949 by extensive work by Schüepp, who discussed earlier types of scattering coefficient and introduced new ones himself.[116] He demonstrated the possibility of determining the Ångström parameters by measurements of solar radiation with a Michelson actinometer in three bands, namely the full range from 300 nm to 5 μm, the range 300 to 625 nm through a Schott RG2 filter, and the range 300 to 525 nm through a Schott RG1 filter. Schüepp's formula included precipitable water content as one variable, and his coefficients were somewhat different from Ångström's. With decadic instead of exponential values, the atmospheric attenuation is $1/\log (k_1 + k_2 + k_3)$, where the Rayleigh coefficient k_1 is proportional to $m/\lambda^{4 \cdot 05}$; k_2 represents aerosol absorption. Then

$$k_2 = m_h B/(2\lambda)^a.$$

The other constants are m, the absolute air mass, m_h, the relative air mass, and B, the turbidity coefficient. a is correlated with size of the acrosol particles and represents the wavelength dependence of k_2. Further conditions apply to k_2 in their treatment: measuring wavelengths in μm, let

$$\lambda' = (0 \cdot 5 \mid 18B).$$

Then if $0 \cdot 3 < \lambda < \lambda'$, k_2 is as stated above. But if $\lambda > \lambda'$,

$$k_2 = m_h B/(2\lambda')^a$$

and is constant. Finally, k_3 represents selective absorption by carbon dioxide, ozone, water vapour and oxygen.

One other method among many based on the same general principles may be outlined. It is due to Linke[131] and was used before

Ångström and Schüepp made their contributions. Linke's turbidity factor T is the relation between the extinction of the real atmosphere under consideration to that of a pure, dry atmosphere of unit thickness ($m=1$). At any given wavelength and air mass m, the extinction factor of the solar radiation is given by:

$$K = \exp\left(-mM-mD-W-A\right)$$

where M is the Rayleigh scattering coefficient for $m=1$, D is the aerosol scattering coefficient, W is the absorption coefficient for water vapour, and A is the absorption coefficient for ozone, carbon dioxide, etc.

If $\bar{K}_R$ is the factor for unit mass of pure dry air, and $\bar{K}$ is that for total extinction for unit mass of the air in question, then

$$\bar{K} = \bar{K}_R \, T.$$

$\bar{K}_R$ is not the same as k_R used above in the discussion of Rayleigh scattering alone. In their fifteen-year-long survey of diffuse (D) and global (G) radiation at Uccle (Brussels), Dogniaux and Doyen[132] quoted IGY values of $\bar{K}_R$ varying from 0·105 at $m=0\cdot5$ to 0·060 at $m=8$. The values of T normally used in these calculations are 1·9 for mountain areas, 2·75 in flat land, and 3·75 in urban areas. In the Uccle survey T was greater than 4 during most of the year. Statistical treatment of the observations gave the empirical relation

$$D/G = (0\cdot051+0\cdot018m)T-0\cdot0016\delta-0\cdot0895$$

where δ is the solar declination in degrees.

With the importance of aerosols well appreciated, assumptions need to be made about their concentration and distribution if standardized daylight distributions are to be calculated from the spectrum of extra-terrestrial sunlight. Moon and Gates both made such assumptions (Chapters 8 and 12 respectively). Elterman proposed a standard atmosphere on similar lines.[133] Foitzik and Lenz calculated extinction and scattering on the assumption of a Gaussian size-distribution of aerosol particles.[134, 135] The problem has been approached in many ways other than those already described, much of the later work being concerned with atmospheric pollution. There have been studies of the effects of aerosols on sky luminance,[136, 137, 273] and of the importance of multiple scattering in such calculations.[138, 139] Aerosols are also a major factor in polarization[145] and in the transfer of radiation between atmosphere and ocean.[486] Other aspects include the origin and quantities of the material;[173, 454] its effects on measurement of the solar constant[165] and in causing unusual celestial appearances;[265, 451–453] and the analysis of earlier observations made by the SAO.[184]

The difficulty of calculating atmospheric transmission is not likely to decrease as pollution by aerosols increases. Comparative measurements at Washington, D.C. of incident solar radiation have shown marked losses during this century, evidently due to increased scattering especially in the infra-red[140]

POLARIZATION OF SCATTERED LIGHT

Both molecular and aerosol scattering cause polarization of the scattered light. It is dependent on the wavelength of the light, the solar altitude, and the angular distance of the observing direction from the sun. Some of the early views and experiments were related by Tyndall in his chapters on the sky, its colour and its simulation.[141] (See Chapter 15.) The Rayleigh scattering theory gives a reasonable prediction of the polarization, as shown for example by Gehrels in a series of observations on clear skies at a height above 2 km.[142] Tousey and Hulbert discussed the dependence of Rayleigh scattering on polarization factors,[122] and Penndorf gave tables for calculation over the range 200 nm to 20 μm in the work previously quoted.[118] There are numerous examples of extensive measurements of skylight polarization and their comparison with calculated values, usually for Rayleigh scattering. Foitzik and Lenz made such experiments,[143, 144] covering a wide spectral range, and Bener used the wavelengths 347, 488 and 533·5 nm in his work at the Weissfluh. In the latter case the amount of polarization was related to angular distance from the sun in good agreement with the theory of a Rayleigh atmosphere. Maximum polarization occurred near 90° from the sun, and minima at different positions near 0° and 180° according to wavelength and h.[145] Nowak used the wavelengths 449, 624 and 844 nm at Mainz, the Jungfraujoch and Maui in Hawaii, with results suggesting the existence of aerosol layers of different particle size.[146] An unusual type of investigation by

Knestrick and Curcio on the radiance of the horizon sky included some polarization measurements (Chapter 10). Gehrels' work covers a number of aspects of polarization; a full account of scattering which includes polarization is found in van de Hulst's book.[147]

In spectral power distribution measurements of the kind discussed in this book, polarization does not usually affect the total radiation flux. It may have slight disturbing effects because of reflection at optical surfaces in measuring instruments. Some few observers have gone to great trouble to eliminate the effects on their measurements (for example Götz and Schönmann, Chapter 8). Most have ignored polarization, some have shown it to produce errors that could be neglected, some have attributed differences between observers to this cause.

In an entirely different context, the presence of polarization in natural daylight is essential to one theory of the origin of optical activity in organic substances, which is typical of living systems.[148]

Fig. 1 gives an approximate picture of the processes discussed in Chapters 2 to 4. Only direct sunlight is depicted. Scattered sunlight is much more variable, as further details of spectroradiometry will show

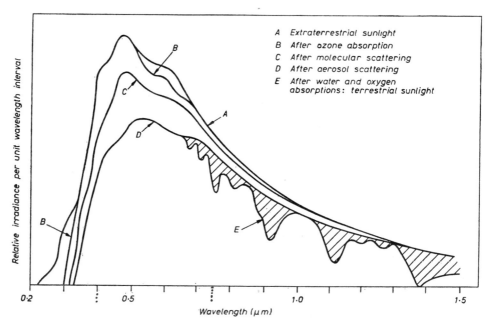

Fig. 1. Successive processes affecting sunlight during penetration of the atmosphere.

in succeeding chapters. It is evident from these curves and the value of the solar constant that the maximum illuminance at sea level from the sun in a clear sky is of the order of 100 kilolux.

Chapter 5 discusses the radiant power content of curve A in Fig. 1. The rest of the book is largely concerned with the ultra-violet and visible regions of curves D and E.

5

The Solar Constant

The world's a scene of changes, and to be Constant, in Nature were inconstancy.

ABRAHAM COWLEY, *Inconstancy*.

The solar constant is defined as the total power input to a surface normal to the sun's direction, outside the atmosphere and at the earth's mean distance from the sun. No symbol has been universally adopted for this constant, but following some other writers, S (without suffix) is used in this book. Others use E_0 or E_{eo}.

The importance of measuring the surface density of solar radiation was recognized before spectroradiometry was established, and it is appropriate to consider the subject before the more detailed discussion of the daylight spectrum. References to individual measurements occur in later chapters.

At first only radiation flux at the earth's surface was measured, a difficult process with the rather insensitive instruments then available, and a determination often vitiated by atmospheric conditions. One of the first instruments was John Herschel's 'actinometer' made about 1826, in principle a water calorimeter. Pouillet, who introduced the idea of the solar constant, also developed a water calorimeter or 'pyrheliometer' for these early experiments, one of the values of S obtained about 1837 being 1·76 cal/cm² min.[149] This unit, the langley per minute (L/min), has been largely superseded by the SI unit kW/m². The conversion factor varies slightly with different definitions of the calorie. Here 1 L/min is taken to be 698 W/m². The langley is still used as a unit in the testing of fading (Chapter 19). Herschel proposed a unit called the 'actine', equal to the power in a beam of radiation which would melt its own cross-sectional area of ice at the rate of 1 μm per minute. This equals 8×10^{-3} L/min or 5·6 W/m². A similar description specified the sun's total power output, stated to be sufficient to melt continuously a column of ice 45·3 miles in diameter falling into the sun with the speed of light.[150] The power required for this dramatic operation would be 4.15×10^{20} MW in modern units, an estimate close to the present accepted value of 3.8×10^{20} MW.

The development of more sensitive thermopile instruments, improvement in spectroradiometry, and increased knowledge of the solar spectrum and atmospheric effects, made it possible to measure the spectral power distribution of direct sunlight more and more accurately, though on a relative basis. The distribution could then be extrapolated to the conditions outside the atmosphere, or air mass $m=0$, by means of the exponential absorption law described in Chapter 4. This was first used extensively by Langley at the SAO in Washington, D.C.

The extrapolated spectral curve, measured on the earth and therefore incomplete at both ends because of the limits imposed by atmospheric absorptions, could then be extended by whatever knowledge of the solar spectrum was available, or if necessary was assumed. A complete spectral power distribution resulted, but not on an absolute scale. The latter requirement was met by simultaneous measurement of the incomplete spectral curve and of the total power received by a non-selective energy detector, the pyrheliometer. Most of these instruments are based on thermocouples, some on calorimetry or on the distillation of measured amounts of a volatile liquid. They need calibration against an absolute standard, and need periodic checks due to some instability.[151] The standard of radiant flux has suffered many small but disturbing changes since precise measurements of S became common. Descriptions of several types of pyrheliometer are available.[152, 153]

The solar constant has in principle become of importance in meteorology, where it is the most fundamental of all the data employed. Its possible variations are perhaps of more interest than the absolute value. In satellite or spacecraft engineering it is of practical use in designing environments for men and instruments exposed to the sun without atmospheric protection. This protective effect relieved earth-bound men from much concern about the solar constant, and even the recognition of 'environmental engineering' as a distinct technology linking architecture, lighting and heating or cooling will increase interest only in the radiation received at earth level, a very different quantity from the value of S.

In physical terms S is unsatisfactory for several reasons. The flux received varies by $\pm 3\cdot 5$ per cent annually because of the earth's ellipti-cal orbit, and there are possibly smaller changes in S related to solar activity in the sunspot cycle. This was treated at length by the SAO[154] Other causes of uncertainty lie in the extrapolation to $m=0$, the correction for the extremes of the spectrum which amount to additions of about 10 per cent to the power between 350 nm and 2·4 μm, and particularly in the past the unsatisfactory state of international standards of radiant flux.

The mean value of S is probably within 1 per cent of 1·94 L/min or approximately 1·35 kW/m², a value adopted by the CIE in 1972 from NASA investigations described below,[155] and used in this book. At sea level with the sun overhead in a clear sky ($m=1$), the radiant flux would be reduced to about 86 per cent in the absence of atmospheric water, to 78 per cent at normal humidity (10 mm of precipitable water), to 75 per cent for 30 mm of water. Averaged over the hemisphere of earth facing the sun, the power received is much less, mainly as a result of reflection from clouds.

For the first half of this century the SAO had spent a great amount of effort on the determination of the solar constant. At the time when the interest in S passed to other organizations, the final Smithsonian mean, based on many hundreds of determinations by Abbot and his col-leagues, and corrected for the scale of radiation,[156] was 1·835 between 346 nm and 2·4 μm and 1·934 for the complete spectrum. Other values based on the SAO work, with later additions and corrections, were those of Moon who found $S=1\cdot 896$ L/min (Chapter 8), and Nicolet who found 1·98 L/min.[157] (Chapter 9.) The determination of S and the corrections applicable to earlier measurements were discussed in detail by Johnson,[156] who decided on $2\cdot 00\pm 0\cdot 04$ L/min as the most prob-able value from the information available in 1953. For this purpose he constructed a new spectral irradiance curve by combining

  (1) the 1940 Moon values (Chapter 8) for the spectral region above 600 nm,
  (2) the 1959 Dunkelman and Scolnik values (Chapter 9) for 318 to 600 nm, and
  (3) information from rocket flights for ultra-violet wavelengths less than 318 nm.[158]

This spectral irradiance is given in Table 3 together with several other attempts at standardized values, including that by Allen who modified

Johnson's curve mainly in the violet and near ultra-violet regions.[42] Allen's value for S was $1 \cdot 98 \pm 0 \cdot 02$ L/min. Thekaekara re-examined the position in 1965,[159] and also made a literature survey.[160] He did not attempt to amend Johnson's curve, which was also used by Gates in his calculations of solar radiation for various atmospheric conditions (Chapter 12). It is perhaps not unexpected to find that the Johnson curve, when converted to a chromaticity value, gives a point close to the full radiator locus, but not very close to the equivalent temperature of 5755 K. The details are given in Table 3 for Moon's basic curve ($m = 0$).

DETERMINATIONS AT HIGH ALTITUDES

Up to this point the conclusions were that a more certain standard of radiant flux was necessary, and that more detailed spectral measurements outside the atmosphere would be valuable. Before such developments occurred, Stair and Ellis made a determination of conventional type at 3400 m on Mauna Loa in Hawaii,[161] repeating similar observations made eleven years before in New Mexico at a somewhat lower altitude (Chapter 9). The longer wavelength half of the visible spectrum and the infra-red region were not measured on either occasion, but in this case Johnson's selected data were used to complete the calculations. The 1955 value of S was $2 \cdot 05$ L/min, the 1966 value $1 \cdot 95$ L/min with an estimated error of ± 5 per cent. The latter agrees well enough with all later measurements except those of Kondratiev (Chapter 11) who found $S = 2 \cdot 016$ L/min by instrumented balloon flights up to 30 km altitude. Stair and Ellis also made measurements with a filter radiometer. These were consistent with the spectroradiometry, and a similar instrument with more filters was operated by Stair as part of the Goddard projects described in Chapter 11.

The Goddard Space Flight Center in Greenbelt, Maryland organized a most extensive measuring programme on solar irradiance under the American NASA.[88] The aim was to work with many instruments simultaneously in level aircraft flights at $11 \cdot 6$ km altitude, above most of the disturbing atmospheric effects. Of the five instruments measuring total radiation, two were of thermopile principle and three used substitution or compensation for incident power by measured currents. They were flown five times in 1967, mostly over the eastern Pacific ocean, each time for $2\frac{1}{2}$ hours in a path perpendicular to the sun's direction. Four of the instruments yielded over four hundred determinations. Each set was examined separately and extrapolated to zero air mass, though the logarithm of signal versus m was not a straight line as it is for single wavelengths. The values of S differed by only 1 per cent overall. The final mean was 1·936 L/min with an estimated error of $\pm 0\cdot041$, or $1\cdot351 \pm 0\cdot028$ kW/m². The water content above the observing level was very small but important, and varied between 0·010 and 0·040 mm. Aerosol was assumed to be absent, though its effects have been reported by Kondratiev at much greater altitudes (Chapter 4).

This determination of S was only part of the Goddard project. Measurements of spectral irradiance were made as well and are described in Chapter 11. Unlike the procedure in other similar investigations, the spectral measurements were not used to determine S. The spectral power distribution, extrapolated to $m=0$, was in fact corrected to agree with the total radiation measurements. The small correction, $-0\cdot1$ per cent, was an indication of the self-consistency of the whole project.

In the year before, another aircraft project was begun by the Jet Propulsion Laboratory at Pasadena, California, sponsored by NASA, with seven flights between 12 and 15 km furnishing many measurements for the extrapolation method.[162] Pyrheliometers of the thermopile type, calibrated by Eppley–Ångström electrical compensation pyrheliometers, were used to record total flux, and (on a multi-channel reader) the flux in broad-band glass filters for separate spectral regions. There was close agreement between the calculated values of S from the different experiments, with a mean of 1·95 L/min (1·360 kW/m²). In October 1967 one flight of an X-15 rocket aircraft was made to 82 km altitude over Nevada. From this many separate observations of solar flux were recovered, with a mean $S=1\cdot361$ kW/m².[163]

A rather low value of 1·912 L/min was determined by Murcray *et al.* using Eppley pyrheliometers with thin silica windows carried in balloons to a height of 31 km over New Mexico.[164] The extrapolation methods and corrections applied to the ends of the visible spectrum were compared by Kondratiev and Nikolsky with their own treatment of measurements from twenty balloon flights in the U.S.S.R. made to a height of 33 km in 1962 to 1967. Actinometers were used in this case and the authors selected $1\cdot943 \pm 0\cdot008$ L/min as the probable maximum value of S.[165] They reviewed earlier determinations including one in 1969 by the spacecraft Mariner 6 at 20 000 km (1·94 L/min). Measurements in 1962 gave abnormally low results, apparently due to the many nuclear bomb tests in that year producing high level aerosols and uncertainty in the extrapolation of measurements made in the atmosphere.

The results of the Goddard programme were revised by selecting measurements from four of the radiometers carried in the aircraft flights, together with data from other projects mentioned above, and from cavity radiometers carried in Mariner 6 and 7 spacecraft. The final mean value of S proposed by NASA and widely adopted in the U.S.A., including standardization by the American Society for Testing and Materials (ASTM), was $1·353 \pm 0·021$ kW/m² or $1·94 \pm 0·03$ L/min.[166, 167] The corresponding solar temperature is 5762 K, slightly different from that calculated in Chapter 2 because of small differences in the constants used.

Other spectrophotometry done in the same Goddard flights, but not used in the calculation of S, is discussed in Chapter 11, namely the Leiss monochromator observations of Webb et al.[168] and those of Arvesen et al.[169] which gave $S = 1·390$ kW/m², reduced to $1·355$ kW/m² if corrected for lamp calibration.[170]

Another publication of the NASA material added a table of values of S as it varies through the year by changes of the earth–sun distance. The mean value and its associated spectral power distribution are for a solar distance of $1·496 \times 10^8$ km. Solar constants for the other planets are included.[171]

Labs and Neckel discussed the value of S in an extensive account of their work on the radiant exitance of the solar photosphere (Chapter 11). In an earlier paper[172] recording their measurements between 328·8 nm and 1·248 μm they arrived at a value of 1·94 L/min, but in the later review[50] their spectroradiometry was considered separately from the pyrheliometry and gave $S = 1·958$ L/min. This became 1·947 L/min by change of the radiation constant c_2 in the 1968 International Practical Temperature Scale,[173] and in yet another weighting of recent measurements[174] Labs and Neckel decided on 1·95 L/min $\pm 1\%$ (1·361 kW/m²), corresponding to a solar temperature of 5770 K, and they maintained this value in a later discussion.[170] An attempt by Smith and Gottlieb to correlate solar irradiation data from many sources led to $S=1\cdot358$ kW/m². [175] This suggests uncertainty in the fourth significant figure, often quoted in values of S.

Another recent assessment of available data was made by Makarova and Kharitonov. [176] Their weighted value for S was $2\cdot03$ L/min of which $1\cdot89$ was contributed between 340 nm and $2\cdot3$ μm. Values proposed for S in a later review were $1\cdot94$ L/min from high level measurements and $1\cdot97$ L/min as the spectroscopic equivalent of a selected spectral power distribution. [711] Their spectral distribution tables are reviewed in Chapter 11.

Labs and Neckel gave reasons for discarding many earlier determinations. Apart from the slight possibility of establishing secular changes in S from historical records, there can be little reliance placed on the repeated correction of old measurements. It is noteworthy that the original low Smithsonian values of S were followed by a phase when values near $2\cdot0$ L/min were usually obtained, but recent determinations have tended towards the earlier ones. It is doubtful if these changes imply a real maximum in S about 1950 to 1960. Variations of this magnitude within a few years seem impossible if due only to changes in the sun. Changing atmospheric conditions are sufficient to conceal small changes of S measured over short periods of time.

THE SUN AS A VARIABLE STAR

The above discussion shows that there are great difficulties in determining the 'correct' value of S even if it is in fact constant over long periods of time. Many attempts have been made to demonstrate periodic fluctuations in the sun's irradiance as from a variable star. From the long series of observations at SAO, Abbot deduced a variation with a period of $8\frac{1}{4}$ years. Much later the analysis suggested periodicities of $22\frac{3}{4}$ years and submultiples of this time, with a maximum change in solar irradiance of only $0\cdot2$ per cent, or $0\cdot002$ in stellar magnitude. [177] His results were not widely accepted. By a different approach, the analysis of existing photometry of the planets Uranus and Neptune, and with the assumption that the changes reflect variations in the sun's radiance, R. Albrecht *et al.* found the mean light curve for the sun to vary by $0\cdot003$ magnitude in the 11-year cycle. [178] There were also variations of 26 to 29 day periods with magnitude variations similar to those found by Abbot.

An examination by Balasubrahmanyan and Venkatesan of some other photometry of Jupiter, Saturn, Uranus and Neptune showed much larger changes in planetary luminance, with minima and maxima at sunspot minima and maxima.[179] These 20 per cent variations could not result from similar changes in S but they might be due to fluctuations in the extreme solar ultra-violet radiation, causing albedo changes in the cloud-covered planets.

There is some evidence that much larger systematic changes occur than those established with difficulty by Abbot and Albrecht. The procedure is to seek a relation between solar irradiance and the number of sunspots at the time, as calculated by the empirical Wolf number R which at present is given by $R = 0.6 (10g + f)$ where f is the total number of spots visible and g is the number of groups of spots. R varies from zero to 300 or more.[180] From 205 measurements of S made at SAO in 1915 to 1917, and grouped according to R values, Ångström deduced the relation $S = 1.903 + 0.011 R^{\frac{1}{2}} + 0.0006 R$. This has a maximum of S at 1.954 L/min for $R \sim 80$ though the curve does not fit the higher measured values of S, of which the maximum occurs at $R = 170$.[181]

Measurements of S had also been tested at SAO for correlation with R, with somewhat inconclusive results, though Abbot[182] quoted Aldrich's 141 values found over ten years, which showed a fairly steady increase in S with increase of R from 5 to 177. The increase of S was only 0.25 per cent, of the same order as in Abbot's own work on periodicities, whereas the differences found by other observers have been at least ten times as great.

Many measurements of global radiation were made by Bossolasco *et al.* at lowland and Alpine sites up to 3000 m in altitude, mostly in Europe during 1956 to 1963, and at times when the skies were at their clearest.[183] The work revealed large variations in S at any value of R, but the envelope of all the points is assumed to show the nearest approach to the true value of S for any selected value of R: an example from the second paper is seen in Fig. 2, which records the measurements made at Krippenstein in Upper Austria (2050 m). From all the observations the maximum of S was 1.99 L/min at $R = 160$. Sometimes an apparent maximum in S was observed at different sites on the same day; at some sites no marked maximum occurred at all, only a decrease at high values of R. Bossolasco *et al.* showed that the long term variation of R, rising to a smooth maximum in the middle of the 11-year solar cycle, has some resemblance to the change of intensity of the Fe XIV line in the spectrum of the corona. There are some grounds for associating the increase of S at low R with increased solar radiance surrounding sunspots, but the decrease of S at high R is less easily explained.

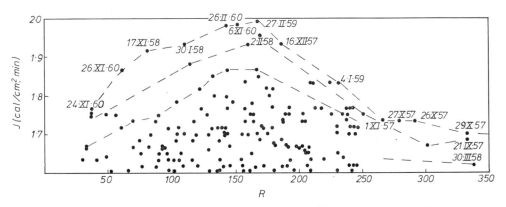

FIG. 2. Dependence of solar radiation on Wolf sunspot number R at Krippenstein. (Bossolasco *et al.*)

Kondratiev and Nikolsky were convinced from their balloon measurements that variations in S were real. They considered 'the most reliable maximum value' of S to be 1·94 L/min at $R=80$ to 100.

In the three investigations quoted the maximum of S from the envelope curves varied by 10 per cent, which from the meteorological standpoint would be catastrophic if truly representative of lasting changes outside the atmosphere. Kondratiev and Nikolsky thought that any possible decrease from the maximum could not exceed 2·5 per cent, but this still corresponds to a decrease of 1·5 K in earth temperature. The very wide scatter among the points within the envelopes shows how difficult it is to obtain, from within the atmosphere, reliable measurements of S over a range of R values with constant conditions throughout. Some calculations revealing former periodic changes in aerosols add another uncertainty to the search.[184] Many observations suggest, however, that there are periodic fluctuations in S within an accessible time scale.

There have been many other estimates of longer periodicities in solar activity. For example, Vitinsky suggested a period of 80 to 90 years from a study of changes in R in successive 11-year cycles,[185] while Cohen and Lintz deduced a period of 179 years,[186] and according to Dilke and Gough internal instability of the sun might cause changes of a few per cent in S over a few million years.[187] This last possibility and the inconsistent solar neutrino measurements are part of the evidence for a modified solar structure mentioned in Chapter 2.

There have been many suggestions concerning possible changes in climate resulting from changes in S, or in the photochemically active part of the solar spectrum, over periods of time long compared with the historical era but short in the earth's evolution. However, local changes could also have large effects, for example, variations in the axial position and distance of the earth from the sun, and changing concentrations of carbon dioxide, water and aerosols in the atmosphere.

This account of the solar constant omits much of the detail found in many other publications. Several aspects are discussed in a book edited by Drummond and Thekaekara,[188] including periodic variations in the emission at wavelengths shorter than 300 nm, demonstrated by rocket and satellite observations. Following the Convair 990 flights at 11·6 km, the Goddard Center plans to operate U2 aircraft with spectro-radiometric equipment at 20 km height, and eventually to have long term observations made in satellites or the 'space shuttle'.[189] More reliable information on the solar constant and its variations may thus be expected.

6

The Literature on Daylight

Look, the world tempts our eye
And we would know it all!
We map the starry sky,
We mine this earthen ball . . .
We scrutinize the dates
Of long-past human things . . .
We search out dead men's words,
    and works of dead men's hands.
MATTHEW ARNOLD, *Empedocles on Etna.*

We come now to the more detailed study of the solar spectrum as observed from the earth. Considered strictly as visible daylight the solar spectrum should be limited to the wavelengths between 380 and 780 nm, according to former CIE publications on colorimetry, or extended to 360 to 830 nm to conform with the latest extension of the range by the CIE. This narrow range is inadequate for many purposes. It is desirable to include all the radiation of shorter wavelength which penetrates the atmosphere, to about 290 nm. In the infra-red region the radiation received at the earth has no wavelength limit but considerable structure, and is of much less interest in the use and applications of daylight than it is to meteorologists and spectroscopists.

The interests of this book are mainly in the visible and near ultraviolet regions of the spectrum. Even so there is a large literature directly concerned with this part of the subject. This profusion arises from several different objectives among those working in the field, and to the nature of the phenomenon itself. Light emitted from the sun has a complex spectrum, and the earth's atmosphere absorbs and scatters this light in a variable, selective and often unpredictable manner.

Four main types of spectrum have been investigated. The first is that of the ideal photosphere of the sun before absorption processes occur. This study is largely a recent development, of importance in astrophysics, and needing the extrapolation technique to arrive at its final results (Chapters 4 and 11). The measurement is one of radiant flux in unit solid angle leaving unit area of the sun, or radiant exitance. It has mostly been conducted at high level sites.

The second type of spectrum is also of the sun's light, in the condition in which it arrives at the earth's surface, or by extrapolation or direct measurement at different depths of penetration through the atmosphere, or as it is before meeting the atmosphere. This measurement is treated as an irradiance, or the rate at which radiant flux falls on unit area. If light only is considered, an illuminance is measured. Supposing the numerical values apply to a spectral band 1 nm wide, then the units might be watts/m² nm or lumens/m² nm respectively. The exitance of the sun has to take account of the polar distribution of flux, with a unit such as watts/m² sr nm. In this context exitance and irradiance are generally denoted by the symbols I_λ and H_λ respectively, instead of the CIE terms M_λ and E_λ. Extraterrestrial values are distinguished by the symbols $I_{0\lambda}$ and $H_{0\lambda}$.

The third type of spectrum is that of indirect sunlight which has been scattered and reflected from the air and clouds. It is by far the most variable since it is never uniform over the sky, and is subject to atmospheric changes having known effects but which are unpredictable in incidence and amount. Variable cloud cover is one obvious source of change, high level air currents carrying suspended dust another less obvious one. Though some measurements are reported as values of sky luminance or radiance, it is most convenient to treat them as spectral illuminance or irradiance. Some writers use H_λ to denote skylight and S_λ for direct sunlight. Both could correctly be denoted by S_λ, the CIE symbol for a relative spectral energy distribution of a radiant quantity. S_λ is used in this sense in later chapters while S is used for the solar constant.

Where continuously emitting sources like the sun or lamps are concerned, it is more suitable to use the term 'spectral power distribution', though in most cases there is no confusion when the former description 'spectral energy distribution' is used. For the sky considered as a large area source of light, when spectral measurements are not made, the light distribution is expressed in luminance units (candelas/m²).

The fourth type of spectrum reported is less precise, but more practically useful than the others. It is the distribution of global radiation or the sum of sunlight and skylight over the whole hemisphere, and ideally received for measurement on a perfectly diffusing horizontal surface. Though subject to the sum of the variations affecting its components, there is often compensation and a surprisingly stable spectral composition during large changes of solar altitude.

The aims of daylight spectrum measurements have altered since the techniques were first established. Initially a means of determining the solar constant and the sun's effective temperature, the spectral power distribution became a source of meteorological information on ozone, water vapour, and the energy balance of the earth. Recently it has become in addition a pattern for the design of artificial daylight sources for technological purposes. Standardized distributions in the spectrum have been important for this last development, with measurements of daylight colour becoming prominent. The latest phase includes measurements of the spectrum at high altitudes on mountains, in aircraft, and by rockets to circumvent the effects of the atmosphere and study sunlight outside the atmosphere in the interests of astrophysics. This is a return with improved methods and equipment to some of the aims of the early work.

On the theoretical side daylight spectra have been used for a thorough examination of the Rayleigh scattering principle, and the modifications necessary on account of the larger particles in the atmosphere, the aerosols which make extrapolation so difficult.

Much of the work on record exists as isolated pieces of information difficult to relate to other work. This is due largely to a lack of knowledge of the meteorological factors at the time of observation, some being difficult to obtain and, if known, difficult to apply as quantitative corrections to observed spectra. Besides this reason there is often no information given on other factors which influence the spectrum, like the geometrical relations of the sun to the observing station and the direction of viewing. Patterns of change have been established, however, and the amount of residual random variation seems to be reasonable apart from a few unexplained discrepancies.

Most measurements have been made in relative terms. Absolute measurements are needed for the values of solar constant or irradiation at the earth's surface, but these can be derived from relative spectral measurements as previously described.

The ultra-violet part of the daylight spectrum has received much attention, sometimes with a quantitative link to the visible spectrum, sometimes without. Its variations caused by changes of absorption and scattering are more marked than is the case for the visible spectrum. In Bener's work on this spectral region the results are mostly recorded as values of irradiation in narrow wavelength bands for many different variables. Some of the measurements have arisen from interest in therapy.

Methods of measurement have changed from the days when the photographic emulsion, the visual spectrophotometer and the bolometer were the only sensitive detectors for the energy in narrow spectral bands. New and even more sensitive and stable detectors have been invented to make higher resolution possible. Spectrographs have changed less, but can now produce an amount of detail in the spectrum which is almost embarrassing. Of particular importance are the improvements in the stability and calibration of reference light sources. Automatic recording and computer processing of data have increased the speed of operation and produced greater numbers of observations. They have also helped to reduce the disturbing effects of rapid changes in daylight. Very high speed spectroscopy is now well developed, but it has not been applied to the less exacting requirements of daylight spectroscopy though it might have value in the measurement of spectral curves during brief high altitude flights by balloon, rocket or aircraft.

Speed is not an overriding condition. Some of the early work, namely Abbot's, has held its value and prestige by virtue of extremely careful techniques and, it might be added, of apparently unlimited time for their operation. The seven volumes of the *Annals of the SAO* form a notable part of the literature, with all aspects of the work fully discussed, not merely the measurements and calculations of *S*. Later chapters of this book refer to other studies showing exceptional skill and persistence, but none on the scale of the Smithsonian work.

Data are presented in the literature in five main forms, of which only the first four concern spectral distributions. The first type includes spectral power distributions in the form of curves, or tables of values of H_λ (and rarely I_λ), and constitutes the bulk of the material till the last few years. The scales are nearly always linear as to wavelength, and linear or logarithmic as to power (or energy), The second group is of chromaticity determinations by calculation from spectral distribution curves or by direct colorimetry, given numerically or plotted on charts. From these values are obtained the third type of result, namely colour temperature or, more accurately, correlated colour temperature determinations. These may also be found by direct observation by meter. The fourth group is based on the first, and consists of tables and curves of weighted means or calculated values which have become standard by continued use, sometimes through lack of later revised data.

The fifth type of information is not discussed in detail below because it is expressed as total power (or energy) and not in spectral terms. It refers to solar or global radiant power, or light flux, incident on the earth's surface, and variations by day, month, season and location. It has long been available in meteorology, and has values for technologies affected by power incident from outside, for example agriculture and architecture, besides giving valuable evidence on the earth's atmosphere, though not on the daylight spectrum. The Kew Observatory has used pyrheliometers since 1907 and made sunshine records by burning traces on paper since 1855.[190] A large amount of information was collected during the International Geophysical Year of 1957–58, for example at Halley Bay in Antarctica.[191] Numerous attempts have been made to derive empirical formulae for the distribution of luminance in the sky and illuminance on the earth.

In succeeding chapters the literature is surveyed in chronological periods, determined to some extent by changes in technique and objective. The first period up to 1921 covers the establishment of reproducible methods, the second to 1949 shows the increasing interest in the ultraviolet region of the spectrum, and in meteorological questions. Measurements made outside the atmosphere stimulated this work further in the third period to 1959, which also saw the end of the Smithsonian project on the solar constant. The present period from 1960 has seen a shift of emphasis to the artificial source and its relation to natural daylight. There has been much interest in high altitude spectral measurements, with the facilities of space research used mainly in regions other than the visible and near ultra-violet.

Some books on aspects of daylight are included in the references (17, 41, 97, 107, 148, 272, 449, 574, 575, 673), and special mention is due to those of Kondratiev[84] and Coulson,[192] the latter emphasizing instruments and methods of measurement.

The divisions of the whole period are not strictly observed when it is more appropriate or convenient to displace items somewhat. Dates in the text are generally those when the work was done, not of publication which is sometimes delayed by several years.

Except in a few of the Figures, spectral distribution curves are presented on linear wavelength scales, and with a constant waveband in each curve. That is, $\Delta\lambda$ is constant for the values recorded on the ordinates. In the published literature the convention $\Delta\lambda/\lambda=$constant is sometimes used, and this has the effect of displacing curve maxima towards longer wavelengths, as in some CIE publications. Other less common methods for representing spectra have been described.[193, 194] Scales of irradiance are normally linear, occasionally logarithmic (Figs. 25–32).

Some astronomical units differ from those given above. $I_{0\lambda}$ may be recorded in erg/cm^2 sr s Hz when the bandwidth is measured by frequency; radiant flux in a vertical column from an extended source of light is measured in rayleighs (Appendix 1).

A confusing variety of descriptive terms and symbols is to be found in the literature: some of these have been mentioned already. Some standardization has been introduced in this book but no attempt made to alter all the symbols used for coefficients of absorption, turbidity, etc.

Appendix 1 outlines some principles of colorimetry used in the daylight spectrum work, and includes a number of standard formulae, units and definitions concerned with radiation, light and colour.

7

Experimental Work
1879 to 1921

If it be well considered the praise of ancient authors proceeds not from the reverence of the dead, but from the competition and mutual envy of the living.

THOMAS HOBBES, *The Leviathan*, conclusion.

Although the spectra of sun, sky and some artificial sources of light had become familiar to physicists by 1868, when Ångström published the first extensive and accurate wavelength atlas of Fraunhofer lines, the measurement of the energy emitted in different regions of wavelength had to await the development of sensitive detectors. The thermocouple and early thermopiles were of little use in the visible part of the spectrum, even for direct sunlight. Photography at this time was unable to record more than the blue and ultra-violet wavelengths, and even here did not give equal blackening for equal energy. Spectrophotometry by eye was therefore widely used in the early days but accurate measurements depended on knowing the energy distribution in some comparison source of light; or on simply taking some arbitrary source as standard because of its convenience, or perhaps because of its assumed black-body distribution. Nichols described how these difficulties were attacked, if not solved.[195] In his own measurements he made use of the acetylene flame, taking a portable gas generator in travels from the U.S.A. to the European Alps. Others used Hefner and Carcel oil lamps, gas lamps, gas mantles and carbon arcs. At first with misgivings, they used the vacuum carbon filament lamp, which appeared in 1879.

The literature of this period includes a variety of measurements on sunlight and skylight which cannot be converted into modern units with any accuracy owing to the primitive experimental conditions. This material is fascinating to read, and shows the difficulties facing the pioneers.

The astronomer Vogel was one of the first to compare the spectral distribution of sunlight with that of an artificial light source.[196] In this case the standard was a petroleum lamp flame which could be reproduced to any acceptable degree of accuracy. Vogel used a polarizing spectrophotometer[197] for visual matches between the test source and the standard at each of the selected wavelengths. Besides sunlight he measured clear and overcast skies, the moon, the then unreliable electric light and, rather surprisingly in view of the small amount of light, a number of stars including Capella. The spectral distribution values of this star were close to those for sunlight from 426 to 633 nm. Another comparison of more present interest was between moonlight and various rocks illuminated by sunlight (Chapter 15).

At about the same time Pickering was developing photometry at the Massachusetts Institute of Technology.[39] His standard both for spectral distribution and for the unit of light flux was a screened-off section of the gas flame of an Argand burner emitting 0·67 cd from an area of about 25 mm^2. A spectroscope with a double slit allowed the side-by-side comparison of spectra from the reference standard, and from the test lamp placed at a suitable distance to produce visual equality of brightness at each of four points in the spectrum. From the information given these were at about 455, 518, 589 and 656 nm. Pickering tried to determine the candle power and 'intrinsic brilliancy' of sun and moon, made measurements on the spectra and intensities of gaslight, lime light and electric light, and incidentally estimated the sun's temperature to be 22 000 °C. The comparison of sunlight and moonlight is described in Chapter 15. The apparatus was somewhat unusual at that time in using a grating instead of a prism.

Crova has already been mentioned in Chapter 4 on account of his interest in scattering. His other publications mostly concerned methods of photometry in which he used a quartz plate between crossed Nicol prisms to make the matching fields approximately equal in hue when the sources were different.[198] He gave one set of figures comparing the spectral distribution of sunlight with that of the Carcel lamp, his usual standard.[199] Intensities at the wavelengths chosen (480 to 740 nm by 20 nm intervals) were determined by decreasing the intensity of the selected narrow spectral band by crossed Nicol prisms until the lines of a graticule at the entrance slit became invisible in that spectral band.

Working at Cornell University in 1889, Nichols employed an Edison filament '16 candle' lamp run at 100 V, a block of magnesium carbonate as a diffusing source, and a spectrophotometer in which reference and test source filled separate halves of a single slit, with Nicol prisms to adjust intensities to equality.[200] With Franklin, he made comparisons between candle, oil and gas flames, the lime light and the carbon arc, measuring at ten wavelengths between 450 and 753 nm.[201] They compared the values for clear and overcast skies with their standard at these wavelengths, and normalized their curves at the D lines. Natura- ally these curves showed little agreement except in general shape with those of Vogel who used a petroleum lamp standard. The primitive nature of these experiments was shown by the evident gratification of Nichols and Franklin in the approximate agreement of the following ratios, recalculated from other work:

Vogel, daylight/petroleum flame   =22·22 at 464 nm

Pickering, sunlight/gaslight   =23·77 at 455 nm

Nichols and Franklin, daylight/incan-
    descent lamp   =21·37 at 450 nm.

This type of work was continued by Fräulein Köttgen at the Physio- logical Institute in Berlin.[202] The spectrophotometer was the one designed by König.[203] Electricity supplies were unreliable, therefore the comparison source had to be either a Triplex gas lamp or a Hefner standard lamp. Measurements were made at 20 nm intervals between 430 and 690 nm on the blue sky, clouded sky, and sunlight with or without clouds. The results, shown as smooth curves normalized at 590 nm, were compared with those of Vogel and Crova. For the latter the erratic data already mentioned[199] were quoted, and as Köttgen remarked, these implied that the sunlight appeared green. In Fig. 3, some of the data discussed are redrawn for comparison, and several sets of points can be represented by a single curve.

At this time a development was occurring elsewhere which made possible the exact calibration and intercomparison of these varied sources. It was that famous, virtually non-selective detector of radia- tion, the bolometer.

Langley invented the bolometer about 1880 and before long the

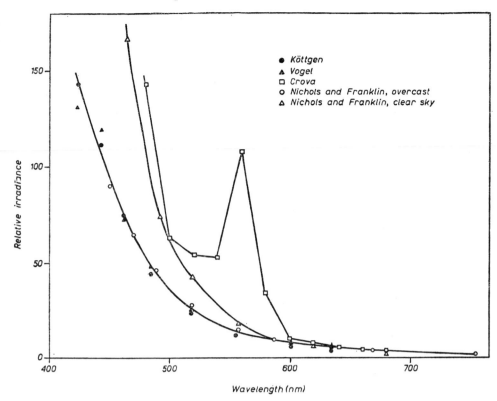

FIG. 3. Comparison of some early solar spectra. Reference sources were oil flames or (by Nichols and Franklin) an incandescent lamp.

measurement situation was transformed. This sensitive Wheatstone bridge arrangement had one absorbing arm of minute thermal capacity which could detect the radiation from a very narrow region of the spectrum, instead of from the wide bands previously necessary to operate thermopiles. Even grating-dispersed spectra could be measured in spite of their relatively low energy content compared with prismatic spectra. Langley described his measurements thus: 'With this apparatus the experiments on the diffraction spectrum were resumed—the first entirely unquestionable evidence of measurable heat, in a width so small as to be properly described as linear, having been obtained on October 7, 1880.'[204] He realized the added advantage of knowing the exact wavelength at any point of the diffraction spectrum. Only one stage of dispersion could be used, but at the time this was no great disadvantage. More troublesome was the fact that artificial sources were very inadequately provided with energy in the blue and violet regions of the spectrum, a difficulty persisting to this day.

Langley's work with the bolometer on the solar spectrum began while he was at the Allegheny Observatory in Pennsylvania. This is not so well known as his later work at the SAO in Washington. His achievements best-known to physicists are in this field of radiation measurement, but he became an important pioneer in aeronautics during the last ten years or so of his life. His reputation was undeservedly damaged by the failure of his full-size aeroplane following successful flights of powered models, and his achievements were overshadowed by the success of the Wright brothers in 1903.

Langley measured the solar spectrum from the near ultra-violet to about 1 μm, using different solar altitudes to calculate the atmospheric absorption for extrapolation to outside the atmosphere. He was interested in the absorption bands in the infra-red, and in extending observations further into this unknown region of the spectrum. With a prism spectroscope on Mount Whitney, California (4000 m) he penetrated beyond the Ω absorption band of water at 1·85 μm, and a few years later to the limit of the carbon dioxide and water absorptions near 14 μm, a range not examined again for about sixty years. He estimated about 60 per cent of the solar radiation at sea level to be in the infra-red, and very little in the ultra-violet. Present accepted values of irradiation at ground level comprise roughly 3 to 6 per cent of the total in the ultra-violet and 40 per cent in the infra-red (Table 9). Fig. 4 reproduces parts of Langley's normal (grating) spectrum of sunlight, and the prismatic spectrum.

Measurements of total radiation by 'actinometers', and allowances for the unrecorded parts of the spectrum, provided Langley with a value of the solar constant by what became the classical method. This, in 1881, gave 2·84 L/min, which like Langley's estimate of infra-red radiation in sunlight was much too high, but nevertheless a creditable attempt at this stage of development of spectroradiometry.

More accurate and frequent determination of the solar constant became the main concern of the SAO during most of its existence in Washington. Langley had planned and founded this observatory in 1890, continued there his infra-red exploration by bolometer till about 1900, and from about 1902 began the long-continued effort on measuring the solar spectrum over the greatest possible spectral range, and in

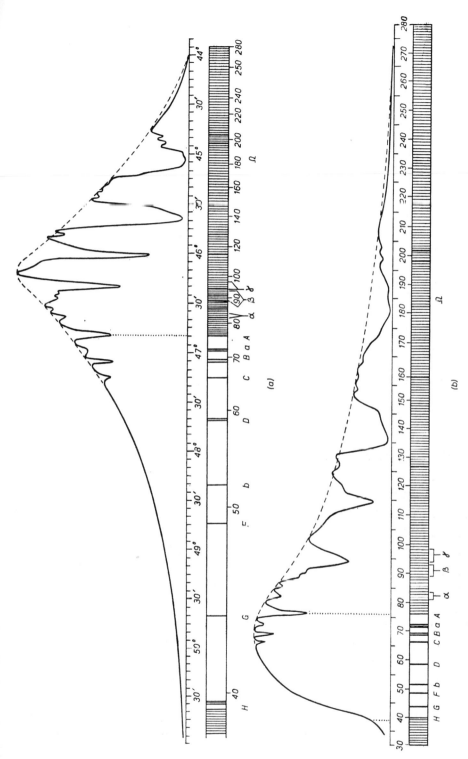

Fig. 4. Solar spectra (after Langley). (a) Prismatic. (b) The 'normal spectrum' obtained by a grating.

total by pyrheliometer. Abbot followed Langley in 1907, extended and elaborated the programme, and established a lasting reputation for the Smithsonian measurements.

American predominance in this field of experiment was assisted by Nichols, Ives, Taylor, Aldrich and others whose names will appear later. There were also contributions of value from several European countries and other continents, though with short-term projects most usual. One institution deserving a special mention was the Astrophysical Observatory at Potsdam, where Vogel had once worked, and where Wilsing now made measurements as detailed as Abbot's, though much less extended in time.

The apparatus constructed and the measurements made at the SAO were described in great detail in the seven volumes of the *Annals*. The work in Washington was amplified and cross-checked at other observatories set up by the parent body, and some others of outside origins who cooperated. Up to 1920 the varied sites included Mount Wilson (1727 and 1790 m) and Mount Whitney (4420 m) in California, Mount Harqua Hala (1770 m) in Arizona, Calama and Montezuma (2740 m) in Chile, and Bassour (1160 m) in Algeria. A summary of the earlier volumes, taking the work up to 1930, was given in volume 5 (1932). In addition to measurements on the whole available spectral range, and on the atmospheric extinction function necessary for the determination of S, reports were made on the dependence of S on sunspot frequency, on ozone determinations, on the spectra of stars, and on fringe subjects like radiation exchange applied to bodily comfort, and the development of a successful solar cooker. Early in the work the variability of S was suspected, particularly as a result of simultaneous measurements in 1911 and 1912 at Bassour and Mount Wilson.[205] Whether random or systematic, this possibility was an objective of measurement and a cause for speculation and controversy for a long time.

On the major subject of the spectral distribution of sunlight, many years' observations were at hand for comparison by the end of the period now under consideration. In 1923 Abbot, Fowle, and Aldrich, who had done most of the work at Washington, summarized the details with critical comments on their own methods, equipment and results, and produced a weighted mean of the 1920–22 measurements at Mount Wilson.[206] They regarded these as their best data for sunlight outside the atmosphere, and rejected much earlier work because of doubts about an imperfect prism used in one case, about the differences between glass and rock salt prisms in another case, about scattered light and the measurement of absorption coefficients in another. Some uncertainty remains about the absolute power standard applicable to these values though various assumptions have been made subsequently. The relative values, as published, were widely accepted till about 1950, and are quoted in Table 2. Interpolated values

TABLE 2. Smithsonian weighted mean for the extraterrestrial solar spectrum (1920–1922)

Wavelength (nm)	Irradiance (arbitrary units)	Wavelength (nm)	Irradiance (arbitrary units)	Wavelength (nm)	Irradiance (arbitrary units)
341·5	262	598·0	485	1377	94
350·4	281	623·8	457	1452	83
360·0	297	685·8	431	1528	74
370·9	318	722·2	409	1603	68
385·3	301	764·4	366	1670	62
397·4	340	812·0	323	1738	57
412·7	480	863·4	283	1870	46
430·7	479	922·0	211	2000	36
451·6	548	986·1	189	2123	25
475·3	566	1062	169	2242	16
502·6	546	1146	143	2348	20
534·8	506	1225	126		
574·2	489	1302	109		

prepared by Gibson and published by Hardy[207] have been adjusted to a reasonable level for insertion in Table 3 and comparison with other absolute measurements discussed in later chapters. While the interpolated values fit the weighted mean well up to about 600 nm, above this wavelength the weighted mean values diverge upwards by about 10 per cent in relative power values; the source of the interpolated values is therefore not entirely clear. The Table includes chromaticities derived from spectral distributions where these are wide enough for the calculation.

Before discussing the better-known investigators we consider some others who made smaller but valuable contributions.

Abney was one of the founders of colorimetry. In 1883, soon after the carbon filament lamp appeared, he was trying to standardize it as a source of 'white light', with advantages over the carbon arc.[208] Ten years later he was using his 'colour patch' apparatus to determine the dominant wavelength of the light from the sun, the sky and clouds,

TABLE 3. Solar irradiance $H_{0\lambda}$ at mean earth's distance, in W/m² μm

nm	Abbot et al.[206] 1920 (a)	Moon[264] 1940 (b)	Nicolet[297] 1951	Stair[275] 1951	Stair et al.[277] 1953	Johnson[156] 1954	Stair and Johnson[279] 1955	Allen[42] 1958	Dunkelman and Scolnik[28] 1959	Sitnik[369] 1965	Stair and Ellis[161] 1968	Thekaekara et al.[88] 1968	Makarova and Kharitonov[176] 1969	Labs and Neckel[173] 1970 (c)	Webb et al.[168] 1970	Arvesen et al.[169] 1971	NASA-Thekaekara[167,171] 1971
220		0	13			30		30				57·5	54·4	46		55	57·5
250		450	56			64						70·4	104	73		68	70·4
280		616	152			240		240				222	243	225		210	222
290		726	358			520						482	511			460	482
300	1140	796	414	1045		610	514	550			540	514	585	521		560	514
310	1212	856	614	~1705		760			760	1180	670	686	663	598	829	660	689
320	1178	916	719			850		750	850	1220	750	819	768	738	1075	750	830
330	1214	976	877		914	1150	1209		1160	1270	930	1037	959	859	1139	1040	1059
340	1395	1046	864			1110	1471	970	1110	1330	920	1050	952	897	993	1000	1074
350	1738	1121	914			1180	1403	1050	1180	1370	1010	1074	972	922	1061	1040	1093
360	1843	1202	942	1164	1262	1160		1090	1160	1750	1040	1055	966	1000	1083	980	1068
370	1841	1304	941		~1430	1330		1120	1330	2050	1150	1173	1115	1046	1141	1150	1181
380	1955	1728	929			1230		1150	1240	2250	1150	1117	1068	993	1043	1110	1120
390	2082	1766	988			1120		1230	1140		1080	1099	1095	1040	1047	1100	1098
400	2128	1788	1546	1879	1960	1540		1530	1540	2500	1500	1433	1510	1383	1524	1540	1429
410	2156	1939	1750			1940		1760	1940		1780	1759	1780	1666	1697	1830	1751
420	2158	2036	1760			1920	~1980	1860	1920	2670	1770	1758	1870	1680	1678	1850	1747
430	2130	2096	1643			1780	~2110		1780		1570	1651	1760	1666	1533	1670	1639
440	2103	2119	1911		~2305	2030		2030	2020	2750	1740	1823	2020	1800	1746	1920	1810
450	2061	2127	2095			2200					1920	2020	2200	1967	1918	2060	2006
460	2012	2103	2161			2160	2154	2150	~2150	2640	1900	2080	2200	1996	1937	2110	2066
470	1965	2061	2150			2170					1900	2045	2200	1988	1899	2070	2033
480	1932	2000	2180	1868	1923	2160	2229	2140	~1950	2530	1950	2085	2220	1938	1882	2100	2074
490	1917	1954	2023			1990					1830	1959	2110	1922	1822	1960	1950
500	1900	1912	2077			1980		2060		2390	1900	1946	2210	1929	1811	1960	1942
510	1893	1894	2074	1794		1960					1920	1882	2150	1867	1780	1970	1882
520	1873	1878	1934	1928	2005	1870				2280	1910	1833	2030	1845	1697	1850	1833
530	1868	1861	2016			1950					1950	1842	2060	1888	1739	1920	1842
540	1851	1841	2019			1980		1950	1880	2180		1783	2050	1887	1724	1890	1783
550	1809	1819	1930			1950						1725	2030	1849	1719	1880	1725
560	1769	1795	2008			1900			1870	2070		1695	1970	1835	1705	1830	1695
570	1729	1762	1890			1870						1705	1940	1831	1688	1850	1712
580		1727	1984			1870						1705	1930	1821	1716	1880	1715
590		1690	1922			1840		1830				1685	1880	1786	1685	1830	1700
600		1653	1898			1810				1970		1646	1880	1752	1638	1790	1666
610			1840			1770						1611	1810	1723	1589	1780	1635
620			1797			1740						1576	1770	1682	1547	1720	1602
630			1755			1700				1880		1542	1730	1648	1516	1700	1570

	A	MO	Ñ	J		S	T	MK-1	LNH	W	A1	NT
700	1513	1405	1483	1440	1460	1610	1369	1540	1433	1373	1500	1369
710	1459	1371	1446	1410		~1530	1344	1460	1401	1376	1430	1344
720	1414	1337	1413	1370			1314		1371	1367	1390	1314
730	1372	1304	1381	1340		~1450	1290	1380	1336	1347	1370	1290
740	1334	1270	1324	1300		1410	1260		1300	1295	1320	1260
750	1296	1236	1305	1270		~1380	1235		1275	1261	1300	1235
760			1282		1280			1310	1251		1280	1211
770			1243						1222		1220	1185
780			1213			~1330			1195		1210	1159
790			1180						1172		1190	1134
800		1097	1154	1127	1130	1280	1107	1240	1149	1156	1160	1109
830			1091					1190	1063		1060	1036
850		976	1046	1003	890	1120	988	1050	996	1015	966	990
900		871	943	895	720	986	889	917	909	963	893	891
1000		706	768	725		796	746	735	730	788	741	748
2000		105	119	108	108	112	103		117		119	103
5000		4·06	4	4·2	4·1		3·8		3·8		4·1	3·79
10000					0·23		0·25	0·251	0·26		0·3	0·24
x	0·3204	0·3179	0·3239	0·3189		0·3041	0·3153	0·3187	0·3237	0·3208	0·3221	0·3171
y	0·3301	0·3297	0·3345	0·3263		0·3152	0·3254	0·3379	0·3337	0·3292	0·3294	0·3263
Δy	0·000	+0·0015	+0·001	−0·003		+0·001	−0·0002	+0·003	+0·0002	−0·0015	−0·003	−0·001
K	6090	6200	5903	6188		7163	6386	6165	5915	6078	6006	6285
Code in Fig. 75	A	MO	Ñ	J		S	T	MK-1	LNH	W	A1	NT

(a) Gibson's interpolated values [207] × 19 (arbitrary factor).
(b) For $m = 0$.
(c) Corrected to IPTS 1968; interpolated linearly from published table.[174]

Δy Distance of chromaticity point above (+) or below (−) Planckian locus.
K CCT of chromaticity point.
~ Interpolated values.

certainly one of the earliest attempts to use this method of specification (see Appendix 1). His values were close to 480 nm for skylight, to be confirmed by many later workers. Clouds gave 486·4 nm, direct sunlight 488·5 nm, and a paraffin candle 588 nm with very little white light needed to desaturate the spectral colour and give a visual match.[209]

Zettwuch, like Crova and others of this period, was much interested in the blue colour of the sky.[210] He used a double-slit spectroscope as a spectrophotometer for measurements made in Rome, recognizing the alterations to the shape of the spectral curve caused by atmospheric dust and water vapour. Zettwuch explored these shapes in terms of inverse powers of wavelength, ideally the fourth power for Rayleigh scattering. In practice he found the exponent to vary through the visible spectrum, and to depend on the part of the sky measured. He observed that the most saturated blue colour occurred at 90° in altitude from the sun, not necessarily at the zenith.

An example of a solar constant determination made independently of the SAO was provided by Millochau. He constructed a simple thermocouple actinometer for measuring solar radiation and used it at a height of 4810 m on the summit of Mont Blanc.[211] Solar temperatures were estimated near to present accepted values, and the estimates of S were 2·55 L/min in 1906 and 2·38 L/min in 1907. By this time Smithsonian values had converged on about 2·0 L/min, which all subsequent work has confirmed to within a few per cent. Millochau did not make any spectral distribution measurements.

Better-known physicists working on light sources and daylight in the early years of this century were Nichols and Ives. Nichols' 1905 paper on the distribution of energy in the visible spectrum described the intercomparison of several reference standards, the determination of source temperatures by the Wien distribution formula, and compared some of the other sunlight measurements already discussed. His own measurements were made at six wavelengths between 390 and 725 nm by means of a portable Lummer–Brodhun spectrophotometer, with visual matches on the standard acetylene flame. He first examined the Hefner lamp, oil and gas flames, carbon arcs and selectively emitting oxides like the Welsbach mantle and the lime light, not without difficulties in the short wavelength region, even with the bolometer. He decided that existing solar spectra agreed in showing the maximum energy close to the peak of the spectral sensitivity curve for the human eye.[195]

This conclusion was accepted by Ives,[212] who showed a full radiator distribution for 5000 K well placed among the curves obtained by other workers, which he recalculated to eliminate the effects of different reference sources. An exception was Nichols and Franklin's curve which differed seriously from the others (Fig. 5). When challenged on

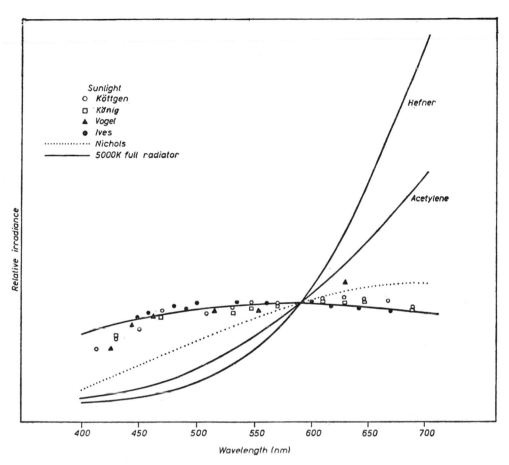

FIG. 5. Sunlight spectra compared with the 5000 K full radiator and other sources. Normalized at 590 nm. (After Ives.)

this, Nichols admitted that stray light probably affected his observations, which had not been made under the best conditions. He continued his measurements on daylight in Switzerland and Austria with the portable spectrophotometer and an acetylene lamp at a reputed colour temperature of 2330 K.

Nichols found the pre-sunrise 'dawn curve' of skylight to approach most nearly to a Rayleigh scattering distribution, with an energy ratio of 8·9 between wavelengths 415 and 725 nm instead of the calculated 9·3. (This example, being based solely on wavelengths, implies an equal energy spectrum for the incident light.) Later in the day it was found that the spectral distribution of skylight often showed a peak at 425 nm. These observations were made to obtain evidence on the cause of the blue sky.[213, 214] Some of the curves are given in Fig. 6. Subsequently Nichols became well known for his research on the luminescence of inorganic phosphors, where his spectroscopic techniques were valuable and used effectively.

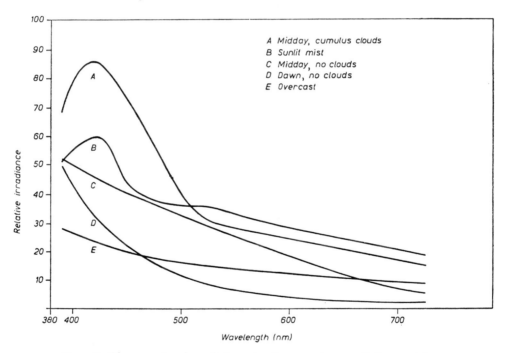

FIG. 6. Sky spectra in relation to the spectrum of the acetylene flame (Nichols). Curves C and D are on the same scale of ordinates, as are B and E.

Ives was more concerned with artificial daylight sources and worked on measuring the colours of illuminants[215] both by colorimeter[212] and spectrophotometer. The two methods agreed well provided that König's primary sensation curves were used, not Abney's. By substitution methods with incandescent and acetylene lamps Ives demonstrated the close similarity in colour between clear noon summer sunlight (considered as average daylight), and that of a full radiator at 5000 K. This was his choice of a distribution for artificial daylight (Chapter 18).

Following Cornu's persistence in tracing the limit of the solar spectrum transmitted by the atmosphere, Fabry and Buisson made use of his results to deduce the presence of ozone (Chapter 3). In a later paper these authors, working at Marseille, compared solar intensities registered photographically at different zenith angles in order to estimate atmospheric absorption at wavelengths between 289·8 and 314·3 nm.[55] They used a carbon arc, a full radiator at 3750 K, as their reference source. By the usual extrapolation method they separated absorption by ozone from the other absorptions, and deduced the concentration of ozone which was found to vary irregularly from day to day. So began one of the standard meteorological measurements.

Extrapolation to find extraterrestrial spectral distributions was also the method used by Müller and Kron, who visited Tenerife in 1910.[216] They took a visual spectrophotometer, standardized filament lamps (probably tantalum) and storage batteries to three sites at 100, 1950 and 3260 m elevation. Observations at eleven wavelengths between 430 and 679 nm gave results shown in Fig. 7, and provided estimates of photosphere temperature between 5900 and 6900 K calculated by the Wien displacement law and the Planck distribution.

Wilsing was a colleague of Müller and Kron at the Potsdam Astrophysical Observatory. He published some very lengthy and much quoted measurements on the daylight spectrum between 400 and 700 nm.[217] His prism spectograph was calibrated by a tantalum filament lamp as a secondary standard, and a ceramic tube furnace at 1300 K as the primary full radiator reference. The extraterrestrial spectrum was compared with that of Abbot at the SAO, but at this time the weighted mean was not yet in existence. Later Wilsing used a bolometer instead of photography in the work, and extended the range examined to 2·34 μm including profiles of the water absorption bands.[218] In the visible spectrum the measurements were at only nine points between 451 and 660 nm and its reproduction was on a small scale, showing no structure (Fig. 7). From 660 nm to 23·4 μm nearly one hundred and fifty points were measured. Wilsing estimated solar temperature in various ways: the photographic work gave 6064 K for the photosphere, the bolometer work gave 'true temperatures' of

5400 K for the upper region of the photosphere, and more than 7000 K for the lowest emitting layer, while the equivalent full radiator temperature for all the radiation became 5900 K. This last value approaches the present estimate even more closely if corrected for Wilsing's use of $c_2 = 1 \cdot 42 \times 10^{-2}$ m K.

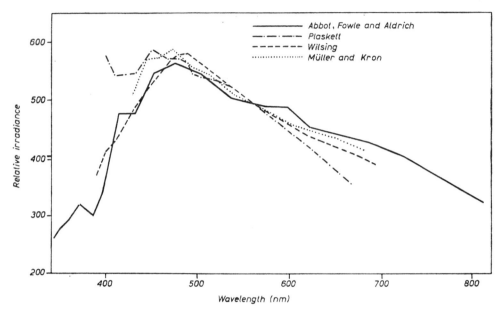

FIG. 7. Comparison of solar spectra, 1912 to 1923, normalized at 560 nm.

The last contribution to be included in the review of this period is that of Plaskett, who worked at Victoria, B.C. with a single prism spectrograph provided with a neutral wedge to give convenient optical attenuation.[219] The light was taken from the central part of the telescopic image of the sun, and the sixteen wavelengths selected were chosen with the aid of Rowland's solar spectrum atlas to avoid Fraunhofer absorptions. Plaskett's work thus had some features of the latest work on the solar radiant exitance (Chapter 11). The primary reference standard was a full radiator of 1373 K, the working standard a carbon arc calibrated by a filament lamp of known temperature when run at a series of different currents. Plaskett took exceptional care in determining the spectra of his standard sources, but the extrapolation to the spectrum outside the atmosphere was based on some inferior linear plots of log I_λ versus m. The photographs of the spectrum were made with five values of zenith angle z between 69° and 28° (or altitude h between 21° and 62°). The wavelength range was 400 to 668·7 nm, and the observations used for the tabulated data all occurred on one day in July 1921. These (relative) values constituted in effect a measurement of the solar continuum, and therefore differed from the irradiance curves of Abbot, Müller and Kron, and of Wilsing chiefly by the lack of the pronounced maximum near 470 nm (Fig. 7). Recent work by Labs and Neckel (Chapter 11) showed Plaskett's measurements to be remarkably consistent with those arising from their much improved techniques. Plaskett's curve approximated to a full radiator curve for 6700 K.

Spectroscopists in this period had much difficulty and inconvenience with their reference sources, and not always adequate knowledge of their spectral power distribution. For example, true temperature, colour temperature and radiation temperature were not often distinguished. The low temperature of their standards was a disadvantage in comparison with the high colour temperature of the daylight being measured: that of the Hefner lamp, for instance, was only 1830 K.[220] Our reliance on tungsten filament lamp standards is a comparatively recent development. In 1914 Jones advanced the claims of a cylindrical acetylene flame in a special burner as a standard, writing: 'Our standards burn on the average 6 hours per day or 36 hours per week. At this rate an electric standard would not last long and would require much more attention to keep it in proper condition.'[221]

For the primary standard of white light it was suggested by Jones that noon sunlight should be adopted, being sufficiently constant in colour for colorimetric purposes if skylight were excluded.[222] In his work described above, Nichols took the contrary view: 'Distribution of energy . . . is so seriously affected by varying atmospheric absorption as to render the sun worthless as a source of comparison in spectrometric work.'[195]

However limited their techniques, the investigators of this period initiated almost all the methods used in later projects. They made measurements of colour, of atmospheric absorption, of the solar constant and the photosphere temperature, and of the ozone content of the atmosphere. They also realized the importance of the ultra-violet and infra-red extensions of the visible spectrum, and the necessity of substitute sources for daylight. One missing feature is the fine structure in the visible daylight spectrum due to Fraunhofer and other absorptions. These were recorded in the SAO work, but not shown in the definitive tables. Other authors had too few points of measurement, while Plaskett took action to avoid measurement at absorption lines or bands.

In this period methods of work were more varied and far from present-day research patterns. There is an entertaining air of adventure in some of the accounts of visits to distant parts which makes them worth reading, however lengthy they may seem in comparison with modern impersonal reports. The following extract from the SAO *Annals* conveys some of the feeling of that epoch.

After a mishap to an instrument on the way up Mount Whitney in 1909, 'Director Abbot carried the pyrheliometer in his hand all the way down the mountain and succeeded in getting it back to Mount Wilson in perfect condition, so that its constant was determined satisfactorily after his return.' In the absence of radio the standard time service in the town was useless to these mountaineers and therefore: 'The time was determined in 1909 by all the watches of the company. In 1910 a vertical iron rod was set up in Portland cement on the first day at the summit, and at noon by the corrected watch time the position of its shadow was marked. On each succeeding day the watch time when the shadow crossed the mark was noted, and on the first opportunity after returning to town the watch correction was determined from the standard time service. From all these observations the watch corrections were deduced.'[224]

8

The Daylight Spectrum
1922 to 1949

It is a great mistake to think that scientific truths differ essentially
from those of every day. They differ only in their greater extension and
precision. . . . The scientist multiplies man's contacts with nature, but
it is impossible for him to modify in any way the essential character of
these contacts. He sees how certain phenomena are produced, though
they escape us, but he is inhibited as much as we are from inquiring
into why they occur.

ANATOLE FRANCE, *The Garden of Epicurus.*

In this period the bulk of the experimental work took place in the
U.S.A. and Germany, or under the auspices of organizations in these
countries. The SAO continued its programmes, still deeply involved
with the measurement of the solar constant. New observing stations
were set up for this purpose at Mount Katherine in Sinai (2620 m),
Table Mountain in California (2280 m), Burns Mountain in New
Mexico (2440 m), and there was collaboration with a National
Geographic Society station on Mount Brukkaros in S.W. Africa
(1520 m). Ultra-violet and infra-red measurements were pursued,
cloud reflectance studied by instrumented balloon flights to 25 km,
investigations made into the dependence of atmospheric ozone content
on sunspot frequency, stellar spectra observed and some attempts at
weather prediction made. No new data for the daylight spectrum were
produced, but all the solar constant values were reviewed and corrected in the 1942 *Annals* (volume 6), a formidable task requiring the
preparation of a typewritten table described as being 15 inches wide
and 200 feet long. There was even more of this to come in the 1954
Annals (volume 7), but in volume 6 the main theme was the dependence
of the weather on variations in *S*. Abbot's belief in this connection
failed to convince many physicists and meteorologists. It is now considered that there may be some correlation between weather changes

and large changes in the very short wavelength solar emission, which amounts to so small a fraction of the total radiation that changes in S would be scarcely appreciable.[225] But previously, in 1938, Abbot was faced with the very hostile comment on different grounds. After the publication of volume 5 of the *Annals* (1932) a close examination of the solar constant work resulted in the suggestion that the observed variations 'are mainly due to the defects in the methods of determining the solar constant.'[226] It was concluded that there was no direct evidence for variations in the constant, and that 'the computations fail to indicate any dependence of weather changes on the observed variations of the solar constant.' Abbot and his colleagues had analysed the variations into thirty harmonics, the main one with a period of $8\frac{1}{4}$ years, but these have never been generally accepted. Some further criticism on this aspect is discussed in Chapter 9.

We now consider a number of smaller projects, commencing with those where the ultra-violet part of the daylight spectrum was the objective of measurement. Some of these were confined to the ultra-violet region, others extended into the visible spectrum to provide a useful quantitative connection. Later the measurements on the visible spectrum are reviewed. A general improvement in methods of measurement is noticeable, an extension of the spectral range up to about 290 nm, and a great increase in attention to structural detail in the spectra. Then, as now, reference sources for the ultra-violet wavelengths were of comparatively low output and their spectral emissivity often uncertain.

THE ULTRA-VIOLET SPECTRUM OF SUN AND SKY

Work was directed towards the structure of this spectral region rather than to its total irradiance or its minimum wavelength. One of the first to explore this field was Hoelper, who made photographic records with a quartz spectrograph at Riezlern in the Austrian Allgäu (1150 m). Tables of intensities at a number of wavelengths below 320 nm showed that the ratio for, say, 314·3 nm to 305·2 nm was dependent on how far the spectrum could be detected into the ultra-violet region.[227] A longer spectrum produced a lower ratio, that is, a flatter curve. From this Hoelper concluded that atmospheric turbidity, which shortens the spectrum, acts in the same sense as an increase of air mass. He found that the measurements made by Fabry and Buisson fitted the same pattern.

Dorno used wide-band filters and argon-filled photocells in his work, which included no spectral distribution curves. At the Davos Observatory which he had founded in 1907,[228, 229] he investigated the distribution of the shorter ultra-violet radiation over the sky, publishing charts of equal radiance as detected by a cadmium cell sensitive between 240 and 315 nm, with a peak near 275 nm. In addition to the work at Davos (1590 m), similar information was collected at Friedrichshafen (400 m) and at Muottas Muraigl (2450 m).

Potassium photocells, with their maximum sensitivity in the far violet or ultra-violet, were much used at this time to measure horizontal illuminance, as by Kalitin at Pavlovsk near Leningrad;[230] or of iso-illuminance curves, as by Aurén for the seasonal variations over Norway and Sweden.[231] In 1931 Kalitin published results for several years more, with analysis of their dependence on solar altitude, on type and amount of cloud cover, and on snow cover.[232]

Some aspects of the expedition to Kings Bay in Spitzbergen in July and August 1929 have been mentioned in Chapter 4.[130] Here Götz made many observations on the ultra-violet spectrum similar to his work at the Lichtklimatische Observatorium at Arosa. Ultra-violet radiation was measured by an acetone–methylene blue actinometer (see Appendix 1), and no spectral curves obtained. For measurements of ozone Götz described and used three methods; they were the calculation from the shortest observable wavelength, the use of two wavelengths of absorption in the normal way, and the Cabannes–Dufay method. The latter is based on the assumption that the ozone layer is so high in the atmosphere that only direct sunshine penetrates it, and that the scattering of light as seen on the earth all occurs below the ozone level. By measuring the ultra-violet spectral irradiance at the zenith, $S_{1\lambda}$ and $S_{2\lambda}$ for solar altitudes h_1 and h_2 ($h_1 > h_2$), and plotting log (S_2/S_1) against wavelength, then the normal increase of S_λ with h should produce a curve of the shape of the Huggins absorption band of ozone, declining to a steady value near 300 nm. The magnitude of this band is related to the ozone content. The method fails if the ozone layer is at a particularly low level, as was the case on one day during the Spitzbergen expedition. The curves for log (S_2/S_1) versus λ showed instead a minimum at 310 nm, or a nearly unaltered ordinate from 313 to 331 nm for the case where $h_1 = 9°$ and $h_2 = 1°$. Light scattering must have occurred above the ozone layer to produce this result.

Important work on the ultra-violet spectrum was begun by Pettit some years earlier, and continued at intervals for about fifteen years. A thermopile receptor was used and assumed to give signals proportional to energy absorbed, like the bolometer. Pettit first reported several years' observations at Mount Wilson (1727 m) where his equipment included an unusual pair of wavelength filters.[233] They were quartz lenses in his 'solar radiometer'. One was chemically silvered and transmitted the narrow band 320 ± 10 nm. The other had an evaporated gold film transmitting green light from 410 to 600 nm with a broad maximum at 500 nm. Galvanometer deflections were recorded photographically, each from 1 minute's exposure of the thermopile to direct sunlight transmitted by one of the filters. Ratios were determined for the irradiance at 320 and 500 nm and their variation over several years appeared to correlate well with sunspot frequency. After corrections made by Allen at a much later date,[42] the correlation seemed much less convincing.

Pettit found no connection between amount of ultra-violet radiation and either atmospheric transmission or ozone content, but it was fairly constant over the sky except near the sun direction (Fig. 8). By the use of crossed spectrographs Pettit made a contribution to the shortest wavelength problem (Table 1), and began in this first period to extend the spectral resolution in his visible spectrum measurements. At Tucson, Arizona (760 m) he used a quartz double monochromator to cover the spectrum at twenty-two wavelengths between 290 and 700 nm. Later work at Mount Wilson, though of no great resolution in the visible spectrum, gave some spectral power distributions which were compared with similar curves from a grating monochromator measured in 10 nm bands to eliminate Fraunhofer structure.[234] Comparisons for $m=0$ were then made between the spectral power distributions with and without Fraunhofer absorptions. These referred to total sunlight, and to the centre of the sun's disc (Fig. 9). The severe reduction of ultra-violet irradiance compared with that of full radiators, and some reduction in the infra-red region, are evident in the figure. The measurements may be compared with those of Peyturaux, Sitnik, and Labs and Neckel (Chapter 11).

In several months' photographic recording at the meteorological

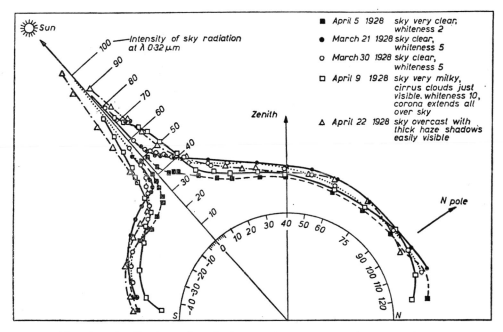

FIG. 8. Distribution of radiation at 320 nm over the noon sky at Pasadena, California. (Pettit.)

Institute in Berlin, Schlömer examined the ultra-violet distribution over the sky during the clearest possible sky conditions.[235] The spectral resolution was only moderate, for after initial trials at many wavelengths to find the detailed shape of the spectral power distribution, six wavelengths were selected between 322·6 and 394·7 nm for most of the measurements, and were found to give sufficient information. Seven points in the sky selected for measurement were on the vertical plane through the sun, and included the zenith, the horizons and the 'dark point' assumed to be at 90° from the sun. Solar altitude h was another variable.

For each location in the sky a spectral distribution curve was given, and the variation of each of the six wavelengths was shown over the vertical plane, providing more information than in previous wide-band measurements. Apart from seasonal variations, Schlömer found no dependence on h of the ratio of total ultra-violet power to the fraction between 356·7 to 394·7 nm (measured as curve areas). The ratio varied somewhat with the direction of measurement in the sun's vertical plane. These observations are probably related to Hoelper's on the shape of

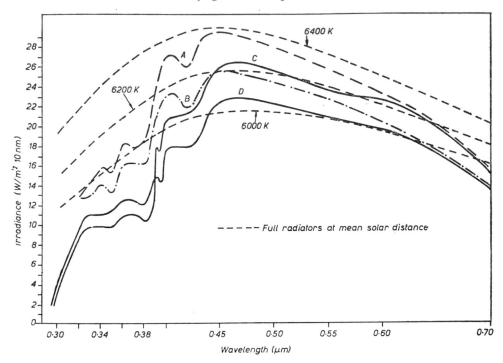

FIG. 9. Extraterrestrial sunlight spectra (after Pettit). Curves A and B
are the envelopes of measurements between the Fraunhofer lines,
and show the relation of the intensity at the centre of the solar disc
(A) to that of the mean disc (B). Curves C (centre of disc) and D
(mean) represent integrated measurements including line absorptions.

the ultra-violet distribution curve, and to the Cabannes–Dufay data of
Götz, but comparison is not very easy.

Schlömer's curves for zenith sky showed peak irradiance at the
342·8 nm wavelength in most cases. For other directions of observation the maximum was often at 373·3 nm but never at 394·7 nm.
Accuracy of wavelength and width of band measured on the plates
come into question here because of the complexity of the spectrum
revealed in later measurements, and the consequent importance of the
choice of measurement points. Compare Fig. 76 which, though for
extraterrestrial ultra-violet distribution, shows how an arbitrary selection of wavelengths may give differing impressions of the general trend
of a spectrum: in Chapter 13 discrepancies in calculation arise from
the same cause.

More ultra-violet measurements came from Africa, where in 1938

Büttner travelled from the Mediterranean to the equator making measurements on total radiation and meteorological observations.[236] With a chemical actinometer (see Appendix 1) he also measured the ultra-violet irradiance at wavelengths below 330 nm. The instrument used had a peak sensitivity at 310 to 320 nm. Büttner measured global flux and the sky component in this wavelength band. The former showed a steady increase with h at a given locality. For longer wavelength ultra-violet measurements he relied on a galvanometer and selenium cell with a glass filter of maximum transmittance at 364 nm. Here again irradiance increased with h at all localities.

On this same journey Orth measured total irradiance with a photocell of peak sensitivity at 580 nm, with a rough spectral distribution provided by the response through eleven filters centred on bands between 365 and 730 nm.[237] The total irradiance within the limits of the cell sensitivity was calculated in cal/cm² min, and in no case did the value exceed 0·601 recorded earlier in Heligoland. In other words, the African sun produced no specially high light intensities, just as Götz had shown in Spitzbergen that high latitudes did not necessarily provide their reputed high levels of ultra-violet radiation.

These African investigations were two of a number in a single sponsored expedition which also studied geographical, botanical and geophysical matters in Africa.

The period under review saw the publication of an atlas of the ultra-violet daylight spectrum including new measurements of wavelengths and intensities for over six hundred Fraunhofer lines of wavelengths less than 340 nm.[238] The famous Utrecht atlas[16] preceded this less extensive one.

A CIE report combined information in the work of Abbot, Fabry and Buisson, and others in the form of ratios of diffuse to direct radiation at 300, 334 and 400 nm for $h=10°$, 30°, 60° and 90°,[239] a form of presentation extended later by Hinzpeter (Chapter 9). For the purpose of selecting lamps for artificial ultra-violet irradiation the 1931 report tabulated the expected annual dose of ultra-violet at sea level, allowing for atmospheric effects, a 60 per cent loss by cloud cover, and irradiation during half the total time: a few of the values are given in Table 4; their equivalents in units of irradiation have been added, but these do not include the reductions for cloudiness and half time in order to compare them with other calculated values for $m=2$.

TABLE 4. Annual dose of ultra-violet radiation[239] and irradiance equivalents.

Wavelength nm	Dose J/m² yr μm	H_λ W/m²	Moon[264] $m=2$	NASA[367] $m=2$
400	$2{\cdot}68 \times 10^9$	425	470	530
370	$1{\cdot}72 \times 10^9$	273	279	320
314·3	$3{\cdot}2 \ \times 10^8$	51	~27	0
299·7	$4{\cdot}68 \times 10^6$	$7{\cdot}4 \times 10^{-1}$	$\sim 7 \times 10^{-2}$	
294·6	$5{\cdot}9 \ \times 10^4$	$9 \ \times 10^{-3}$	2×10^{-4}	
293·1	$9{\cdot}6 \ \times 10^3$	$1{\cdot}5 \times 10^{-3}$		

THE VISIBLE SPECTRUM, WITH OR WITHOUT ULTRA-VIOLET

Priest was mainly interested in artificial light sources and their comparison with sunlight.[240] Here we consider his measurements of the colour temperature of sunlight. This method of describing light sources began about the time of Nichols and Ives (Chapter 7). Experimentally Priest used a rotatory dispersion photometer, an instrument with a quartz plate cut perpendicularly to the optic axis and inserted between the Nicol prisms (or two plates between three Nicols). Compare Crova's photometer.[198] Choice of angular settings allowed the production of spectral distributions close to Planckian ones over the range 1800 to 24 000 K, starting with a light source that might be a carbon filament, or vacuum tungsten, or gas-filled tungsten lamp. With this photometer he measured sun, sky and global irradiance at Washington, D.C. During one day sunlight varied from 4510 to 5300 K, skylight and global radiation from 6000 to 24 000 K. Priest used a 'spectral centroid' scale for an alternative method of ranking colour temperatures of sources. Though having some similarity to the dominant wavelength method of description, this used the spectral distribution and not merely the chromaticity value. The spectral centroid λ_c was defined by

$$\lambda_c = \frac{\int V(\lambda) \, H_\lambda \, \lambda \, d\lambda}{\int V(\lambda) \, H_\lambda \, d\lambda}$$

where $V(\lambda)$ is the spectral luminous efficiency (then called the 'relative visibility'). Later Priest introduced his mired method of describing light sources.[241] This gave a more uniform perceptual scale than the colour temperature scale to which it was related: the mired value is

10^6 divided by colour temperature in K. The paper shows its close correlation with the value of λ_c.

A photographic method was used by Cunliffe in studies of light sources used in the fading of textiles.[242] Spectrograms of sunlight were measured at eight wavelengths between 370 and 720 nm, and the values used to establish diurnal and seasonal variations, with a good deal of emphasis on meteorological conditions. The primary reference standard was an acetylene flame at a colour temperature of 2360 K, while a tungsten lamp at 2700 K served as a working standard. Cunliffe also measured the spectra of carbon arcs, incandescent lamps with various daylight filters, and a few examples of skylight. Like other workers in other places, he concluded that the local noon summer sunlight, in this case at Didsbury near Manchester, was a reasonably constant standard. The resolution in his spectral curves is naturally rather low. The paper is one of the few published in this country on daylight spectrum measurements; it includes useful summaries of earlier work.

Dorno's work on the ultra-violet spectrum at Davos was supplemented in another report[243] by tables of daily, annual and secular variations of solar irradiance in wide bands, extracted from over twenty years' observations. This did not include information on detailed spectral distribution, except for a table of minimum wavelengths recorded. The therapeutic use of ultra-violet radiation and its artificial production were discussed, these being a major concern of Dorno's.

Similar work on sunlight was done at sanatoria in Ohio (<300 m) and in Colorado (2000 m) by Greider and Downes.[244, 245] They determined the proportion of 'antirachitic' ultra-violet radiation (290 to 310 nm) in winter and summer, comparing the outputs of artificial sources used medically. Wide bands were selected by filters at 290 to 310 nm, 310 to 650 nm, 650 nm to $1·4\ \mu$m. and $1·4$ to $12\ \mu$m, and measured by a quartz spectroradiometer with a thermopile detector. Though the mountain site gave an increase in each band over the values for Ohio, the short wavelength band showed the largest increase. Some curves showed measurements at much closer spectral intervals for comparison with a calculated distribution for a zenith sun at Mount Wilson, where the power in the 290 to 310 nm band was ten times as great. The conclusion on artificial sources was that an arc with cerium-cored carbons could provide a higher proportion of its radiation in the ultra-violet region than was found in sunshine. See Chapter 19.

The observations made at Utrecht by Ornstein, Eymers, Vermeulen and Postma utilized a novel form of photometer, at any rate in this field of work, namely a disappearing filament instrument like a pyrometer.[246] Visual matches were made at fifteen wavelengths between 450 and 680 nm selected by a prism monochromator. At each wavelength the background was provided by the image of a nearly horizontal magnesium oxide screen exposed to the sky, or to sun and sky together. Against this background was seen an incandescent fila-ment, for which the necessary current to cause its disappearance was recorded. Previous calibration of the filament temperature against current gave the required factors for calculating spectral power dis-tribution in absolute terms. The main variables were solar altitude and the cloud conditions. The authors developed sorting and averaging procedures for their seven hundred spectral distributions, with particu-lar regard to the variations of (power) ordinates from the mean. In one such method they determined four parameters α, β, γ and δ vary-ing with h, and plotted these separately against h. The best fitting spectral curve could then be stated in the form of an equation; for example, for a cloudless sky,

$$H_\lambda = \alpha + \gamma[\tfrac{1}{3}(x-\beta)^3 - (x-\beta)\delta^2]$$

where x is the wavelength in units of 10 nm and α, β, γ, δ were to be read off from the curves. Though many single spectral curves showed erratic shapes, all the definitive ones were quite smooth with peaks near 500 nm. The authors also calculated illuminance values by hour of day and month, for different h and cloud conditions, and for skylight and global radiation.

Measurements to take advantage of high altitude atmosphere were made by Herzing in 1935–36.[247] The site was at 800 m in the Taunus mountains (near Frankfurt am Main). Small scale curves over the spectral range 300 nm to 2·5 μm, and for various solar altitudes, resulted from this work whose chief purpose was to determine the infra-red absorption bands of water vapour, and hence a method of quantitative estimation. The bands were those near 900 nm ($\rho\sigma\tau$) and at 1·2, 1·4 and 1·85 μm (Φ, Ψ, Ω). Herzing used a quartz double mono-chromator and thermopile, recording the galvanometer deflections.

Hess worked with the same monochromator as Herzing, and at the same observatory.[248] His detector was a gas-filled photocell of peak sensitivity at 350 nm and wide range, 290 nm to 1·1 μm, and photo-currents were recorded continuously from galvanometer deflections. Spectral power distribution curves for skylight at a number of points related to the sun's position were measured from 310 nm to 1 μm. The angular distance from the sun was significant with regard to the content of ultra-violet radiation. This reached its highest relative amount at 90° from the sun, as in Fig. 10. Hess listed persistent peaks in these

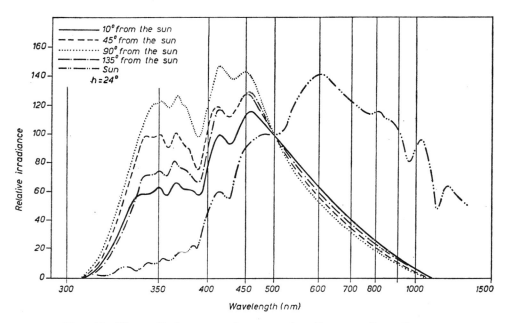

FIG. 10. Sky radiation at various angular distances from the sun, normalized at 500 nm. (Hess.)

curves at 347, 358, 374, 386 and 428 nm, and expressed changes in the irradiation by ratios for selected wavelengths and different angular distances from the sun. This work has been much quoted, but is somewhat inaccessible in this country.

Reiner again used the Leiss monochromator at the Taunus observatory, this time with a selenium barrier layer cell as detector, and recorded galvanometer deflections.[249] Since the cell had a peak sensitivity at 590 nm, a blue filter was included to reduce the range of the correction factors. Reiner was primarily concerned with the extraterrestrial solar spectrum, but his work is more valuable for his detailed study of absorption bands between 370 and 600 nm. Forty-three bands were tabulated and shown in a curve, and spectra at different air masses compared, as in Fig. 11.

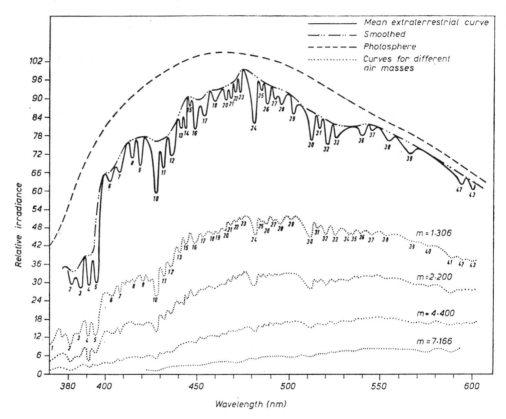

FIG. 11. Extraterrestrial and other sunlight spectra with the main absorptions. (Reiner.)

In 1941 Bullrich measured daily variations of sky luminance at Frankfurt, for the zenith and other directions, and as a global luminance.[250] Colorimetry of the sky by three calibrated colour filters, and comparisons with a blue pigment scale (Chapter 15), would have been more useful if a standard colour triangle had been used. The white point was not defined, but dominant wavelengths and saturation were specified, respectively between 484 and 493 nm, and 50 to 63 per cent. In Chapter 9 further experiments by Bullrich on these lines are discussed.

Colour measurements were becoming more common. In 1942–43 Reesinck made a great many measurements by colour filters and a selenium cell on the colour of the twilight zenith sky at Arnhem.[251] Absolute brightness measurements were also made, but as only three spectral regions were used, near 416, 529 and 604 nm, the results are not of much interest here. In 1944 further measurements using a different pair of blue and red filters (about 472 and 648 nm) gave the distributions of these bands in relation to solar altitude and angular distance of observation direction from the sun.[252] Reesinck's tentative conclusions were that the sky colour near the sun was mainly determined by large particle scattering, and that at large angular distances Rayleigh scattering was more important.

Herrmann used a double monochromator with an integrating sphere input, a gas-filled alkali photocell, an amplifier and galvanometer to record spectra of skylight and global radiation at medium values of solar altitude.[253] He confirmed Albrecht's 1935 results to the effect that the global curve was largely independent of h in flat country when the sky was clear. Albrecht's calculations are described later in this chapter. Herrmann's curves covered the range 315 to 800 nm and showed little structure. They were reproduced, together with some due to Hess, in a review of daylight for photographic purposes by Nagel.[254] This author adopted Herrmann's skylight values as a standard distribution, and for direct sunlight the calculated values due to Moon. Herrmann's curve for a clear sky approximated to the 10° curve given by Hess (Fig. 10). It was now becoming evident that serious variations occurred in the ultra-violet spectrum according to angle of view and h. Herrmann's work took place at Giessen and included measurements on sources for ultra-violet skin therapy, with calculations of exposure times, dosages and costs.

Taylor accomplished many useful measurements in the course of long attention to the daylight spectrum. The initial work was colorimetric. Sunlight and skylight at Cleveland, Ohio were studied over a whole year. By matching with combinations of tungsten lamps and blue filter glasses, and the use of Priest's spectral centroid method, he arrived at equivalent colour temperatures for the daylight conditions, and these served as a basis for the selection of artificial daylight sources.[255] More importance was then attached to spectral distribution, and Taylor and Kerr produced an extensive series of curves for sunlight, skylight and their combinations, and included the earlier colour temperature values in the paper.[256] To correct for variations of light during a spectral determination, they made check measurements at 560 nm by the spectrograph in use, or by an extra photocell with a narrow-band pass filter, or by a light meter. The main detector was a caesium photocell, and another innovation was the modification of the calibrated 1 kW tungsten reference lamp by a blue daylight filter to approximate the spectral distribution of test source and reference.

The published curves showed only major irregularities although the measurement interval was 10 or 20 nm between 400 and 700 nm. Some of these curves, plotted on a basis of equal lumens, are reproduced in

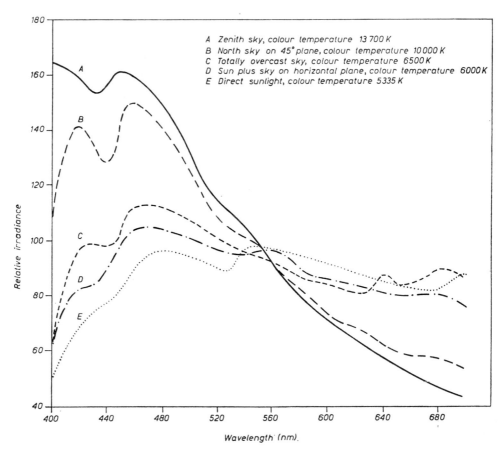

FIG. 12. Average spectral distribution curves for types of daylight, on the basis of 'equal illumination'. (Taylor and Kerr.)

Fig. 12. In some cases they have been difficult to reconcile with later work. Colour temperatures ranged from 4975 to 60 000 K, and since the chromaticities were not far from the Planckian locus, these values would be scarcely altered by later reformulations of the iso-temperature lines. Most of the points were on the upper (greenish) side of the locus, a feature of daylight that was to be confirmed by nearly all later measurements. Another observation was that the quality of combined sunlight and skylight received on a horizontal plane varied normally only over 6000 to 7000 K, with a mean of 6500 K, while north skylight was much less constant in quality.

In 1949 Taylor published many small scale curves for extraterrestrial sunlight and its resultant at sea level, with wide-band energy values between 300 nm and 2·6 nm.[257] His work provided some of the earliest daylight distributions to be used as acceptable standards.

CALCULATIONS OF DAYLIGHT SPECTRA

We now turn to those authors of the 1922 to 1949 period who were predominantly calculators of spectral distributions rather than recorders of experimental observations. The distinction is not always sharp, for some of them made their own measurements on daylight. The calculations in this and subsequent chapters are largely concerned with the scattered component of daylight, though in some cases global radiation is formulated as well.

Some estimates of daylight at the earth's surface were prepared by Kimball for the CIE in 1928,[258] starting from the SAO weighted mean for sunlight corrected by the available atmospheric transmission coefficients, and combining this with skylight values in ratios measured by himself. The skylight was derived from Priest's colour temperature measurements.[240] For each of the days June 21, September 21 and November 21, and for latitude 41° N, a diagram showed the variations at fifteen wavelengths between 397 and 764 nm with distance of the sun from the meridian. The curves reflected the variations between sunlight and skylight with time of day and season.

About this time Berlage, a meteorologist in Java, developed formulae for the spectral distribution of sunlight and diffuse skylight falling on a horizontal surface.[259] Replacing frequencies by wavelengths, and sec z by m (air mass), with the atmospheric transmittance denoted by q_λ and extraterrestrial sunlight by $H_{0\lambda}$, these formulae become:

for sunlight

$$S_\lambda = \frac{1}{m} H_{0\lambda} q_\lambda{}^m$$

for skylight

$$H_\lambda = \frac{1}{2m} H(1_{0\lambda} - q_\lambda{}^m)/(1 - 1 \cdot 4 \ln q_\lambda).$$

The foregoing expression for diffuse radiation was used by Albrecht to analyse his own measurements of sky radiation, excluding direct sunlight, which he had made at the Potsdam Observatory with the aid of a Moll–Gorczynski solarimeter (a type of thermopile instrument). Horizontal colour filters over the solarimeter admitted radiation from three-quarters of the sky hemisphere.[260] A few ultra-violet measurements down to 330 nm were made in addition to those on radiation though Schott GG7, OG1, RG2, 5 and 7 filters. Histograms from the observations allowed only approximate spectral energy distribution curves to be drawn, nevertheless Albrecht was able to subtract from them the calculated molecular scattering curve, given by Berlage's formula, leaving residual differences to be attributed to scattering by dust particles. These differences were smooth curves with a peak in the region between 530 and 600 nm. Other graphical results showed radiation balance in the visible and near infra-red range. Unfortunately the curves in this paper are badly reproduced and extremely difficult to read. Albrecht produced a formula for total irradiance in terms of solar altitude, when $25° < h < 70°$, namely

$$I + D = 0 \cdot 33 \sin h \, (3 \cdot 0 - \sqrt{\operatorname{cosec} h})$$

where I and D are his symbols for sunlight and diffuse radiation respectively. Chapter 9 contains further references to Albrecht's work.

Calculations of a less empirical nature were made by Elvegard and Sjöstedt in 1941.[261] They found that the Smithsonian weighted mean data could be represented reasonably well by a Wien-type radiation equation with a Priest mired value of 167·5 (or colour temperature 5970 K) and $c_2 = 1 \cdot 433 \times 10^{-2}$ m K. This equation was

$$S_{0\lambda} = 409\lambda^{-5} \exp(-2 \cdot 400/\lambda)$$

which differed from the experimental data by 10·8 per cent at 410 nm, 8 per cent or less from 420 to 500 nm, and less than 3 per cent from 510 to 720 nm. A similar equation was due to Kastrow[262] who adjusted the numerical constant to give values close to those of the SAO weighted mean:

$$S_{0\lambda} = 189\cdot7\lambda^{-2\cdot3} \exp(-0\cdot0327/\lambda^4).$$

Elvegard and Sjöstedt represented sunlight at the earth's surface, for $h=45°$ and the Ångström coefficient $\beta=0\cdot10$, in terms of the weighted mean $(S_{0\lambda})$:

$$S_\lambda = S_{0\lambda} \exp[-1\cdot413(0\cdot0085/\lambda^4+0\cdot10/\lambda^{-1\cdot3})].$$

The Wien-type approximation to this was, with a mired value of 206·2 or colour temperature 4850 K,

$$S_\lambda = 760\lambda^{-5} \exp(-2\cdot955/\lambda).$$

The last two equations agreed within 10 to 20 per cent from 410 to 430 nm, to 7 per cent or less up to 500 nm, and by 5 per cent or less to 720 nm. In the four equations above the wavelengths are to be measured in μm.

The work of Götz and Schönmann at the Lichtklimatische Observatorium at Arosa (1860 m) closely followed Herrmann's in time, but used a photographic method with tungsten lamp calibration.[263] Forty wavelengths between 310 and 530 nm measured on the spectrograms gave information of the variation of the zenith sky spectrum with solar altitude, and allowed the calculation of extraterrestrial sunlight distributions. Each spectrogram was made twice, with the instrument axis respectively in the vertical plane containing the sun, and perpendicular to it, the sum of two measurements at a given wavelength being independent of polarization effects. Rayleigh scattering formulae gave poor agreement with observation except under very clear atmospheric conditions, and extra terms for aerosol scattering were required, as well as an arbitrary choice of the value of N_D to obtain the agreement shown in Fig. 13 ($N_D=2\cdot7\times10^6/cm^2$).

The equation used was of the form:

$$S_\lambda = S_{0\lambda} \frac{0\cdot434}{m-1} \frac{q_1q_2-q_1{}^mq_2{}^m}{-\log q_1q_2} (\Gamma_1 kn+\Gamma_2 N_D) d\omega$$

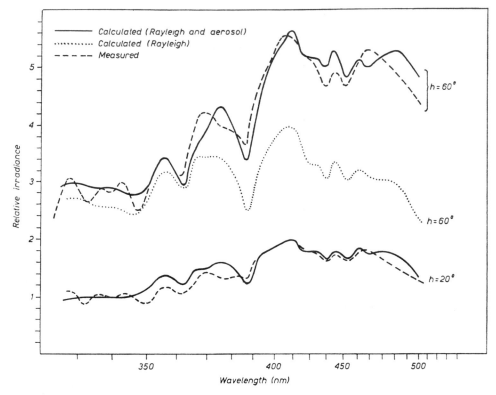

FIG. 13. Observed skylight spectrum, and calculated distributions for Rayleigh scattering with and without aerosol at $N_D = 2\cdot7 \times 10^6/\text{cm}^2$. (Götz and Schönmann.)

where m is the air mass, q_1, q_2 are transmittance coefficients for Rayleigh and aerosol scattering respectively, Γ_1 is the Rayleigh scattering function for a volume element, of which there are n above the observation point, Γ_2 is the scattering function for a dust particle, N_D is the number of dust particles in a vertical column of atmosphere above unit area at the measuring site, k is a correction factor for multiple scattering, and $d\omega$ is the solid angle of zenith sky illuminating the observing plane. The experimental curves for the spectral distribution at Arosa showed very high ultra-violet components, with a peak of emission near 330 nm sometimes equalling the level of the normal peak at 410 nm.

The best-known standard data for sunlight, founded on experimental values for the spectrum and calculation of scattering and absorption effects in the atmosphere, are those of Moon.[264] Relying on the observations of Abbot (not only the weighted mean), Wilsing, Pettit, and Fabry and Buisson, he proposed in 1940 a spectral distribution for sunlight outside the atmosphere, extending from 295 nm to 2·13 μm. The section from 1·25 to 2·13 μm was assumed to follow the shape of the full radiator emission at 6000 K. By corrections for reasonable amounts of water vapour (20 mm), ozone (2·8 mm) and dust particles (300/cm³), this distribution gave rise to a set of curves for air masses 1 to 5, of which the one for $m=2$ ($h=30°$) was recommended for general use and has in fact been widely used. It is equivalent to an illuminant of correlated colour temperature 5089 K and a chromaticity appreciably above the full radiator locus (Table 5 and Fig. 72). In the visible spectrum the curves show the oxygen absorptions at 688 and 759 nm, and water vapour near 723 nm, also near 810 and 930 nm in the near infra-red. The absolute scale for $m=0$ corresponds to 1·322 kW/m², or roughly 2 per cent below present estimates.

Another set of spectral distributions appeared in 1940. They were calculated by K. S. Gibson[265] from the SAO data of Abbot *et al.*, as smoothed by Gibson,[207] and are known as the Abbot–Gibson distributions. Tables of spectral distributions were given by Nickerson.[266] They were based on the assumption that the spectral distribution from overcast sky or clouds was close to that of the extraterrestrial sun, and that the blue sky was equivalent to this same distribution corrected by a $1/\lambda^4$ factor. Different proportions of these distributions added together gave a range of illuminants varying from a correlated colour temperature of about 6100 K for the unaltered Abbot curve, to ∞ K for the sky curve. The most commonly used one was the 15 per cent sky/85 per cent sunlight mixture, of correlated colour temperature 7400 K, which became an accepted standard for colorimetric calculations, especially in the U.S.A. The chromaticities of the Abbot–Gibson distributions lie on a straight line close to the full radiator locus, but cutting it at two points which makes all the chromaticities at less than 18 000 K lie slightly below the locus. Another limitation in these largely theoretical spectral distributions is that they do not depend on air mass or any atmospheric absorption process except Rayleigh scattering. Unlike Moon's curves, they show strong absorptions of Ca_{II} in the HK lines, and in the FeMg lines near 517 nm, but no water vapour or oxygen bands.

TABLE 5. (a) Solar irradiance H_λ at earth's surface for $m=2$, in W/m² μm; and (b) solar radiance, centre of disc, $I_{0\lambda}$ in MW/m² μm sr.

nm	Moon[264] 1940 (a)	Brandhorst et al.[367]–NASA 1975 (a)	Kok[340] 1972 (a)	Sitnik[368] 1965 (b)	Labs and Neckel[50] 1968 (b)	Makarova and Kharitonov[176] 1969 (b)	Harvard–Smithsonian reference atmosphere[380, 381] 1971 (b)
195							1·50
230							19·82
280					32·0		
290					33·4		
300	0·08	0·0	0·4		34·6		36·84
310	11	1·0	10		35·6	25·0	
320	54	50	48		36·4	27·1	37·91
330	101	105	108	25·0	37·0	28·1	
340	151	177	135	25·3	37·5	27·8	38·39
350	188	217	178	26·1	37·8	27·2	
360	233	255	217	26·9	37·9	29·3	
370	279	320	288	28·1	46·8	33·5	
380	336	344	321	36·0	47·3	38·4	48·07
390	397	371	351	42·0	47·5	43·0	
400	470	530	507	46·5	47·1	46·4	48·49
410	672	693	568			48·1	
420	733	737	636	50·05	46·2	49·7	47·23
430	787	737	568			50·5	
440	911	867	738	52·3	45·1	50·2	45·70
450	1006	1022	884			49·6	
460	1080	1089	933	52·8	43·7	48·2	44·02
470	1138	1107	952			47·0	
480	1183	1167	1027	50·05	42·2	46·6	42·23
490	1210	1133	961			45·9	
500	1215	1165	1012	47·5	40·7	45·5	40·77
510	1206	1147	1030			44·5	
520	1199	1136	966	44·45	39·1	43·5	
530	1188	1159	1051			42·3	
540	1198	1140	1047	41·9	37·5	41·0	
550	1190	1120	1067			39·5	
560	1182	1111	1055	39·65	35·9	38·0	
570	1178	1134	1056			36·9	
580	1168	1147	1067	37·4	34·2	35·8	
590	1161	1148	1054			35·0	
600	1167	1135	1051	35·4	32·5	34·2	32·2
610	1168	1137	1062			33·3	
620	1165	1137	1058	33·6	30·9	32·7	

Empirical formulae were advanced by Moon and Spencer to represent some of the functions used in colorimetric calculations.[267] Polynomial expressions gave approximations to Planckian distributions and the $\bar{x}$, $\bar{y}$ and $\bar{z}$ functions of the CIE 1931 Standard Colorimetric Observer.[268] A third paper proposed three- or seven-term polynomials to fit the Taylor and Kerr daylight curves, and Moon's sun-

TABLE 5 (*continued*)

nm	Moon[264] 1940 (a)	Brandhorst et al.[367]–NASA 1975 (a)	Kok[340] 1972 (a)	Sitnik[368] 1965 (b)	Labs and Neckel[50] 1968 (b)	Makarova and Kharitonov[176] 1969 (b)	Harvard–Smithsonian reference atmosphere[380, 381] 1971 (b)
630	1176	1137	1024			31·8	
640	1175	1141	1027	32·2	29·5	31·2	
650	1173	1140	1009			30·5	
660	1166	1136	979	30·9	28·1	29·9	
670	1160	1127	1019			29·1	
680	1149	1119	992	29·5	26·7	28·6	
690	978	1114	897			28·2	
700	1108	1102	917	28·15	25·4	27·8	25·0
710	1070	1089	894				
720	832	1071	792	~26·8	24·2	26·3	
730	965	1058	816				
740	1041	1040	871	~25·5	23·0	24·8	
750	867	1026	851	24·8			
760	566	~755	653	~24·2	21·8	23·2	
770	968	997	804				
780	907	981		~23·0	20·7	21·9	
790	923	965					
800	857	950		21·85	19·4	20·9	19·4
830	863	898					
850	839			18·95	17·4	18·3	
900	480	348		16·65	15·5	15·9	
950	487	199		14·65	13·8	14·1	
1000	630			13·1	12·17	12·6	12·17
2000	2.7			1·77	1·84		1·83
5000					0·0568		0·0562
10000					0·00369	0·00376	0·00366
x	0·3431	0·3435	0·3494	0·2977	0·3037	0·2992	
y	0·3567	0·3534	0·3610	0·3081	0·3130	0·3105	
Δy	+0·0065	+0·003	+0·0055	+0·0005	−0·001	+0·0015	
K	5089	5065	4930	7734	7205	7576	
Code in Fig. 75	M2	B	K2	SI	LNI	MKI	

Δy Distance of chromaticity point above (+) or below (−) Planckian locus.
K CCT of chromaticity point.
~ Interpolated values.

light curves.[269] For the latter, with $m=2$, their three-term expression gave the spectral emission as:

$$H_\lambda = 1510·8 - 515\lambda - 0·07915\lambda^{-10}$$

where H_λ is in W/m² μm and λ is in μm.

9

The Daylight Spectrum
1950 to 1959

The desire of knowledge, like the thirst of riches, increases ever with
the acquisition of it.

LAURENCE STERNE, *Tristram Shandy*, Vol. 2, chap. 3.

At the beginning of this period the highly variable nature of daylight
was well appreciated, and much effort was therefore expended on
calculations from the known spectrum in attempts to formalize the
effects of scattering and absorption, and so produce theoretical distributions of real use. Most of these were of limited application. The
SAO weighted mean of 1920 to 1922 for extraterrestrial sunlight, and
Moon's curves for terrestrial sunlight, continued to be regarded as
fundamental and were on the whole the best available data.

In this decade there was activity in the application of knowledge
about the spectrum of daylight, in the development of better artificial
sources and in meteorology. The uncertainty in the solar constant
became a matter of concern since the performance of space vehicles,
manned or unmanned, was critically dependent on the quantity and
quality of incident radiation outside the atmosphere. More work was
done on the ultra-violet end of the spectrum and on the solar constant,
and a few investigations on the visible spectrum in different parts of
the world.

Experimental methods continued to improve with the added power
of the photomultiplier and photoconductive detectors, though the
bolometer was by no means discarded and in improved forms has
continued in use.[270] The grating spectrograph or monochromator
was not commonly used, and quartz prism instruments were still the
usual equipment.

The SAO ceased operations in Washington, D.C. in 1955, and was
transferred to the control of Harvard University at Cambridge,

Massachusetts. Its subsequent work became wider in scope, with much emphasis on work concerning meteors and artificial satellites. The final volume of the *Annals* (volume 7, 1954) contained solar constant data for 1939 to 1952, articles on instruments, the measurement of ozone, and related subjects. One remarkable statistic showed that between 1921 and 1948, more than 16 000 determinations of S were made at one station (Montezuma). The succeeding publication, *Smithsonian Contributions to Astrophysics*, contained another critical survey of previous work on the solar constant. In a comparison of the simultaneous SAO observations at Montezuma and Table Mountain over nearly thirty years, Sterne and Dieter failed to find any common periodicities.[271] Abbot defended his position again and would not accept the implication of erroneous observations, but reiterated his belief in the great significance of periodic variations in S and their value in long-range weather forecasting.[182] His views on the connection between S and the sunspot frequency have been discussed in Chapter 5. The Washington era of the SAO has been surveyed by Jones in a book more historical than technical.[272]

An examination by A. K. Ångström of observations made at SAO between 1940 and 1952 and recorded in volume 7 of the *Annals* led to conclusions about periodic aerosol variations. The measurements, made at Montezuma and Table Mountain with Abbot's pyranometer, were of solar radiation from an annulus of sky with the inside circle at 3·5° and the outer at 13·5° from the edge of the sun's disc. For values of m near 2 constant annual patterns appeared with strong maxima in January for the Chilean site and in April for the Californian, and minima at about half these levels roughly six months later. Ångström gave equations for the relation between β and the circumsolar radiation.[184] The values of β were of the order of 0·01, somewhat lower than those found at Mauna Loa in Hawaii (3350 m) at a much later date,[273] and discussed in Chapter 12.

PROJECTS UNDER THE NBS

The National Bureau of Standards in Washington, D.C. now became prominent by the researches of Stair in the period 1950 to 1956, beginning with measurements of solar ultra-violet radiation at Washington, then on Chalk Mountain near Climax, Colorado (3410 m), where favourable climatic conditions were to be expected.[274, 275] In the first phase of the work the combination of a quartz double mono-chromator, a photocell with peak sensitivity at about 375 nm (RCA 935), a light modulator, amplifier and recorder had proved adequate to record fine structure of ultra-violet Fraunhofer lines in the solar spectrum. A siderostat reflected the light on to the entrance slit without other optical components. Calibration of the energy scale was effected by a tungsten-in-fused-silica lamp of known colour temperature, but using spectral emissivity data for tungsten in the region 300 to 430 nm. This range was extended to 535 nm at the mountain station. Stair discussed the possible polarization error due to reflections in the siderostat: there was no significant dependence of the effect on wave-length but intensity changes of ± 5 per cent were observed between morning and afternoon measurements at $m = 3$.

The rapid spectral scan in 2 to 3 minutes, with eight successive amplifier sensitivity ranges for different wavelength ranges, dispensed with the need for monitoring the total light flux or some fixed wave-length band to provide a correction for fluctuating irradiation. Re-peated observations were made daily during the three week operation in September 1951, and from these the material from four days was selected for the calculations. Atmospheric transmittance was deter-mined by Bouguer lines at twenty-five selected wavelengths (Fig. 14), and irradiance values extrapolated to $m = 0$. The tabulated results also included values for $m = 1, 2$ and 3.

With a different double monochromator and an RCA 1P28 photo-multiplier, Stair and Johnston next measured ultra-violet radiation from the moon when near full.[276] Comparison with the Colorado observations revealed a general absorption by the declining reflectance from 400 to 320 nm. This was interrupted by a minimum due to a selective absorption at 380 to 390 nm, a possible indication of the surface material (see Chapter 15). Atmospheric transmittance was calculated again, to give a straight line relation between the logarithms of wavelength and transmittance between 325 and 550 nm.

The next use of this equipment, with the original photocell, took place at Sacramento Peak, near Sunspot, New Mexico (2800 m). Spectral irradiance measurements were within 5 per cent of those found at Chalk Mountain, and the ozone content was the same 2·1 mm. A lead sulphide photoconductive detector extended the range to 2·5 μm.

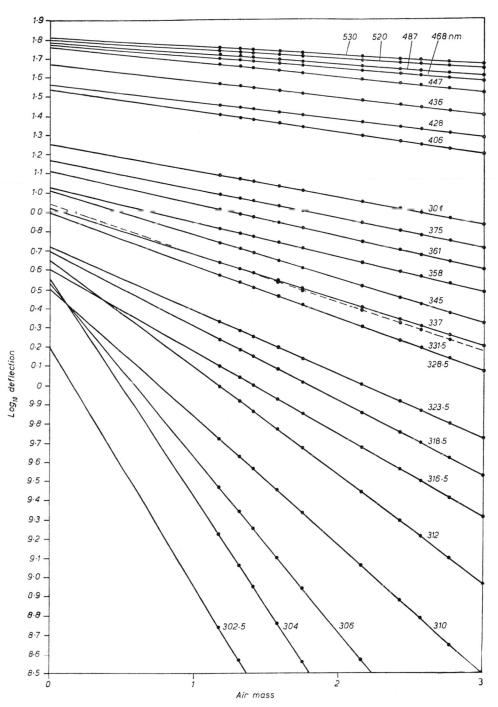

FIG. 14. Determination of solar irradiance outside the atmosphere by extrapolation from observations at separate wavelengths (after Stair).

The resulting value of the solar constant, an estimate only, was 'slightly exceeding 2·00 langleys per minute.'[277]

A separate publication by Stair described a method of filter radiometry based on his earlier measurements of the solar spectrum. It was used at Sacramento Peak and elsewhere to determine total ozone. A titanium photocell and four filters of different transmittances, all low at 300 nm and higher at 340 nm, provided the measurements. Each filter had a set of curves relating transmittance to ozone concentration and air mass, from which interpolation of the observations, after some simple calculation, gave readings of the ozone content.[278]

The final work in the project also occurred at Sunspot, and was directed towards a redetermination of the solar constant, once more with the photocell detector from 299 to 535 mm, and the lead sulphide cell beyond.[279] Polarization corrections were eliminated by using no

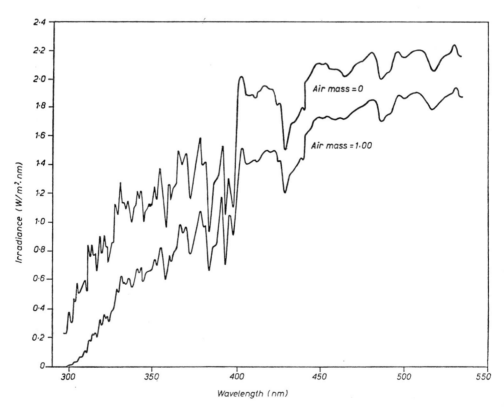

FIG. 15. Solar irradiance, mean of four days' observations at Sunspot, New Mexico. (Stair and Johnston.)

siderostat, but instead pointing the whole monochromator at the sun continuously. Increased accuracy was attempted by more elaborate methods than before in checking the calibration lamp, a tungsten ribbon in a fused silica envelope. Four days' observations were treated as in the Colorado experiments, with special corrections in the infra-red region required by some instability of the detector, and the avoidance of water absorption bands. With a small addition for wavelengths below 300 nm, estimated by extrapolation of the spectral curve to zero irradiance near 200 nm, and a further addition of 0·06 for infra-red beyond 2·5 μm, the value of the solar constant came to 2·05 L/min or 1·43 kW/m². Stair's later determination of S has been discussed in Chapter 5.

In Fig. 15 the wealth of detail obtained in Stair's New Mexico measurements may be compared with that in recent work (Figs. 49 and 77) and in the Utrecht atlas of 1940 (Fig. 76). Fig. 16 shows the transmittance curve from New Mexico: it is almost identical with that for the Colorado observations, and separates scattering (mostly molecular) from ozone absorption (2·2 to 2·3 mm during these observations). Table 3 includes some sets of Stair's spectral power distributions for $m=0$, quoted where the wavelengths of measurement were multiples of 10 nm; that is, interpolation has been avoided except over small intervals.

THE MOUNT LEMMON PROJECT

At this time photographs taken in rocket flights were providing information on the short wavelength ultra-violet spectrum. Difficulties occurred in fitting the new spectral curve to previous solar curves terminating near 300 nm and extrapolated from terrestrial measurements, for example those made by Stair. There were differences in shape and in the absolute energy level needed for solar constant calculations. In another attempt to provide accurate spectral measurements Dunkelman and Scolnik of the Naval Research Laboratory in Washington, D.C. carried out in 1951 a four week programme of observations on Mount Lemmon, near Tucson, Arizona (2440 m). This site was expected to have excellent atmospheric conditions.[28]

The equipment resembled Stair's with a siderostat, double quartz monochromator with automatic drive and recording of the photo-

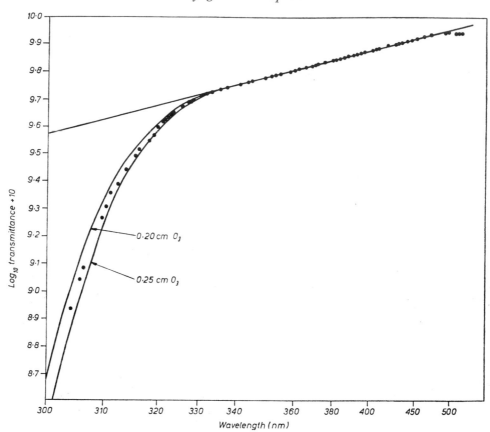

FIG. 16. Atmospheric transmittance at Sunspot, New Mexico, and determination of total ozone. (Stair and Johnston.)

current from a photomultiplier (RCA 1P21). Polarization was eliminated by a block of magnesium carbonate between siderostat mirror and entrance slit, and the angle of view of the sun was limited to 1° to minimize sky radiation accepted. A tungsten-in-fused-silica lamp, calibrated by tungsten emissivity values, provided reference spectral irradiance values on the magnesium carbonate block. The illuminance of this white surface, measured respectively under the tungsten filament standard and sunlight, gave the absolute level for the spectral measurements with an estimated accuracy of ±10 per cent. The energy scale thus found was raised by 9 per cent between the date of the experiments (1951) and their publication (1959) in order to conform with Johnson's review of solar constant work, particularly the scales of radiation.[156]

Scanning the spectrum took 5 minutes and was repeated throughout the day. No monitoring for variations during a single scan was considered necessary. The aim was much more ambitious, namely to obtain constant atmospheric conditions during a whole day, or at least half a day. The authors explained how difficult this was in spite

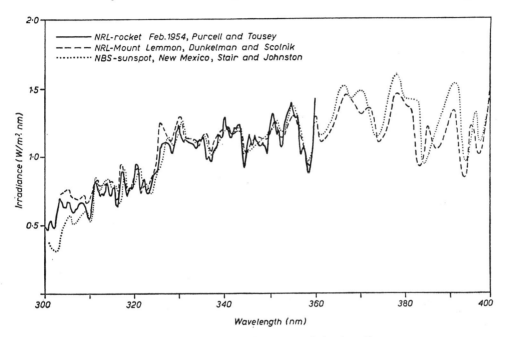

FIG. 17. Comparison of some solar ultra-violet irradiance spectra. (Dunkelman and Scolnik.) Stair and Johnston's curve is as in Fig. 15. The rocket curve is adjusted on the vertical scale to the best fit.

of apparently perfect clarity and stability of the atmosphere. Plots of log photocurrent for a given wavelength versus m sometimes extended to $m=12$, but only on a few occasions were they sufficiently consistent to show acceptable linearity, so permitting confident and accurate extrapolation to $m=0$. The authors gave several plots showing how, under certain atmospheric conditions, the slopes could change and suggest incorrect extrapolation limits. The material selected for the final processing consisted of twenty-six spectral traces taken in one noon to sunset period. They were very closely linear when corrected for atmospheric refraction and curvature at high values of m, and for the curvature of the ozonosphere at wavelengths below 340 nm where ozone absorption occurs.

This investigation produced valuable information, including graphical comparison of earlier work (Figs. 17 and 18). These showed, for example, that Stair and Johnston's values were higher at wavelengths greater than 500 nm, and there were some close similarities of structure below 400 nm. Dunkelman and Scolnik tabulated exponential co-

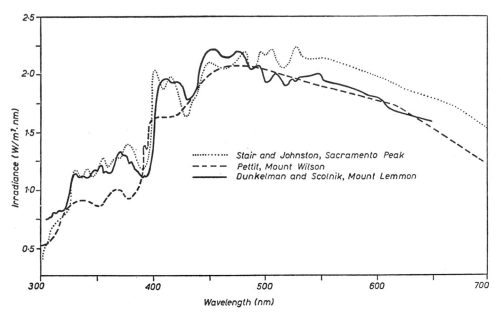

FIG. 18. Comparison of absolute extraterrestrial spectra, integrated over 10 nm intervals below 450 nm. (Dunkelman and Scolnik.) Pettit curve (D of Fig. 9) adjusted to lower power level in accordance with solar values accepted in 1959.

efficients of spectral attenuation for the equivalent atmosphere of the test site between 303 and 650 nm, values of $H_{0\lambda}$ in the same range at the measured wavelengths, and smoothed values at 1 nm intervals from 308 to 445 nm where averaging had been done over 10 nm bands. The best day's data were shown as a plot of log attenuation versus log wavelength (Fig. 19) where the attenuation is seen to be 15 per cent greater than that required for Rayleigh scattering apart from regions of ozone absorption. The authors suggested that this was the nearest approach to pure Rayleigh scattering so far recorded. At 550 nm the attenuation was 30 per cent greater than calculated, or somewhat better than Hulburt and Tousey's determination from aircraft (Chapter 4). The values of $H_{0\lambda}$ appear in Table 3.

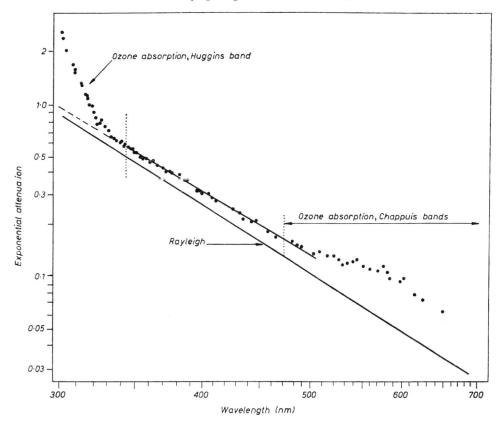

Fɪɢ. 19. Total vertical attenuation of the atmosphere on Oct. 4, 1951 above Mount Lemmon, Arizona, compared with calculation. (Dunkelman and Scolnik.) Equivalent atmosphere 5·96 km thick.

MEASUREMENTS IN EUROPE AND NEW ZEALAND

The other experimental measurements of this decade were less important. They serve to illustrate points of technique or qualitative features in daylight rather than to provide standard values. Schulze made filter measurements at Wyk auf Föhr, an island in the North Sea off North Friesland.[280] Two ultra-violet and five visible spectrum bands, and as detectors a cadmium cell for the shorter ultra-violet and a selenium cell for the rest, gave some global radiation curves of little structure and no great differences between $h=20°$, 30° and 60° except in the 280 to 315 nm region (UV-B). This shorter ultra-violet band increased strongly in relation to the band centred on 510 nm as h increased.

Schulze proposed $h=35°$ as normal for the provision of standard data. His information was obtained for the benefit of the CIE which in 1951 recommended the investigation of 'the most suitable energy distribution to adopt for average natural daylight', and 'the suitability of light sources for reproduction of natural daylight'. The former aim has been achieved (Chapter 12) but the latter is incomplete (Chapter 18).

Another filter investigation in this period was made by V. Hisdal at Hardangerfjord in Norway.[281] Irradiance from the zenith sky in summer was measured by a selenium cell and galvanometer, and nine interference filters with maximum transmittance at 339, 351, 371, 391, 418, 463, 488, 547 and 591 nm (and considerable overlap in the ultraviolet region). Calibration was effected by a standard tungsten lamp, and total irradiance was measured as a check on random variations during a set of observations. Changes of spectral distribution, illustrated by histograms, resulted from changes in h and from cloud variations. A striking feature was the approximate constancy of the proportion of total flux in the 463 nm band except for a slight increase near sunset. Later work by Hisdal is described in Chapter 10.

A study of skylight and total daylight at Wellington, New Zealand, was made by Hull in 1951 to 1952.[282] This is the only known information from that area. The glass constant deviation spectrophotometer with integrating sphere input by way of an annular opening in the top, was described by Cooper and Probine.[283] It used an RCA 1P22 photomultiplier with chopped and amplified output, and a special feature was the compensation for detector fatigue by reflection of a constant beam of light from one face of the spectrograph prism on to the photomultiplier, whose sensitivity could be checked at intervals. Cooper and Probine discussed the error due to departure from cosine law in the integrating sphere for angles of incidence above 60°, and calculating from the observed contributions of sun and sky on a number of occasions, they found that with $h=10·6°$ the colour temperature might be estimated 240 K too high, with $h=14°$ about 130 K too high; with $h>25°$ the errors would be negligible.

In Hull's measurements the scan extended from 420 to 700 nm with a normal bandwidth of 20 nm. Corrections were applied when photometer readings of the total irradiation varied, though constant conditions were chosen when possible. Little structure appeared in the curves, though a double peak was said to be always found at the maximum (470 nm) in measurements at 10 nm intervals. The curves applied to different cloud conditions, some to sunlight alone, and some comparisons were made with full radiators. They showed no unusual features compared with daylight curves in other parts of the world. Colour temperatures were between 5490 and 7400 K, with one day's sun plus sky variation shown to be nearly constant (5500±200 K) for 9 hours except during a period of smoke cover. No seasonal variations were discussed.

The infra-red measurements of Peyturaux were made at a series of fifteen wavelengths, including 679 and 788 nm in the visible region and extending to 2·311 μm.[284] He demonstrated a nearly linear relation between the logarithm of irradiance versus air mass for different wavelengths. Results were given for radiance of the photosphere surface, which he and others studied more extensively later (Chapter 11). On this occasion the fall in radiance from the centre of the disc to the edge was shown to be greater if the wavelength of the light measured was shorter. Peyturaux estimated the temperature of the photosphere as 4530±90 K, and its absorption in the spectral range 600 nm to 1·2 μm was attributed mainly to the ion H$^-$. The equipment used in this case was a quartz double monochromator, a cooled lead sulphide cell detector, and a recorder. The experiments were conducted at the observatory of Haute Provence (600 m altitude).

In 1956 B. Hisdal investigated the light from the overcast sky near Oslo.[285] The spectral irradiance from 320 to 600 nm was measured by a double-pass quartz prism monochromator and a 1P21 photomultiplier, with a white magnesium carbonate block inclined at 45° to the horizontal reflecting the light from the eastern sky. Spectral distribution curves were given for $m=1·2$, 1·5 and 2·0, and compared with those of Taylor and Kerr. Mean colour temperatures were 6450 K for $m=1·2$, 6080 K for $m=1·5$, and the corresponding chromaticity values slightly on the green side of the Planckian locus, like the RD distributions (Chapter 10); colour temperature and chromaticity values would have been altered if the spectral measurements had been extended beyond 600 nm.

On the whole spectrographic instruments have not varied much in principle, but a paper by MacAdam described a novel recording spectroradiometer intended for the measurement of light sources.[286] This constant deviation double-grating monochromator received modulated light from the test source and from a reference tungsten lamp, automatically adjusted them to equality at each moment and recorded the required attenuation of the standard source. The record could be a relative logarithmic spectral power distribution, or by a further circuit an absolute distribution for the test source. All collimation in the instrument was accomplished by systems of parallel baffles which eliminated lenses and their associated light scattering properties. Final concentration of the beam was necessary, and for this a plastic Fresnel lens was used.

The instrument operated at a bandwidth of about 10 nm, and the time of scan was 3·5 minutes. Detectors were a Dumont 1292 photo-multiplier for 350 to 650 nm, and a type 6292 with a yellow filter for 650 nm to 1 μm. A few spectral curves for sunlight and skylight were measured by MacAdam, with the main absorption bands, Fraunhofer and atmospheric, showing on them. A question raised by some of these curves is discussed in Chapter 14.

A very brief communication by Lenz concluded the spectrographic work of this period.[287] Measurement of the zenith sky by a prism monochromator provided a single curve between 300 nm and 1·05 μm with considerable absorption band detail, compared with a smooth curve calculated by Dave and Sekera for a water- and dust-free atmosphere. Reasonable agreement from 330 to 450 nm changed to an increasing excess of the measured curve above the theoretical at longer wavelengths.

COLORIMETRY OF THE SKY

An earlier type of measurement was rejuvenated by a new instrument, a portable colour temperature meter operated by blue filter glasses in front of a tungsten filament lamp held at a controlled temperature.[288] This permitted approximate colour matches between the sky and a comparison field with its chromaticity varying along a line close to the full radiator locus. The instrument included extra green glass filters to allow more exact matching, but these were not used in the measurements to be described here. The same instrument was used again by Collins (Chapter 12).

Observations were made in September 1950 from an aircraft at two different heights, and for various azimuths to the sun and zenith angles.[289] Chromaticities reported were roughly in the range of 6000 K to near infinity at 6·1 km, and well above infinity for 12·2 km. It is not possible to discover the exact chromaticities of these very blue colours, but they appear to be in the same area as the twilight colours measured by Gadsden (Chapter 15). At the higher altitude the sun's colour temperature was found to be 5800 K, confirming estimates from other sources.

CALCULATIONS OF DAYLIGHT DISTRIBUTIONS AND CHROMATICITIES

In the period 1950 to 1959 interest continued to expand in the calculation of atmospheric scattering effects, and there was more research on the daylight spectrum and its equivalent colour. While direct sunlight could be dealt with by experiment, given settled atmospheric conditions during the period of measurement, the quantity and quality of scattered light was difficult to measure or to calculate even under normal atmospheric conditions. Several different attempts are discussed below.

In earlier work Albrecht concluded that global radiation could be expressed as a function of solar altitude h, but that the spectral distribution was nearly constant and independent of h (Chapter 8). He now started from the SAO weighted mean distribution and applied corrections for water absorption in the atmosphere and for reflection from the atmosphere into space.[290] Calculation of reflected and scattered components of sunlight under a clear sky gave a distribution of global radiation which for $h=60°$ agreed moderately well with an attenuation according to $1/\lambda^2$. At $h=15°$, $1/\lambda^{1·3}$ produced closer agreement. In both cases the proportion of radiation between 300 and 400 nm was too high by calculation, as many investigators have found.

Another attempt to apply the principles of scattering, in this case to predict the colour of the sky, was made by Bullrich, de Bary and Möller.[129] Dominant wavelength λ_d, saturation and luminance were calculated for two values of h and a number of points on the vertical plane through the sun. Nicolet's extraterrestrial distribution was assumed, with Rayleigh scattering only, or with the addition of four different degrees of turbidity. The colour data were known to be affected by reflection from the ground, and by secondary scattering which could be as high as 50 per cent in skylight away from the sun direction. The tabulated data did not include corrections for these causes.

As in Bullrich's earlier report (Chapter 8), there was an omission to identify the white point used. This might have been the equi-energy spectrum, which at the present time is defined as achromatic by the CIE (see Appendix 1). In Fig. 20 chromaticity values have been calculated from the data for Rayleigh scattering without turbidity, and $h=60°$, and views of the sky from horizon to horizon in the plane of the sun. The white point is assumed to be Standard Illuminant C to allow comparison with Lenoble's calculation. The chromaticity has to be estimated from λ_d and percentage saturation. It can be seen that the choice of the white point alters chromaticity only slightly for a given value of λ_d in the 480 to 490 nm region. For the horizon sky values Bullrich *et al.* gave $\lambda_d=553$ nm, saturation 4 per cent: here a change from equi-energy spectrum to Standard Illuminant C would alter the correlated colour temperature from 5500 to 6500 K. References to other points in Fig.20 are made later.

Middleton made calculations to discover the colour of the overcast sky.[291] He simplified the problem by investigating the behaviour of upward and downward radiation in a horizontal cloud layer, neglecting absorption in the water. The first important deduction was that the exponential absorption process did not hold in a cloud, where decrease of light flux varied in a linear manner with depth of penetration. The cloud was assumed to be illuminated from above by an extra-terrestrial distribution derived by Middleton from data by Moon, Stair and Dunkelman and Scolnik. Ozone absorption values were those given by Vigroux,[292] while further corrections were based on the water vapour absorption bands. Finally, the reflectances of three types of land surface were introduced (snow, grass and urban), and a selection made of values of h, cloud thickness, and the size and concentration of water drops in the cloud. Insoluble aerosols were not considered. Chromaticity differences of more than 0·02 in x and y were shown to be possible (Fig. 20). In one case the colour temperature of the incident light was altered from 6350 K to 8150 K under the cloud as a result of snow cover on the ground. To observers on the ground such differences would not be noticed without instruments, since no side-by-side comparison would be possible.

The theoretical spectral distribution for an urban ground surface

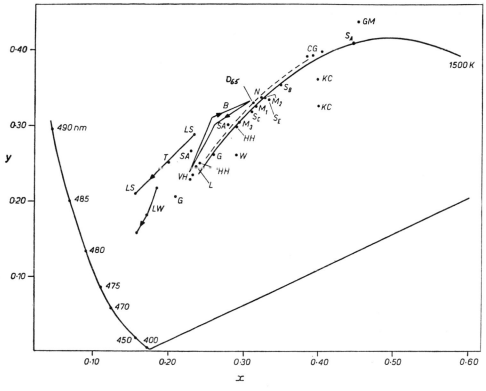

FIG. 20. Chromaticities of standard sources, Planckian radiators and CIE daylight: observed and calculated values for air, sea and sky.

Solid curve: Planckian radiators from 1500 K to ∞.

Broken curve: reconstituted daylight from 4000 K to 25 000 K.

$S_{A, B}$: Standard Illuminants.

GM: gas mantle. 2982 K (Chapter 18).

CG: Condit and Grum, daylight examples. [333]

KC: Knestrick and Curcio, daylight examples (see Fig. 53). [357]

N: Nicolet, extraterrestrial sunlight (Table 3) used by Bullrich and Lenoble. [297]

M_1: Middleton, incident radiation assumed. [291] 6350 K.

M_2: Middleton, calculated overcast sky, urban conditions. 5620 K.

M_3: Middleton, calculated overcast sky, snow cover. 8150 K.

B: Bullrich, calculated sky, 0°–90°–0°, $h = 60°$. [129]

SA: Schimpf and Aschenbrenner, observations on air. [463]

HH: Hendley and Hecht, observed sky range. [507]

W: Wright, sky in eclipse. [430]

G: Gadsden, observed range of twilight sky. [514]

L: Lenoble, calculated sky. [294]

LS: Lenoble, calculated light in sea, increasing h. [474]

LW: Lenoble, calculated light in pure water, increasing h. [474]

T: Tyler, observation on lake water. [478]

VH: Hisdal, zenith sky light measured and calculated. [359]

D_{65}: reconstituted daylight for 6500 K.

S_E: equi-energy spectrum.

differed notably from that given by Taylor and Kerr for overcast sky (Chapter 8), chiefly by a 10 to 15 per cent lower irradiance in the 460 to 520 nm region, and a marked dip at 430 nm where the earlier curve had a slight peak (Fig. 21). Middleton's values were considered by the CIE in 1955[293] as a possible daylight standard like Schulze's in 1951, but no formal adoption of spectral distributions occurred till much later (Chapter 13).

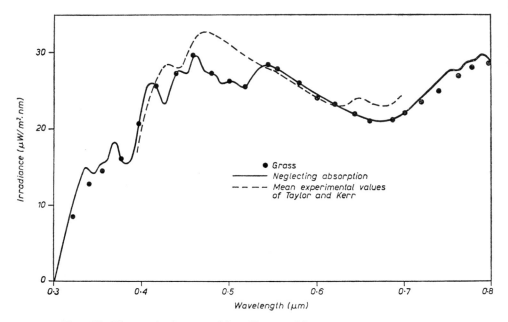

FIG. 21. Theoretical spectral irradiance of heavy overcast over grass, with and without correction for absorption due to polluted water found in clouds (Middleton). Compare curve C of Fig. 12.

Other calculations on the colour of daylight by Lenoble employed Rayleigh–Chandrasekhar scattering functions for a homogeneous atmosphere above a diffusing ground surface, which in this case was assumed to be dark. The distribution of incident radiation was that due to Nicolet,[297] and resulting changes of sun and sky radiation were reported both spectrally and in total.[294] They confirmed what has been frequently mentioned, that global radiation of nearly constant spectral distribution (or chromaticity) was to be expected over a wide range of solar altitudes.

Lenoble's determinations of dominant wavelength and saturation were referred to Standard Illuminant C (the chosen white point). Calculations were made for h from 0° to 90°, for sky alone and for sun plus sky. Where comparable the values agreed with those found by Bullrich. In Fig. 20 a theoretical chromaticity for skylight is given ($x=0.238$, $y=0.245$) which is representative of Lenoble's calculated points for h between 90° and 37°. Both authors showed the shortest dominant wavelength of sky chromaticity to be 478 nm, which, according to Lenoble, changed only to 481 nm at $h=12°$, and then to 570 nm at $h=0°$, whereas Bullrich *et al.* found variations in the same wavelength range by using a fixed value of h and different points of measurement over the sky. Global radiation varied only from 573 to 578 nm on the same scale according to Lenoble's calculations, and for $h=0$ the chromaticity was close to that of Standard Illuminant B (Fig. 20).

Previously Lenoble had shown how to compute the radiation from various directions in the sky, with simplifying assumptions.[295] Her subsequent work concentrated on the ultra-violet part of daylight, to calculate the radiation in the 301·5 to 371·5 nm band for different azimuths from the sun direction.[296] Experimental work followed, and is described in Chapter 10.

MORE STANDARD DISTRIBUTIONS

Nicolet's work in this direction consisted of a new interpretation of the SAO material with a view to a revision of S. Accepting the weighted mean, he used some of the early rocket information to apply different corrections to the ultra-violet end of the spectrum.[297] The infra-red end was modified by then unpublished data from de Jager. All this gave a value of S 2 per cent higher than the SAO value, or 1·98 L/min. This came at a time of numerous attempts at revision of the constant, particularly Johnson's somewhat later survey of the information available in 1953[156] (see Chapter 5). Nicolet's spectral power distribution is included in Table 3.

The Potsdam Observatory was concerned at this time with calculations on radiation in a diffusing atmosphere. After some experimental work like Albrecht's on global radiation, using a solarimeter and Schott colour filters, Hinzpeter began calculations on the model of a vertical stream of radiation impinging on a turbid atmosphere.[298] At first only Rayleigh scattering was assumed, and an equation resulted similar to that devised by Bullrich *et al.* This was subsequently made more complicated by corrections for turbidity, but the simpler form may be expressed as follows, for any given wavelength:

$$H_\lambda = H_{0\lambda} \cos z \frac{[\eta+(1-\eta) \cos z][1-\exp{(-a \sec z)}]}{1+a(1-\eta)(1-R)}$$

$$-H_{0\lambda} \cos z \frac{[a(1-\eta)](1-R) \exp{(-a \sec z)}}{1+a(1-\eta)(1-R)}$$

where z is the zenith angle, a the extinction coefficient of pure air, R the albedo of the earth's surface, $H_{0\lambda}$ the incident radiation flux outside the atmosphere, H_λ the vertical radiation flux on the earth's surface, and η that fraction of direct radiation scattered by air molecules which is received at the earth's surface. For calculation of numerical values Hinzpeter used $R=0.15$, and a spectral distribution of $H_{0\lambda}$ similar to that given by Nicolet but equivalent to $S=1.94$ instead of 1.98 L/min.

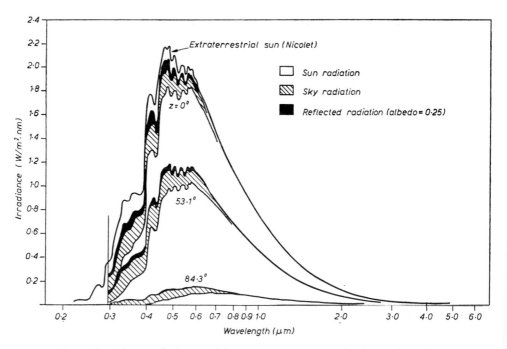

FIG. 22. Solar radiation and its components on a horizontal surface at sea level, for several zenith angles. For each value of z the hatched and solid areas are additional to the direct solar spectrum below. (Deirmendjian and Sekera.)

The spectral curves found for diffuse sky radiation with variable *z* differed from those of some other authors but agreed well in comparison with the calculations of Deirmendjian and Sekera. A further agreement with these authors was seen in Hinzpeter's calculations for the ratio of sun to diffuse radiation versus wavelength.

Calculations of global radiation on a horizontal surface, assisted by Chandrasekhar solutions for multiple Rayleigh scattering, were made by Deirmendjian and Sekera.[299] They also found approximate constancy of spectral distribution for a wide range of zenith distances of the sun. The ratio of direct to scattered sunlight could not in their view be correlated with atmospheric turbidity. Their less detailed curves separated the contributions to irradiance at sea level due to molecular scattering and to earth reflection (Fig. 22). They commented on the scarcity of measured spectra of skylight at the earth's surface, but their calculated curves for this component of daylight were similar in some respects to those of Götz and Schönmann especially in the high levels of ultra-violet irradiance found in the zenith sky at Arosa (Chapter 8). The calculation showed the persistent maximum near 370 nm found by observation. Fig. 23 gives some of the curves: compare Fig. 13.

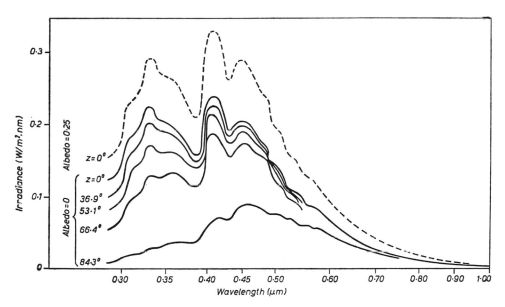

FIG. 23. Spectra of sky radiation calculated from Nicolet distribution and multiple Rayleigh scattering: effects of varied zenith angle and earth's albedo. (Deirmendjian and Sekera.)

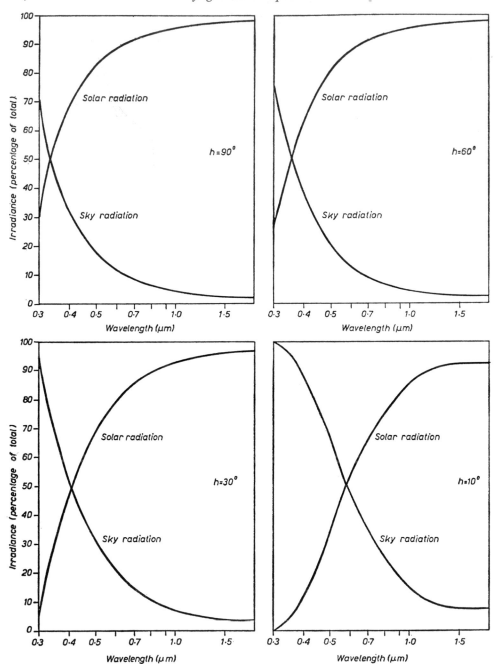

FIG. 24. Graphical representation of Hinzpeter calculations on sunlight and skylight.

Hinzpeter extended his calculations with the addition of Ångström turbidity coefficients and the $\beta/\lambda^{1\cdot3}$ law.[300] Relative values of spectra irradiance now agreed better with experiment than the absolute values which were too high by calculation. In his third paper Hinzpeter modified his model by the addition of a non-selective absorption corresponding to an aerosol affecting only wavelengths in the red and infra-red region. This grey absorption was taken, for example, as 40 per cent above 1·2 μm. To reduce the high calculated values still more it was necessary to assume a further anomalous selective absorption in the ultra-violet part of the spectrum. This was based on some rather doubtful experimental work.

Hinzpeter's extensive tables of his calculations were on a basis of 40 nm bands between 280 nm and 1 μm. They have been much quoted, being one of the last investigations of the kind.

Fig. 24 is an example of the data in their usual representation as ratios between skylight and sunlight.

In 1959 Sekera revised his previous calculations and with Dave made an additional correction for ozone absorption. Comparison was made with some experimental work by Liljequist in Antarctica,[301] but the values for diffuse irradiance agreed only if the atmospheric ozone content was assumed to be 5 mm. Dave and Sekera calculated the total sky and global radiation for different values of albedo, ozone content, and solar altitude.[302] They used Nicolet's solar irradiance values and Vigroux's measurements of ozone absorption bands. Much later, in the period covered by the next chapter, calculations made by Dave on absorption, reflection and transmission of solar radiation included the effects of aerosols, ozone, water vapour and clouds.[303]

10

Spectrophotometry of Daylight from 1960

Often that which has come latest seems to have accomplished the whole matter.

LIVY, *History of Rome*, Book 27, 45.

In the period since 1960 a large amount of experimental work has been done in widely separated parts of the earth. Some of this has been of great precision and elaboration, and concerned in particular with the radiance of the photosphere of the sun, and irradiance at high levels above the earth: this is discussed in Chapter 11. Theoretical distributions and the dissipation of solar radiation in the atmosphere have been the subject of numerous calculations, but more interest has been shown in the standardization for practical purposes of representative spectral distributions for daylight. The colorimetry of the sky and daylight has become much more frequent in this period. First to be considered are a few important researches on the ultra-violet spectrum.

ULTRA-VIOLET MEASUREMENTS AT DAVOS

At Davos, Bener has been carrying out a very extensive project at the Physikalisch-Meteorologische Observatorium, which Dorno founded for the study of climatic factors in this mountain site, renowned for its curative effects.[151] In 1944 and 1945 Bener had made over three thousand measurements of total sky radiation by a solarimeter, with the intention of determining the effects of cloud cover and its type, and of solar altitude.[304] Later the spectral work began with a quartz double monochromator furnished with an integrating sphere input and a photomultiplier detector (EMI 6256). Some thousands of spectra were recorded between 290 and 400 nm with a resolution of 1 to 2 nm, most of them from the hemisphere of sky, some with direct sunlight included, and a few of sunlight alone.[305] Different aspects of the measurements were analysed in successive reports: first a comparison with calculated values based on the methods of Sekera, next the pattern of daily and annual variations in ultra-violet radiation from nearly cloudless skies, and then the effects of clouds expressed as ratios of the radiation with and without clouds. Other major factors recognized throughout were ozone content of the atmosphere and solar altitude, while the winter snow cover produced recognizable effects complicated by the normal changes in ozone concentration. The measured spectral curves showed considerable detail, but most of those reproduced in the reports either gave the variation of irradiation at a single wavelength against other parameters, or were without fine structure by being plotted at fixed wavelength intervals.

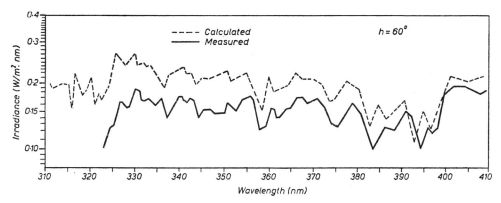

FIG. 25. Measured and calculated spectra of ultra-violet sky radiation for summer conditions. (Bener.)

A separate series of ninety-nine spectra between 330 and 410 nm, a range which allowed ozone absorption to be neglected, was used for comparison with calculated values based on Rayleigh scattering without aerosol addition.[306] Fig. 25 shows an example of the failure to agree which was normally found. Calculations for single wavelengths and for specified solar altitudes showed discrepancies which were nearly always in the sense of lower experimental than calculated H_λ, with differences up to 32 per cent at 330 nm in summer conditions. Better agreement was found with increasing wavelength, especially at increased h though often the calculated value still fell somewhat below the measured one at $h > 40°$. In winter the shapes of the experimental

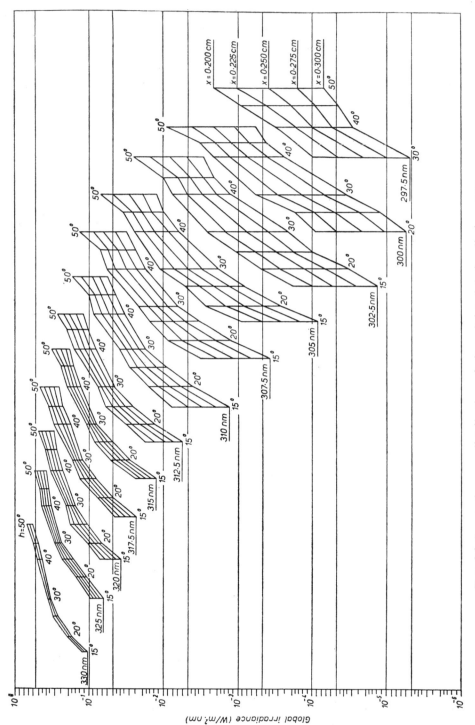

Fig. 26. Spectral power distribution of ultra-violet global radiation at Davos; its dependence on solar altitude *h* and atmospheric ozone content *x*. (Bener.)

curves were close to the calculated ones but their levels were consistently lower by about 26 per cent between 330 and 370 nm. The discrepancies might be due to the extraterrestrial $H_{0\lambda}$ values used (those of Dunkelman and Scolnik), or to incorrect assumptions regarding ground reflection where the albedos chosen were 0·15 (summer) and 0·75 (winter).

The second large section of Bener's work was to prepare tables and curves for the variation of ultra-violet radiation for the daylight hours throughout the year.[307] Sky and global variation and spectral distributions were all represented, and data for the single wavelengths 300, 310, 320 and 330 mm. In all cases the values were reduced to a standard

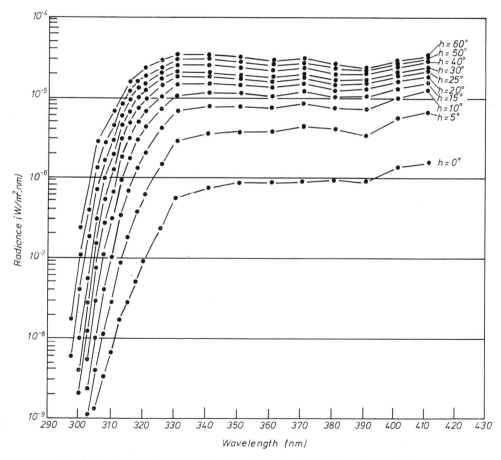

FIG. 27. Spectral power distribution of ultra-violet sky radiation at Davos for variable h and 3·35 mm ozone. (Bener.)

ozone content of 2·5 mm. Fig. 26 shows how this affected the global radiation at different wavelengths and different values of *h*. Among other conclusions Bener showed that diffuse ultra-violet radiation from the sky formed a larger fraction of the global radiation than is the case for the visible part of the spectrum. This diffuse ultra-violet accounted for more than half the total for considerable periods, depending on the time of year, solar altitude and wavelengths in

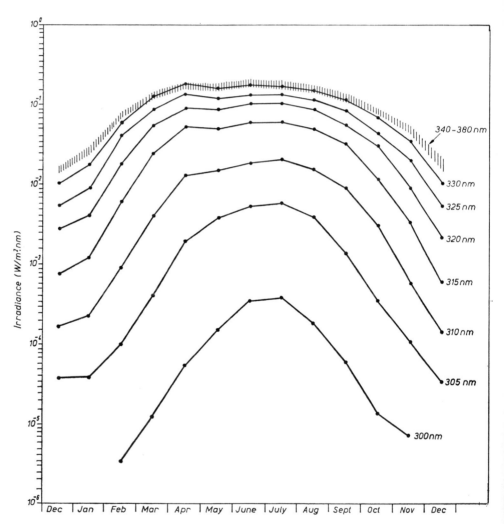

FIG. 28. Annual variation of irradiation at selected wavelengths from the sky at Davos, at 8.00 and 16.00 hours. (Bener.)

question. Examples of the diagrams are given in Figs. 27 and 28. This section of the work referred to cloudless conditions.[308]

Bener analysed his measurements for the effect of clouds which were classified by type and level, comparisons being made at 330 and 370 nm.[309] Clouds affected the ultra-violet sky radiation less than the visible radiation, and could reduce or increase the irradiance. Fig. 29 gives

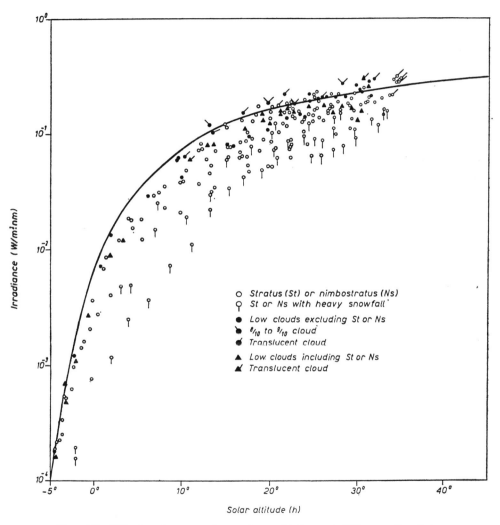

O	Stratus (St) or nimbostratus (Ns)
♀	St or Ns with heavy snowfall
●	Low clouds excluding St or Ns
◗	8/10 to 9/10 cloud
◖	Translucent cloud
▲	Low clouds including St or Ns
◢	Translucent cloud

FIG. 29. Effects of overcast on sky radiation at 330 nm, for 8- to 10-tenths low cloud in winter, and variable solar altitude, at Davos. The curve gives mean values for nearly cloudless skies and snow cover. (Bener.)

mean values for nearly cloudless skies as a function of solar altitude, while the scattered points show how the irradiation is affected by low cloud. In the same paper Bener compared diffuse radiation for cloudless skies in summer conditions to those of winter with snow cover. At 330 nm, in winter, mean irradiance (as shown in the curve in Fig. 29) was higher than in summer by 50 to 75 per cent throughout the range of $h=0°$ to 40°. At 370 nm the winter excess was less, but still 30 to 40 per cent.

A more extensive and varied programme was made possible by the construction of a transportable spectroradiometer for measurement of sky and direct solar radiation at 11 wavelengths between 297·5 and 400 nm.[310] The instrument consisted of a Zeiss double prism monochromator and a cooled 12-stage photomultiplier, with an achromatic telescope to image the sun (or an area of sky) on the photocathode and an automatic drive to keep the image fixed. For low intensity radiometry, photon counting replaced photocurrent measurement. In a three year programme the observations were extended to Biel at 316 m elevation and the summit of the Weissfluh (2818 m) (Fig. 30), besides Davos (1590 m), with measurements of zenith sky radiance I_λ and direct sun irradiance H_λ, and the main variables h, air mass and the known ozone concentration at the time.[311] Polarization effects from skylight were eliminated by adding the components polarized parallel and perpendicular to the entrance slit. The ground albedo was assumed to be 0·1.

By means of the Rayleigh atmosphere calculations of Coulson *et al.*[115] and the Labs and Neckel tables for solar irradiance,[50] some agreement between theory and practice was found, but discrepancies were probably due to the lack of theoretical correction for aerosol effects. A convenient measure was the ratio I_λ/H_λ (in units of sr^{-1}). The theoretical ratio R_{th} was always less than the measured ratio R_m; it was found that R_m/R_{th} increased for lower site elevation, for lower h, and for shorter wavelengths, and was generally less than 2 except for the Biel measurements. On the other hand the ratio of measured to theoretical zenith sky radiance I_m/I_{th} was mostly between 1 and 1·5, except for the Weissfluh results which gave somewhat higher ratios. In these theoretical calculations the wavelength range was from 330 to 400 nm only.

From the measurements of direct solar radiation the extraterrestrial

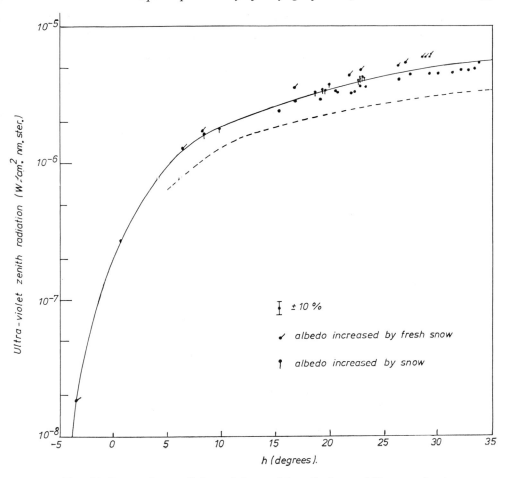

FIG. 30. Dependence of ultra-violet zenith radiation at 340 nm on h, at Weissfluhgipfel. (Bener.) Full curve, mean trend. Dotted curve, theoretical trend for a Rayleigh atmosphere and for an assumed albedo of $A = 0.10$.

spectral values could be calculated, but these were scattered both above and below the Labs and Neckel values, due to the erratic effects of aerosol scattering. From all this work Bener obtained tentative data for standard values of zenith radiance at above 330 nm as a function of optical thickness of the atmosphere.

At the highest site, measurements were made along several angles of elevation and azimuths with respect to the sun, and additionally two visible wavelengths were used, 488 and 533·5 nm.[145] Here there was better agreement, at the longer wavelengths, with the calculated values of direct solar irradiance as a function of *h*, angle of elevation and wavelength. Fig. 30 shows the results for 340 nm. This investigation included polarization measurements already discussed (Chapter 4).

Further study was intended of the very large collection of measurements in these later reports.[311, 145] One summary report covers much of the long-continued work at Davos, with extrapolation by standard data[64, 292, 519] and by the calculations of Dave and Furukawa on scattered radiation in the ozone absorption bands for a Rayleigh atmosphere.[312] Irradiance values are given at wavelengths from 297·5 to 380 nm for the whole sky, the direct sun and its vertical component, and the global radiation. Tables show (1) variation of these quantities with *h* (0° to 90°), ozone content (2·4 to 4·4 mm), and height above sea level (0 to 5 km); (2) the effect of reduction of the normal ozone content by up to 50 per cent at selected values of *h* and northern latitudes; (3) changes with season at various northern latitudes.[85] The report considers cloud effects, and the minimum wavelength observable in daylight (Chapter 3). Fig. 31 is an example of the final results, in this case for sea level.

Since the conclusion of the ultra-violet work at Davos, the observatory's main objective of studying the interaction of radiant energy with the earth's atmosphere has been shown in the construction of a mobile station for automatic measurement and recording of solar radiation. Three identical double monochromators are designed for (1) the centre part of the sun's disc, (2) small areas of sky at distances of 1° to 16° from the sun's centre, and (3) small areas of sky along the sun's vertical. The solar instrument (1) is suitable for high level measurements in a balloon; polarization measurements are included in (2) and (3); and all the spectroradiometers use twelve wavelengths between 370 nm and the near infra-red. In addition there is a pyroheliometer for the whole spectrum of the sun, and seven pyranometers for global radiation in different directions. Observations made at different sites could provide useful practical data, as well as much needed information on problems of atmospheric turbidity.[313]

Calibration was effected by tungsten ribbon lamps in fused silica envelopes. In the later work the primary standard was acquired with calibrations against full radiators at fifteen wavelengths between 294·1 and 500 nm, with a claimed absolute accuracy of 1 per cent. A tungsten halogen lamp was a secondary standard in part of the programme.

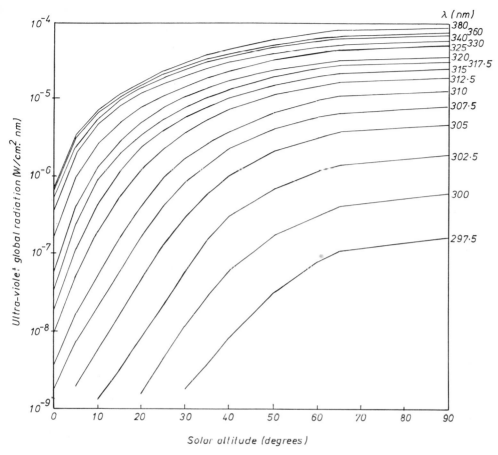

Fᴵɢ. 31. Dependence of ultra-violet global radiation on *h* for different wavelengths at sea level; 3·4 mm ozone, ground not covered with snow. (Bener.)

Calibration, accuracy and sensitivity of the measuring system were discussed at length.[310, 311, 314]

OTHER WORK ON THE ULTRA-VIOLET SPECTRUM

The other studies in the present period are of much smaller scope. In one of these a portable quartz spectrograph with a Ferry focusing and reflecting prism of 30° angle was used by Andreychin to photograph the spectrum of direct sunlight at Sofia (590 m) and other sites at sea level and 2290 m elevation.[315] Daily variations of the irradiance at

330, 354 and 394 nm showed the expected increases at higher levels, and a longer wavelength was observed to persist longer than a shorter one at both ends of the day. Using the same equipment Andreychin and Kechlibarov made some measurements recalling earlier attempts to extend the observable ultra-violet spectrum to the limit (Chapter 3). Photographs of the solar ultra-violet spectrum at different heights up to 2000 m established a minimum wavelength near midday for each site, and a decrease with increased elevation of the site, but no new records were made.[316]

In Chapter 9 Lenoble's theoretical calculations on the ultra-violet irradiation from the sky were mentioned. In that period her experimental work consisted of recording photographically the ultra-violet component from the zenith, and if no clouds were present the measured values agreed with theoretical estimates. This was done in Paris, but by experiments outside the city it was shown that haze was more important in causing variations in irradiation than the coarser aerosol of city dust.[317]

To continue this work a better quartz prism instrument was constructed for the range 297 to 414 nm, with a caesium–antimony photomultiplier of twenty stages and a thin glass entrance window. Automation provided records at predetermined solar altitudes, when an eight-minute scan of the short wavelength spectrum was made.[318] The equipment included a polarimeter, and attenuating filters of evaporated nickel–chromium on silica which had nearly uniform absorption between 300 and 400 nm.

In the spectral measurements made with this spectrograph in Paris and Lille the range was limited to 297 to 340 nm, with h between 5° and 60°. No spectral curves were published by Lenoble, Arceduc and Coquelle.[319] Ratios of flux at chosen wavelengths, for example 376·5 : 329 nm, were compared with calculations of the Rayleigh–Chandrasekhar type. Agreement was moderate. Similarly the daily variation of the single wavelength 340 nm, recorded continuously throughout 1959, was compared with the theoretical curve by suitable adjustment of the ordinate scale. Fig. 32 shows the excess of measured radiation at high solar altitudes: abscissae represent $\cos z$ or $\sin h$ or $1/m$, ordinates calculated or measured luminance.

Ultra-violet measurements were made at Stamford, Connecticut, by Hirt, Schmitt, Searle and Sullivan, beginning with a comparison of the

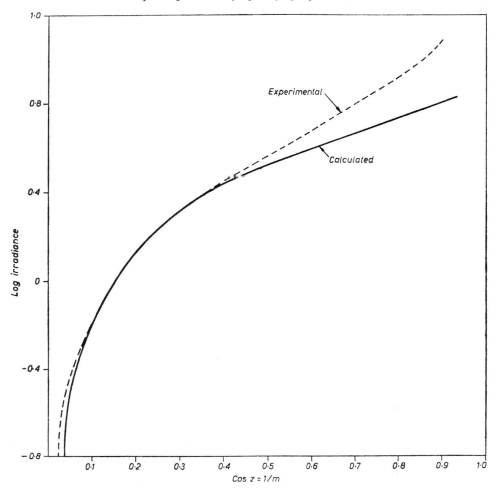

FIG. 32. Diurnal variation of zenith radiation at 340 nm in Paris, compared with calculation by Rayleigh–Chandrasekhar theory. Curves brought to nearest fit on vertical scale. (Lenoble, Arceduc and Coquelle.)

ultra-violet content of sunlight with that in artificial sources used for testing materials liable to photodecomposition.[320] With a grating monochromator and 1P28 photomultiplier the range covered was 266 to 410 nm, the lowest recorded solar wavelength being 293 nm. To provide absolute measurements a ferrioxalate actinometer was also used (see Appendix 1), and a filter allowed a narrow band at 365 nm to be isolated for this purpose. Extensive measurements were made on artificial sources, but only a few solar spectrum curves with moderate structural detail were given. These were measured at 3 nm intervals with an exit slit of 5 nm width. Searle and Hirt later reported measurements on solar ultra-violet and total irradiance made with the same equipment and a heliostat.[321] Skylight was largely avoided by restricting the monochromator angle of view to 2°23′. As others have found before and since, the total amount of ultra-violet radiation in the 290 to 410 nm band was not proportional to total solar power received, measured in this case by a radiation meter, but it was highly dependent on atmospheric conditions. Ratios between 1 per cent and 7 per cent were found for the proportion of ultra-violet to total power. Diagrams gave daily variations of radiation at 305 and 410 nm, and annual variations of the part below 340 nm, besides a few spectral curves for different times of the year. Fig. 33 shows that the variation of ultra-violet irradiance is much greater than that of the total power received.

In order to provide reasonable estimates of the radiation between

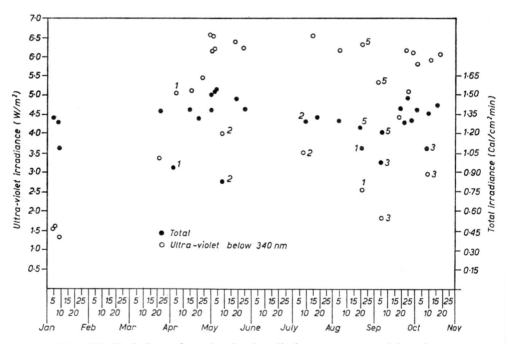

FIG. 33. Variation of total solar irradiation at noon and its ultra-violet component below 340 nm, at Stamford, Connecticut. Observations made with clear sky except those numbered (1) very slight haze, (2) slight haze, (3) haze, (5) partly cloudy but sun clear. (Searle and Hirt.)

280 and 340 nm for different values of h, ozone and aerosol concentrations, ground albedo, elevation of the site and cloudiness, Green and his co-workers developed semi-empirical formulae from the available measurements of direct solar, diffuse, and global ultra-violet. The formulae were intended for use in studying man-made changes affecting normal ultra-violet radiation, and with the skin cancer problem in view extensive tables were produced giving the daily erythemal dose for different months, ozone contents and north latitudes.[322] One calculation showed that reduction of ozone from 3·2 mm to 1·6 mm would result in an increase of 130 per cent in the incident ultra-violet. This appears to be not inconsistent with Bener's estimates for separate wavelengths at $h=60°$ and latitude 40° N.[85] A useful summary of the exploration of the ultra-violet from Ritter in 1807 to the rockets of 1959 was given by Tousey.[323] Other useful surveys were made by Massey[324] and by Tousey.[325] An account of a rocket flight in 1960 to a height of 235 km gave details of emission lines, absorption bands and absolute energy levels between 50 and 260 nm.[326] By 1967 the line spectrum had been extended to 0·6 nm.[44] More recent results from rockets and satellites were reviewed by Hinteregger[327] and Heath.[328]

The 1960 rocket flight confirmed the minute proportion of total solar radiation which is contained in the short ultra-violet region. The power in the Lyman α line of hydrogen, detected by an ionization chamber, was 6×10^{-3} W/m², or 5 millionths of the whole irradiation. A comparable measurement by a platinum photocell, carried in a free balloon to upwards of 15 km over Trinidad, made use of a narrow atmospheric window centred on 210 nm. At this wavelength the irradiation extrapolated to outside the atmosphere was $1·08 \times 10^{16}$ photons/m² s nm.[329] This is about 7 millionths of the total solar irradiation.

EXTENSIVE SAMPLING OF THE DAYLIGHT SPECTRUM

Since 1960 a number of similar projects to measure the terrestrial daylight spectrum have made available a large amount of information on the visible and near ultra-violet regions. The colorimetric aspect has been emphasized, particularly for transforming the original measurements of spectral power distribution into an easily recorded form, though with the inevitable reduction in the information conveyed. The work used for devising new standards of artificial daylight is considered first, then similar work not included in the standardization, and more specialized investigations appear in later chapters.

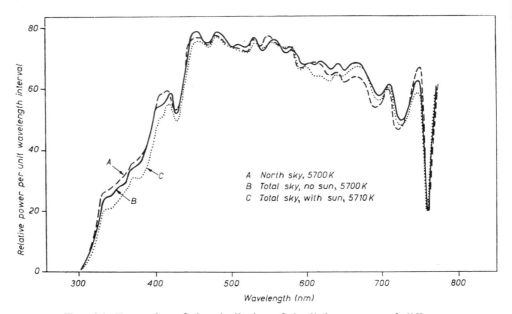

Fig. 34. Examples of the similarity of daylight spectra of different origins but the same CCT. (Henderson and Hodgkiss.)

The former British Standard Specification for artificial daylight fittings (BS 950:1941) was based on the spectral power distribution of Standard Illuminant C (Chapter 18). When the revision of this specification was begun in 1960, no work in this country on the spectrum of daylight could be traced, except that of Cunliffe where insufficient spectral detail was given (Chapter 8). To remedy this lack of information Henderson and Hodgkiss began a programme of measurement at Enfield, Middlesex.[330] During fifteen months from April 1961 they measured the distribution on 274 occasions over the spectral range 300 to 780 nm at 10 nm intervals and with a bandwidth of 1·5 nm. The equipment included a small portable monochromator (Hilger & Watts D292), a grating of 575 lines/mm, two photomultipliers (Mazda 27M3 for 300 to 580 nm, RCA 1P22 with a Chance OY6 filter for 540 to 750 nm), and a standardized tungsten-in-glass lamp at

2854 K for calibration. In a direct view of the north sky the mono-chromator accepted a 6° cone of incident light at 45° elevation; alternatively a horizontal white Vitrolite plate was viewed from above at 45° reflection for global measurements, some including direct sun-light. No correction was made for fluctuations of irradiance during the 10-minute measuring period for each complete spectral curve. Fairly steady sky conditions were chosen as far as possible, with the aim of obtaining representative averages rather than accurate single curves. It was soon established that metamerism of the light source was negligible in amount: in other words, for a given correlated colour temperature determined from the spectral power curve by CIE 2° field standard data, the curves were very similar for north sky, total sky without sun, and total sky with sun. Fig. 34 gives an example. This is another expression of the tendency to uniformity in spectra observed by several earlier authors. It is an important fact in justifying simple procedures for averaging large numbers of curves, as was done at first in this investigation. Later more sophisticated methods were used (Chapter 13). Fig. 35 includes a selection of single spectral curves for different

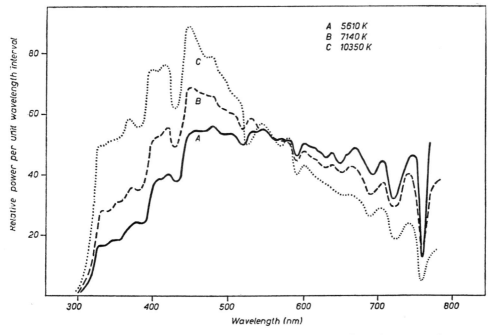

Fig. 35. Examples of daylight for various CCT. (Henderson and Hodgkiss.) Normalized at 560 nm.

colour temperatures; Fig. 36 shows the means for sixteen pairs of spectra in which sunlight was alternately included and excluded, together with full radiator distributions of nearly the same equivalent correlated colour temperature (CCT). This description is used for brevity to denote the calculation of chromaticity from a spectral power distribution followed by determination of the CCT of the chromaticity point, graphically or by calculation (Appendix 1). Note that the full radiator temperature is higher than that of the daylight CCT because daylight has comparatively less blue radiation.

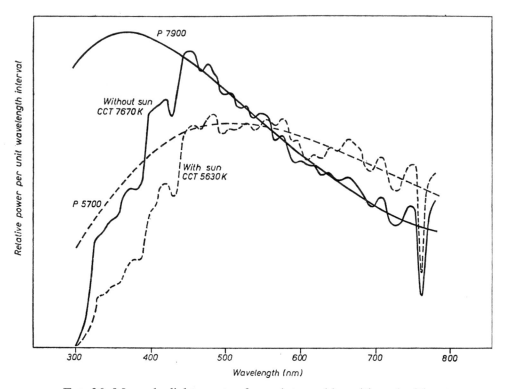

FIG. 36. Mean daylight spectra from sixteen skies with and without sunlight; also full radiator distributions of reasonable fit for 7900 K and 5700 K. (Henderson and Hodgkiss.) Normalized at 560 nm.

The distribution of chromaticities found in this work is seen in Fig. 37, a very similar pattern to that found by other authors of comparable work described below, though there are some minor differences possibly worth closer investigation. The histograms of CCT distribution (Fig. 38) may be compared with others in this chapter.

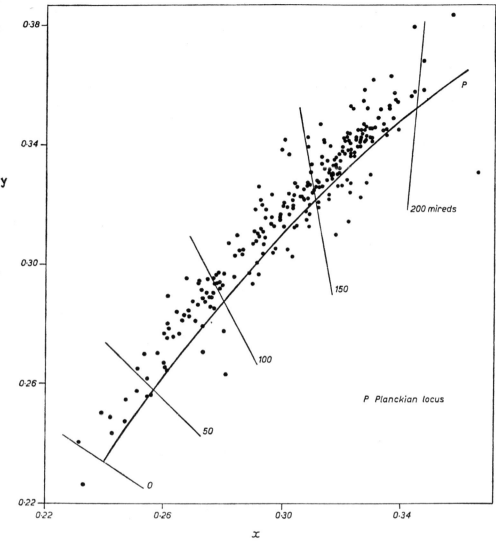

FIG. 37. Distribution of chromaticities of daylight at Enfield, Middlesex, 1961 to 1962. (Henderson and Hodgkiss.)

Similar spectral distributions and chromaticities measured about this
time in Germany were reported by Bodmann and Jantzen; they range
from 6020 to 13 500 K in CCT.[331]

  Means of numbers of curves for the daylight spectrum in selected
areas of CCT were used in two proposals for new standards. They
were not implemented because of the wider scope of action taken later.

They were also complicated by the appearance of a new set of iso-tempcrature lines[332] for the CIE 1960 Uniform Chromaticity Scale. This made recalculation of the CCT values necessary: it has been done in Figs. 34 to 39.

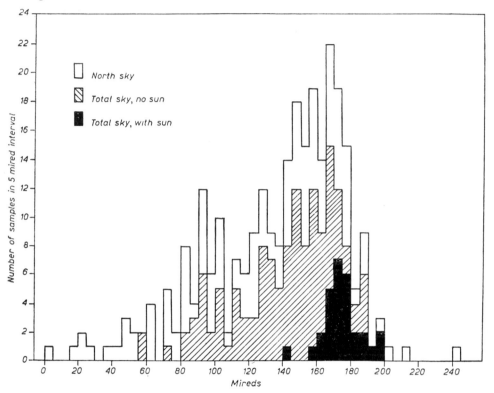

Fig. 38. Distribution of CCT (mireds) of daylight. (Henderson and Hodgkiss.)

The first proposal was based on the means of seventy spectra which gave equivalent CCT in the 5500 to 6500 K range, and ten spectra above 20 000 K. The two representative distributions, of equivalent CCT 5990 and 30 770 K, were used as the limits of mixtures in the Abbot–Gibson manner, with the advantage that the line joining the chromaticity points lay entirely on the green side of the full radiator locus, besides having a completely experimental basis. A check on an experimental curve of equivalent CCT 10 000 K showed this formulation to be a closer fit than the Abbot–Gibson distribution.

The second proposal was for a spectral distribution near 6500 K as a standard of artificial daylight. This was the average of a selection of normalized curves with equivalent CCT in a suitable interval. Details are given in Chapter 13 where the international standardizing procedure is described. Its object was to provide typical spectral power distributions for daylight at any selected CCT, or chromaticity on a defined locus. These are referred to as 'reconstituted daylight' distributions (RD) and they lie on an RD locus nearly parallel to and on the upper side (the 'green side') of the Planckian locus (Figs. 20 and 68). Planckian distributions are referred to as P6500, etc. The RD for 6500 K proved to be very similar to the experimental mean proposed by Henderson and Hodgkiss.

Another feature of these observations was the information they gave on the ultra-violet content of daylight. The 300 to 400 nm fraction of the total irradiance varied widely and with no evident dependence on CCT. This has, of course, been often observed before and has defeated attempts at standardization. Fig. 39 shows the variations graphically and includes a comparable mean line for Winch's measurements (see below). A closer examination of the points in Fig. 39 demonstrated some dependence on time of year, with higher than average proportions of ultra-violet in summer and lower than average in winter. This provides a contrast between the urban site and the Alpine one in Bener's work.

The next two investigations of the daylight spectrum provided further material for the standardization of RD distributions.

In June 1962 Condit and Grum recorded about five hundred daylight spectra at Rochester, N.Y.[333] They had in mind the needs of outdoor photography, in particular the effects of the ultra-violet content of daylight, of solar altitude, and of irradiation on planes other than the horizontal. Their instrument was a Beckmann prism reflectometer adapted as a spectroradiometer, with automatic slit control to provide a constant photomultiplier signal from the reference source, a tungsten lamp at 3000 K. Recording from 300 to 700 nm took 75 s, and resolution was 5 nm or better in the curves which were later measured at forty-six wavelengths selected at peaks or troughs in the spectrum. The traces were automatically corrected by instrumental and lamp calibration functions, while a record of illuminance served as a further correction applied if the level changed by more than 2 per cent during a spectral scan.

Condit and Grum used a barium sulphate coated plate to receive

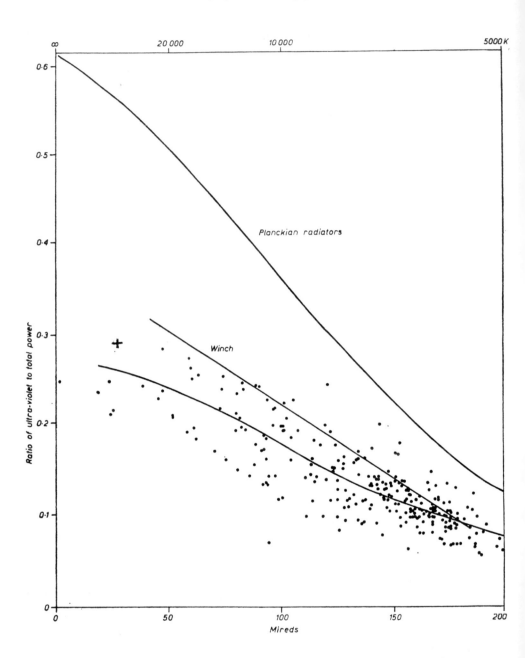

Fig. 39. Ratio of power in spectrum from 300 to 400 nm, to that from 300 to 780 nm, for observations of Fig. 38. The middle line represents the ratio obtained by Winch in South Africa. The upper line is the ratio for Planckian radiators.

the daylight in numerous aspect positions at different solar altitudes. The complete sky was photographed every minute to record cloud conditions. The authors emphasized the solar altitude parameter more than most others have done. Fig. 40 shows some of their curves from the two hundred and fifty spectra selected. Their observations at low

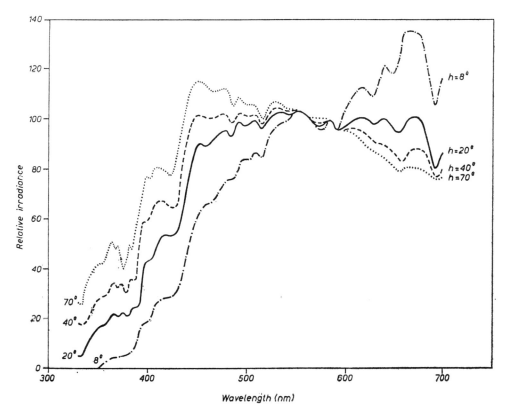

FIG. 40. Daylight spectra from sun and clear sky for several solar altitudes. Irradiation on a plane facing the sun in azimuth but tilted 15° from the vertical away from the sun. Normalized at 560 nm. (Condit and Grum.)

values of h produced some of the lowest equivalent CCT on record (Figs. 20 and 68). They submitted their spectral distributions to characteristic vector analysis (Chapter 13) but the results were described only in general terms.

The third contributor to the collaboration on RD distributions was Budde, who in seven weeks of March and April 1963 recorded ninety-nine spectra of north skylight or total skylight at Ottawa. He used a quartz double monochromator with a bandwidth of 1 to 7 nm in the range 300 to 720 nm. Radiation was accepted from an integrating sphere and detected by a 1P28 photomultiplier. The recorded curves were automatically corrected by a separate signal for irradiance at 560 nm, and in subsequent measurements 10 nm bands were averaged throughout the spectrum. Colorimeter checks were made by a Donaldson six-stimulus visual colorimeter simultaneously with the recordings: compare the observations of Nayatani and Wyszecki from the same site,[334] when the chromaticities of the north sky showed a much wider spread across the full radiator locus than those of Budde (Chapter 12). This work by Budde was not published, and the details are those given by Judd *et al.*[335]

For the purposes of this chapter the RD distributions may be considered as the weighted means of the measurements by Henderson and Hodgkiss, Condit and Grum, and Budde, for selected equivalent CCT. Publication of the investigation and approval of its proposals by the CIE had some effect in encouraging more measurements in other parts of the world, where the characteristics of daylight might perhaps not have agreed with the RD formulations. The new locations of this work were South Africa, India and Japan.

At the first of these Winch, Boshoff, Kok and du Toit obtained over four hundred sets of spectral power data on daylight with and without direct sunshine.[336] This was in the period from September 1964 to March 1965, at Pretoria in Transvaal province at 1400 m altitude. The intention was to test the CIE proposals on standard illuminants of the RD series. Nearly forty years earlier the Mount Brukkaros station, not far from Pretoria in latitude and at a similar elevation, had contributed to the Smithsonian enterprise, otherwise this is one of the very few locations in the southern hemisphere where spectroradiometric work on daylight has been done. The apparatus used was a quartz double monochromator with integrating sphere input, an EMI 9558 tri-alkali cathode photomultiplier with a silica window, a digital voltmeter from which the photocurrent values were observed, and a 500 W tungsten lamp in a fused silica bulb for comparison. As a primary standard a calibrated tungsten–iodine lamp in a silica envelope was used.[337] No correction for sky variations was made during the 4-minute scan over 285 to 775 nm at 5 nm intervals. The maximum resolution was 1 nm at 500 nm.

The equivalent CCT values lay between 5200 and 12 000 K. In order to obtain representative spectral distributions and others at the CIE-recommended standard colour temperatures, the observations were grouped over intervals of 25 or 10 mireds and interpolations made at the exact colour temperatures required. Comparison with RD values for 5500, 6500 and 7500 K revealed some differences (Table 6, Chapter 13) especially in the ultra-violet and violet spectral regions. The values for 6500 K have been plotted on Fig. 67, Chapter 13, to show how the curves differ. Apart from the displacement of the strong oxygen absorption at 759 nm by about 10 nm, the agreement in the visible spectrum is adequate.

According to the measurements of Winch *et al.*, the ultra-violet contribution to daylight is about 30 per cent higher than in the RD formulations. This was plausibly explained by the site and altitude in South Africa. The trend is evident from Fig. 39. In the 300 to 500 nm region the shape of the curves resembled that of others from high altitudes, for example those of Stair, Johnston and Bagg,[277] and even those calculated for $m=0$ by Pettit (Fig. 9), Stair and Johnston (Fig. 15) and Dunkelman and Scolnik (Fig. 17).

Inspection of Winch's curves and tabulated values does not offer any immediate reason for another well-marked feature of his results, namely the closer approach of the chromaticities to the full radiator locus than is the case in most other work of the same kind. Compare Fig. 41 with similar figures in this chapter. The authors suggested that the strong Fraunhofer G absorption near 431 nm might be responsible for the slight green shift found in most daylight chromaticities compared with full radiators. Filling up this absorption band brought the chromaticity across the Planckian locus but with an increase in CCT of 140 K. Perhaps there is some significance in Winch's lower values near 520 to 530 nm and increased values on each side of these iron and magnesium absorptions (see Table 6). One other comparison is available. Fig. 42 is a histogram of the South African results, plotted like Fig. 38. Although the selection of occasions for measurement was not entirely random in either case, the South African conditions appeared to be less variable than those in Enfield, but with some similar features: for example, the wide and fairly even spread of south (or north) sky chromaticities, and the position of the most frequently observed CCT. The predominance of sunshine in the climate of Pretoria is clear from

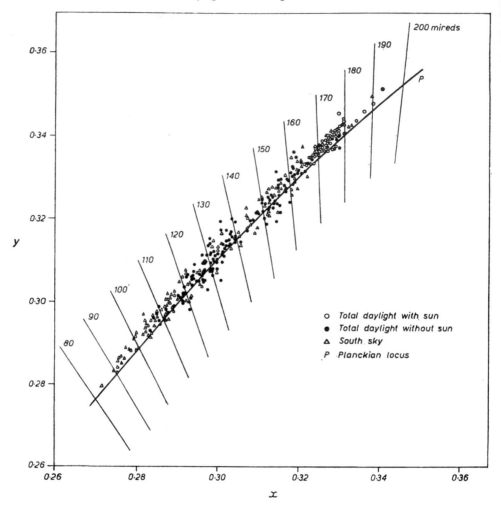

FIG. 41. Distribution of chromaticities of daylight at Pretoria, 1964 to 1965. (Winch, Boshoff, Kok and du Toit.)

this diagram. This work seems to provide the only available comparison of south sky in the southern hemisphere with the frequently measured north sky in the northern hemisphere.

Another set of measurements on the 300 to 400 nm part of the spectrum, made at the same time, was published later by Boshoff and Kok.[338] This included relative values for different cloud conditions, showing the same high levels at about 330 to 350 nm as in the main experiment (Fig. 67). Absolute measurements were reported for a number of conditions, with the largest amount found on a 'very hot summer day' (including direct sunshine). This occasion provided quite remarkable values, for at 370 and 400 nm they were 1080 and 1456 W/m² μm respectively, or within 5 per cent of several of the extraterrestrial irradiation ($H_{0\lambda}$ in Table 3). Below 370 nm the irradiation decreased though still remaining well above Moon's calculated curve for $m=1$, and at 409 and 699 W/m² μm for 320 and 330 nm the values

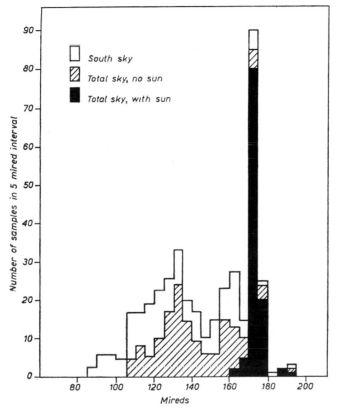

FIG. 42. Distribution of CCT (mireds) of daylight. (Winch *et al.*)

were close to Bener's highest standard levels (Fig. 27); but Bener found a few unaccountably high values in his work.[311]

Kok used the data to obtain mean absolute spectral irradiance for sunlight and for total daylight on a horizontal plane, under cloudless conditions with high atmospheric clarity. Diurnal variation of the attenuation due to atmospheric pollution still appeared in the shorter wavelengths. The results were expressed for $m=1$[339] and later extrapolated to $m=2$ for global values on a horizontal plane, and global and solar irradiance perpendicular to the sun's direction, from 300 to 775 nm.[340] The direct sunlight values are 10 to 20 per cent lower than Moon's for $m=2$ and the calculated CCT is lower at 4930 K. This distribution is given in Table 5; the direct sunlight chromaticities appear in Fig. 75.

Das and Sastri had been developing methods for spectral measurements on the sky at New Delhi, with the intention of comparison with Abbot–Gibson and RD distributions.[341] They first used a quartz monochromator and a 1P22 photomultiplier to produce some measurements on the north-west sky at 40° elevation, under varied cloud conditions. The spectral range was extended from 340 to 700 nm to 300 to 750 nm in the next set of observations, where a 1P28 photomultiplier was used and the spectral curves continuously recorded.[342] For calculation of chromaticity the curves were read at 10 nm intervals. Very marked divergences from the full radiator locus began to appear at higher colour temperatures, while the reverse held at lower colour temperatures when hazy sky conditions had been observed. Sastri and Das later reported their main group of 187 spectral power distributions of the north sky for a year commencing October 1964.[343] The main changes in the experimental methods were the substitution of an aluminized mirror instead of a magnesium oxide screen on the entrance side of the monochromator, and an independent measurement of illuminance to correct for variations in sky irradiance during the 7·5-minute scan.

The chromaticities covered an unusually wide area of the diagram (Fig. 43) with equivalent CCT between 4000 and 20 000 K. The greenish colours were mostly those measured in monsoon conditions when the air was freed of dust by rain. The tendency previously noted by the authors for the points to lie on a line of lower slope than the Planckian locus was marked in the main group of measurements, with higher CCT above, and lower CCT below the locus. The usual preponderance of points on the green side of the locus was absent in this work. No representative locus for the chromaticity distribution was given by Sastri and Das, though they resorted to the characteristic vector method of analysis to derive typical spectral distributions for given colour temperatures (Chapter 13). The shape of the Indian curves was distinctly different from most earlier ones. They showed sharp peaks at 420, 470 and 500 nm with the second often predominant and

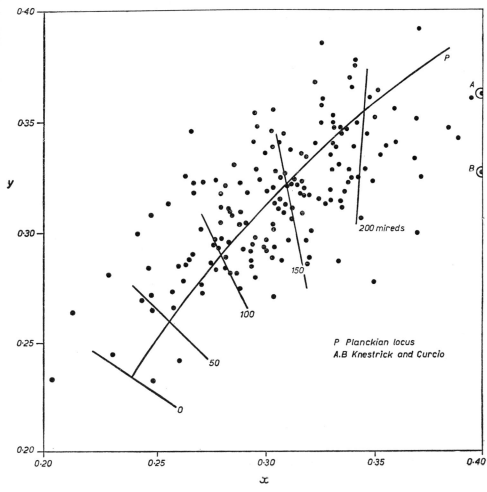

FIG. 43. Distribution of chromaticities of north skies at Delhi, 1964
to 1965. (Sastri and Das.)

separated from the third by a deep absorption, apparently Fraunhofer
F (hydrogen). Above 600 nm they were consistently high in spectral
power, and some unexplained changes of wavelength occurred in the
absorption bands at shorter wavelengths. See Table 6 and Fig. 67,
where the Sastri and Das distribution for 6500 K is included. Like the
D_{65} distribution the Sastri and Das curve is not directly experimental,
but computed by the vector method.

Like Winch *et al.*, the authors considered the possibilities of chromaticity shifts towards the full radiator locus by filling up absorption

Daylight and its Spectrum

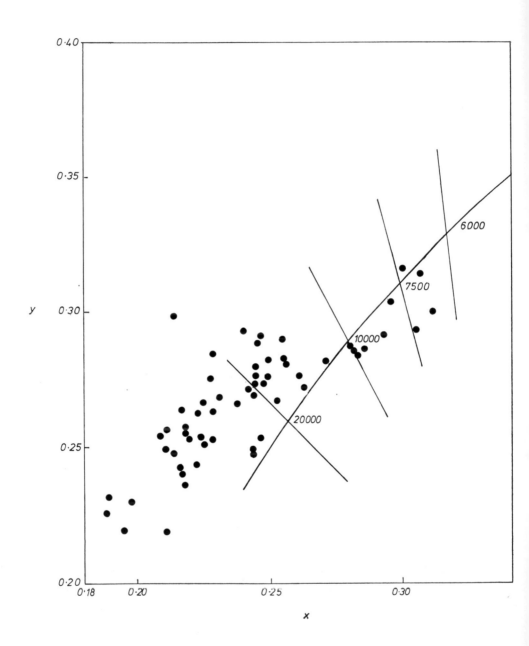

FIG. 44. Distribution of chromaticities of north sky at Bombay. Isotemperature lines are shown for 20 000 K, etc. (Sastri and Manamohanan.)

bands in the observed spectral power distribution, not only the 430 nm
absorption but the prominent ones at 490 and 520 nm in their curves.
The results were inconclusive, and in any case about half the chromaticity points lay on the other (purple) side of the locus.

In late 1966 Sastri and Manamohanan made measurements of the
north sky spectrum at Bombay, providing the first data from a tropical
site.[344] By comparison with their results in Delhi these showed a
preponderance of high CCT, with ten out of sixty chromaticities falling
beyond the ∞ point in spite of the polluted local atmosphere (Fig. 44).
Fig. 45 combines the two sets of CCT and shows the wide difference
between Delhi and Bombay measurements. Specimen distributions for

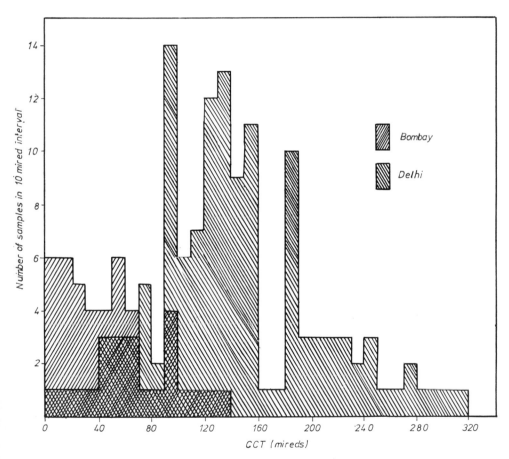

FIG. 45. Distribution of CCT (mireds) of north skylight at Delhi
(Sastri and Das) and Bombay. (Sastri and Manamohanan.)

7500 and 10 000 K resembled those from Delhi in the maxima near 420, 470 and 500 nm, but differed by a small peak near 620 nm and a marked falling off beyond 650 nm. The chromaticity locations of the Indian daylights suggested a similarity to Abbot–Gibson distributions[266] with increased greenness at higher CCT values due to more Rayleigh scattering. The greenness of daylight compared with full radiator chromaticities was also attributed in part to absorption by ozone, oxygen and water vapour at wavelengths longer than 560 nm.[345] A contrary trend in chromaticities appeared in Moon's calculations.[264] Here at $m=0$ he found $x=0.318$, $y=0.330$, a slightly greenish colour at 6200 K. As m increased and CCT decreased the chromaticities diverged further above the Planckian locus, unlike many of the observations quoted above (Fig. 75).

Another unusual feature was a sharp maximum in the Bombay curves at 330 nm where the Delhi ones had little or no emission, perhaps on account of higher aerosol concentration over the continent compared with that beside the Indian ocean. The maximum was much more prominent than any to be seen near 330 nm even in the extra-terrestrial data in Table 3; but on the other hand Arvesen's extended table for $H_{0\lambda}$ does show a peak at 330 nm,[169] and so does a spectrum calculated for Rayleigh scattering (Fig. 23).

The next comparison of local observations with the RD values took place in Japan at Amagasaki.[346] The site was close to Osaka on the southern coast of Honshu. Here Nayatani, Hitani and Minato recorded the spectrum of the north sky at 45° elevation, using a single quartz monochromator and photomultiplier for the range 310 to 720 nm, with a resolution from 1·5 to 18 nm over this spectrum. The light was reflected from a barium sulphate plate, the calibration made by a standardized tungsten–iodine lamp, and corrections made for overall illuminance changes during the 4-minute scan recorded by a barrier-layer cell. Nearly all the measurements were made in winter. One hundred and twelve sets were presented as chromaticity plots which, while lying almost entirely on the green side of the full radiator locus, showed much less scatter than other extensive sets of measurements, except Budde's. This suggests that the set was not representative of all seasons since wider areas of the chart are commonly covered. It may however be significant that, as in the work of Sastri, Das and Manamo-hanan, the higher CCT values were further from the full radiator locus than those of lower CCT, so that a line of lower slope would represent them (Figs. 43, 44, 46).

Some spectral curves were shown compared with corresponding RD values. A somewhat higher ultra-violet content occurred in the Japanese

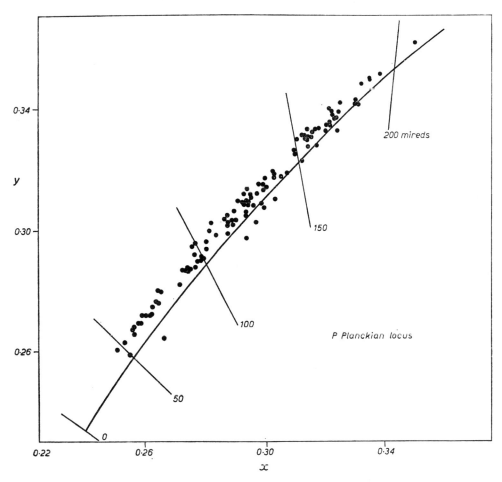

FIG. 46. Distribution of chromaticities of north skies at Amagasaki, 1965 to 1967. (Nayatani, Hitani and Minato.)

measurements, estimated as an excess of 8 per cent between 340 and 380 nm for all the available data, whereas the South African measurements, similarly treated, showed an 18 per cent excess over the RD distributions. Late afternoon measurements revealed an unusual rise of H_λ from 600 nm onwards, not unlike that seen in many of the curves given by Sastri and Das.

Comparison of spectral curves is unsatisfactory by descriptive methods like those frequently used in this book. Nayatani *et al.* made use of a metamerism index proposed by Nimeroff and Yurow[347] in order to give quantitative values to the differences between a measured spectral curve and the RD curve of the same equivalent CCT: these may be considered a metameric pair of light sources (the same colour appearance, different spectral distributions). By the $_{\text{B}}M_t$ index it appeared that over 90 per cent of the Japanese curves had values of 0·014 or less, interpreted by the authors as not likely to introduce 'any

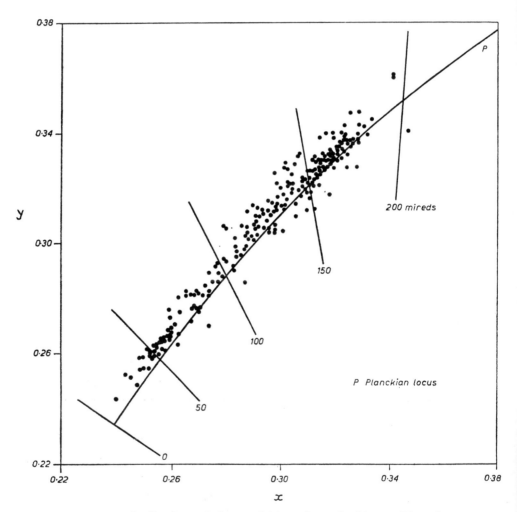

FIG. 47. Distribution of chromaticities of north skies at Nagaoka, 1966 to 1967. (Ando, Ikemori and Sekine.)

larger [large?] observer difference in visual colorimetry'. This work was reported again as the subject of vector analysis[348] (Chapter 13).

Another Japanese investigation took place about this time at Nagaoka near the northern seaboard of Honshu.[349] Ando, Ikemori and Sekine worked at two sites, of 25 and 95 m elevation, with similar equipment and methods to sample the light from the north sky at 45° elevation, chiefly in the middle of the day between June 1966 and February 1967. Grating spectrometers and photomultipliers were used, and auxiliary records of illuminance and colour temperature, or alternatively of H_λ at a fixed wavelength, to allow the exclusion (not correction) of measurements made during large variations of sky conditions during the 2·5- to 3-minute scan. The chromaticity distribution for the 303 spectral power curves showed once more the tendency to a lower slope than the Planckian locus (Fig. 47). Few points were on the lower side of the locus. These workers also used the metamerism index to compare the spectral curves with RD ones, with results very similar to those reported by Nayatani *et al.* Two other methods of comparing spectra of light sources were used. The CIE Test Colour method gave a general colour rendering index R_a of 95 or over. The ASTM Conformity Index[350] yielded a distortion factor of 1 or 2 per cent.

The distribution of CCT in the work, extended up to June 1967, appears in Fig. 48, where the observations on north sky are separated according to clear or overcast conditions. The prominent maximum at 160 to 170 mireds in overcast conditions was not seen in the north (or south) sky sections of other histograms (Figs. 38 and 42).

The work reported above was evidently based on earlier work by Kawakami, Ando and Ikemori, another study of north sky chromaticities.[351] Diagrams showed these points to be represented by a locus still further above the Planckian locus than the RD locus, and once more with a lower slope. Equations to the locus are given in Chapter 13. The spread of chromaticities included a number below the Planckian locus.

An extensive project of recent years was that carried out by Tarrant in 1963 to 1967.[352-354] More complex measuring and data processing equipment was used than in most other work of the same kind. The main instrument was an Optica CF4 grating monochromator with trialkali photomultiplier (EMI 9558A), with light taken from a large ground Vitrolite plate placed at 45° to the horizontal and facing either

south or north. A second monochromator of the same type produced
a continuous recording on 560 nm for correction of irradiance measurements by ratio recording throughout the spectral range, from 300 to
800 nm. At the beginning and end of each day's measurements calibration was made by a standardized tungsten-in-glass projector lamp and
a magnesium oxide diffuser.

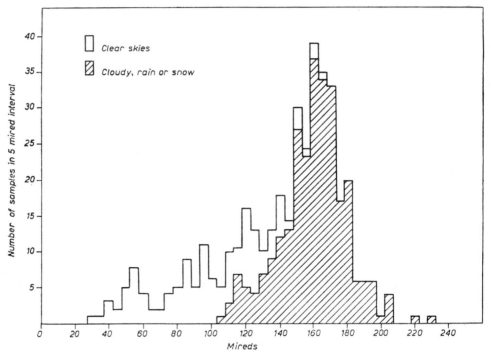

FIG. 48. Distribution of CCT (mireds) of north skylight. (Ando,
Ikemori and Sekine.)

The equipment was driven automatically and produced continuous
spectral curves from which, in the early stages of the work, values of
photocurrent were read off at 10 nm intervals and punched tape prepared for further calculations. In most of the work the tape was
punched automatically at 1 nm intervals. From it a Sirius computer
delivered a table of spectral power values at the same intervals, also
XYZ and *xyz* values. Corrections were incorporated for gradual
changes in zero levels. Fig. 49 is an example of a recorded curve.

These operations on 434 spectra, recorded at a bandwidth of 1·2 nm,
gave an enormous mass of information which might be analysed for

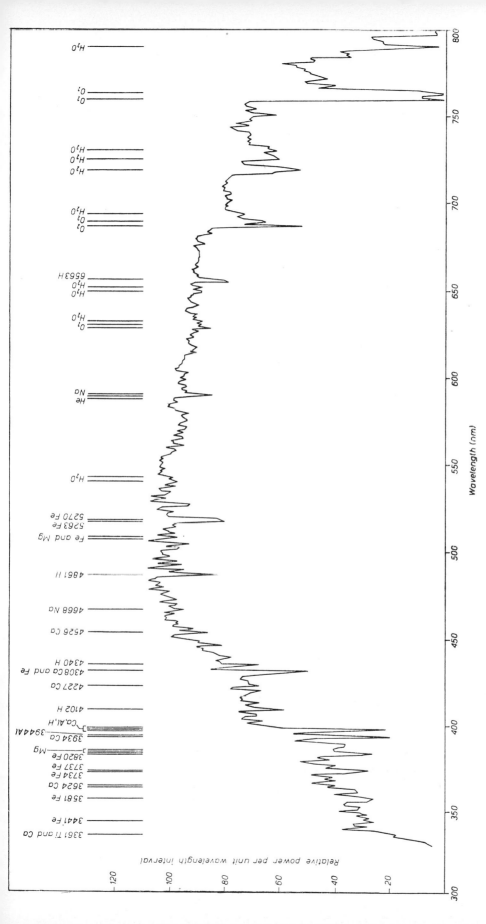

Fig. 49. Individual recorded spectrum for a south-facing observation, CCT 5500 K. Absorption bands are identified. (Tarrant.)

many types of effect. Much of the initial value lay in the chromaticity determinations. Tarrant's diagrams showed a distribution very similar to that of Judd *et al.* (Fig. 68); Fig. 50 gives the main group of north- and south-facing runs together. When separated these fell within CCT limits of 11 000 to 5000 K for north-facing, and 8000 to 4800 K for south-facing. For convenience in handling the material Tarrant followed Winch's method of forming ten-mired groups. Histograms gave

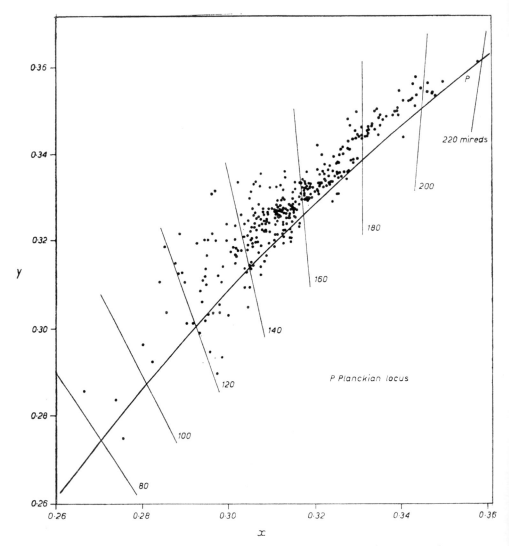

FIG. 50. Distribution of chromaticities of daylight, north- and south-facing, at Putney, London, 1966 to 1967. (Tarrant.)

140 to 160 mireds as the most frequently observed facing north, and 150 to 160 mireds facing south. Fig. 51 shows these combined, and having a rather more contracted distribution on the low mired side than others have found. This figure uses wider mired intervals than others in this chapter, and there is no division into types of sky, etc.

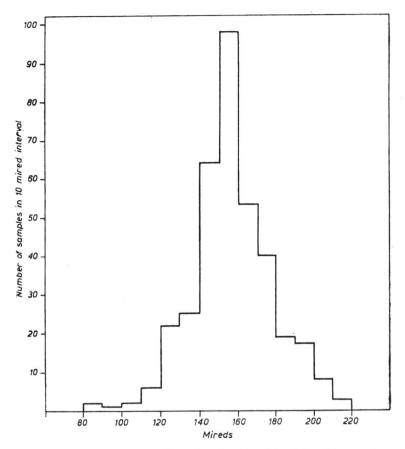

FIG. 51. Distribution of CCT (mireds) of daylight. (Tarrant.)

All the distributions considered in this chapter agree that the most frequently observed CCT is within 150 to 175 mireds, or 5700 to 6700 K. The observations were made in such a way that they might be considered to constitute a random sample if all taken together. There is therefore some reason for thinking that the CCT range mentioned is the most frequently occurring as well as the most frequently observed.

Of all Tarrant's observations 95 per cent were made in Putney,

London, at 15 m altitude, the rest in Saffron Waldon, Essex, at 100 m in a rural site. No great differences in spectral power distribution were found between the two sites, nor between summer and winter daylight at the same CCT. Between north- and south-facing measurements there was some increase of H_λ below 350 nm and above 690 nm for the north-facing ones. Attempts to correlate conditions of wind and cloud with chromaticity variations were not rewarding, but a consistent change to a higher CCT during the morning hours was established, and fair stability followed this later in the day. This was attributed to diffusion resulting from the morning increase of atmospheric pollution: compare this with Lenoble's observations on ultra-violet radiation, and Sitnik's on atmospheric turbidity changes (Chapter 11).

Tarrant's measurements were hampered by lack of detector sensitivity at wavelengths above 700 nm. In the rest of the visible spectrum his curves showed much the same shape as those measured elsewhere. It is well known that the quantitative ultra-violet content of daylight is not related in any fixed way to the visible emission, but consistently high values were found by Winch *et al.*, by Boshoff and Kok, and to a lesser degree by Nayatani *et al*. Tarrant's results agreed fairly closely with Winch's in this respect, which was unexpected for measurements made in a polluted atmosphere. The discrepancy from the RD curves was most noticeable in the 360 to 400 nm region, where possible calibration difficulties are not so severe as in the shorter wavelength range of daylight. Fig. 52, shows some of Tarrant's mean spectral distributions.

A closer examination of the ultra-violet part of the spectrum resulted from a further 178 measurements made in 1971 and 1974 at Weybridge, Surrey.[355, 356] The range of CCT of the spectra was about 6900 to 4900 K (145 to 205 mireds) with the main concentration near 5800 K. Tarrant and Brock studied the relative content of ultra-violet power by an 'ultra-violet ratio' and a 'short ultra-violet ratio', for the ranges 300 to 399 nm and 300 to 329 nm respectively, as a fraction of the power in the full range 300 to 699 nm. Values of the first ratio lay among the points for the same mired range in Fig. 39, though below the line for Winch's values. The north-facing observations between 150 and 160 mireds gave a ratio of 0·0978 compared with 0·0993 (Henderson and Hodgkiss), 0·1236 (Winch *et al.*), and 0·0912 (Judd *et al.*, RD for 6500 K), confirming again the comparatively low level of ultra-violet content in the RD distributions. On fairly cloudless days both ratios agreed approximately with an exponential decrease with increase of air mass, while the short ultra-violet ratio showed some negative correlation with the product of air mass and ozone content at $m=1$, but less than would be expected if cloud and atmospheric conditions had remained constant. Some effects of cloud on the amount of ultra-violet radiation were described.

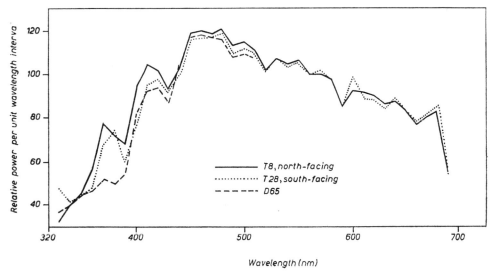

FIG. 52. Mean spectral power distribution of daylight for the 150 to 160 mired group of chromaticities, compared with D_{65} distribution. Normalized at 560 nm. (Tarrant.)

Differences thus persist between several different sets of measurements in the ultra-violet region and from those in the RD distributions (Table 6, Fig. 74), especially below 330 nm.

MEASUREMENTS AT UNUSUAL SITES

A novel project was reported by Knestrick and Curcio of the Naval Research Laboratory in Washington, D.C.[357] In connection with work on atmospheric scattering, they measured the spectral radiance of small areas of horizon sky at varied solar altitudes and azimuths, over land and over water. A glass double monochromator was operated with photomultipliers to cover the spectrum in two stages from 380 to

760 nm, and from 610 nm to 1·0 μm. A ratio of signals from the sky and a standard lamp was recorded, also a red-filtered monitor signal to provide absolute values of radiance by a pyrometer operating at 665 nm. Meteorological range, cloud cover and relative humidity were noted, and no corrections for luminance changes were required owing to the fast scan (1·5 minute). Polarization was measured, but calculation showed its effects to be negligible.

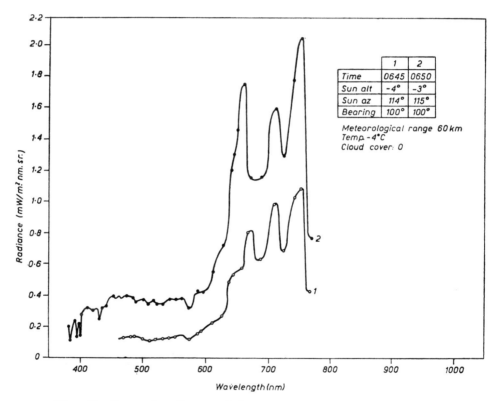

FIG. 53. Spectral radiance of the eastern horizon sky over water before sunrise. (Knestrick and Curcio.)

The forty-five spectral curves were plotted at selected peaks and troughs from the recorded data. Two-thirds of the curves referred to near sunrise or sunset conditions, with solar altitudes between +11° and −5·5°. Many have lower equivalent CCT than in the usual range of daylight, with maximum emission at or above 600 nm, as in Figs. 53 and 54, and with shapes very different from those measured for sunlight or larger areas of sky. In particular they showed a group of strong absorptions near 690 and 760 nm (O_2), 730, 823 and 945 nm (H_2O). No differences appeared between paths over land and water, though the water absorption bands varied with cloud cover and relative humidity. Under an overcast sky the whole of the spectrum up to about 650 nm in Fig. 54, curve (2) was raised with respect to the longer wavelength end. Only in hazy conditions was the shorter wave-

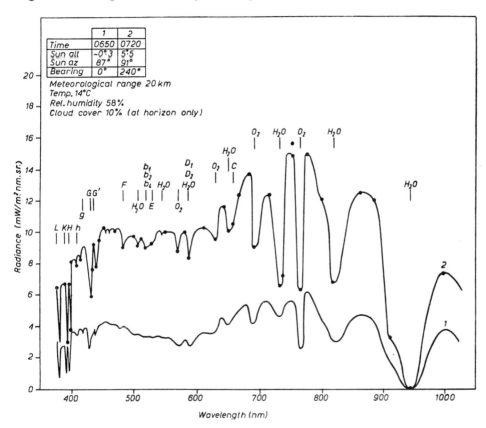

	1	2
Time	0650	0720
Sun alt	-0°.3	5°.5
Sun az	87°	91°
Bearing	0°	240°

Meteorological range 20 km
Temp. 14°C
Rel. humidity 58%
Cloud cover 10% (at horizon only)

FIG. 54. Spectral radiance of the horizon sky; north over land at sunrise, south-west 30 minutes later. Absorption bands identified. (Knestrick and Curcio.)

length end modified so as to produce a prominent maximum near 450 nm, and a minimum near 600 nm. Chromaticity determinations would have been valuable, but were not recorded. Some of the curves were evidently lacking in the usual proportions of radiation in the green region of the spectrum. This has been confirmed by calculations made on some of the curves. For a clear eastern horizon over water, with

$h=-3°$ before sunrise (the upper curve in Fig. 53), the chromaticity is $x=0.400$, $y=0.324$. For a southern horizon over water near sunset, with $h=2°$, the chromaticity is $x=0.398$, $y=0.360$. Both these are below the full radiator locus, representing pink shades, with CCT of 2970 and 3390 K respectively, though the distances from the locus are too far for any similarity between the sky colours and those of the full radiators. A few of the chromaticity plots by Sastri and Das were near this region; the above calculated values have been added to Fig. 43 and are also shown in Fig. 20. Compare the low CCT measurements by Condit and Grum.

Knestrick and Curcio extended their work to the ultra-violet region under similar conditions, using a grating spectrometer of higher dispersion calibrated by a tungsten–halogen standard, with monitoring for intensity by means of a broad band of ultra-violet. Curves show the spectrum of horizon sky radiance for $h=0°$, $1.5°$ and $3°$ in the east, south and north respectively. At noon in May ($h=68°$) there was much less change with different azimuths; at 400 nm the radiance was fifty times as great as for $h=0°$ under clear skies facing east (70 against 1.4 W/m² sr μm). Most of the curves had a sharp peak near 400 nm but another near 410 nm disappeared for $h=0°$ to $3°$. Sunset and sunrise results were similar, and cloud cover reduced the radiance.[358]

Knestrick and Curcio did not measure zenith radiance and no direct comparison is possible with CIE tables for the visible spectrum which show the relative luminance of clear skies according to h and measuring point (Chapter 12). Thus if h is near $0°$ the zenith luminance is not more than one quarter of the horizon luminance. For similar conditions, though at elevated sites, Bener's measurements[145, 311] show the zenith radiances at 330, 347 and 360 nm to be consistently 1.5 to 2.5 times the horizon values found by Knestrick and Curcio, which suggests massive attenuation of ultra-violet over the sea. There is much better agreement for $h=70°$ where the CIE tables require a zenith to horizon factor of about 1.2, and a few of Bener's values are only about twice those found by Knestrick and Curcio at the horizon.

Another unusual investigation was that of V. Hisdal who measured spectra of the zenith sky in Spitzbergen and in Oslo (60 m altitude). The instrument was a small grating monochromator (D292) with a 6° field of view and photomultiplier detectors (RCA 931A for 320 to 600 nm, Philips XP1002 for 300 to 800 nm). Calibration was by a tungsten-in-silica lamp at 2500 K; photocurrents were recorded manually at 10 nm intervals, and corrections were applied for sky variations by a total irradiance measurement with a selenium cell. In all cases the solar altitude was close to 30°.[359]

At Ny-Ålesund, the more northerly of two sites in Spitzbergen, where Götz had worked in 1929, Hisdal recorded fourteen spectra, at Slettebu thirty-four, at Oslo twenty-two. Fairly good agreement with a calculated spectrum was found for the best clear sky measurements at Ny-Ålesund except below 340 nm where the measured curve fell away more steeply than the calculated one. In the theoretical case Rayleigh scattering was assumed, a ground albedo of 0·25, and for extraterrestrial sunlight the distribution given by Gast.[360] An un-expected difference appeared between the clear sky curves for the two Spitzbergen sites, which were only 150 km apart. A marked decrease in the proportion of ultra-violet radiation was found at Slettebu, probably attributable to different wind conditions affecting the aerosol con-centrations. Hisdal made detailed comparisons between clear and overcast skies, and between his own observations at Oslo and others by Taylor and Kerr, and Götz and Schönmann. These showed Hisdal's spectral curve for Ny-Ålesund to be exceptionally high in the ultra-violet region, where the maximum of irradiation at 340 to 360 nm approached that in the visible region at 415 nm.

From chromaticity calculations Hisdal found dominant wavelengths with respect to the equi-energy spectrum to be near 479 nm for clear skies at Ny-Ålesund, Oslo and for the theoretical curve. The first of these gave $x=0·232$, $y=0·227$, close to the theoretical values at $x=0·232$, $y=0·234$, both chromaticities being beyond infinity on the Planckian locus, and having excitation purities of 46 and 44 per cent respectively (VH in Fig. 20).

In measurements at Ny-Ålesund in the summers of 1970 to 1974, the equipment was automated, calibrated by a tungsten–iodine lamp and the total irradiation monitored by a pyranometer. In all, fifty-two spectra were recorded, and all of the radiation from the whole hemi-sphere of sky collected by a Budde-type integrating sphere.[361] At noon, with $h=30°$, the global radiation for clear and overcast skies gave mean CCT in the 6500 to 7000 K region with chromaticities very close to the Planckian locus (Fig. 75). Light from the sky at midnight ($h=10°$) was slightly more blue than at noon.

The CCT of diffuse light from the sky at noon (without sunlight) was 37660 K, with a greenish chromaticity of $x=0.244$, $y=0.249$ and excitation purity 38 per cent, a point very close to the bluest colour matched by Hendley and Hecht (Chapter 15) and shown on Fig. 20 (HH). At 26·6 mireds the ultra-violet ratio of the spectrum was about 0·3 (point+on Fig. 39), while the clear zenith sky with a chromaticity beyond the ∞ point of the full radiator locus gave the very exceptional value of 0·374.[362]

11

High Altitude
Spectral Measurements

Nothing is difficult to mortals; we strive to reach heaven itself in
our folly.

HORACE, *Odes*, Book 1, no. 3.

SOLAR IRRADIANCE

Several spectroradiometric projects of more than usual elaboration
remain to be considered. They were carried out at high altitudes in
order to take advantage of clearer atmospheres and avoid much water
vapour absorption. Some related measurements were made without
spectral resolution for the purpose of direct solar constant determination. They have been described in Chapter 5 and are only briefly
mentioned here.

Kondratiev and his colleagues made measurements up to 3700 m
elevation on Mount Elbrus on atmospheric transmission, on atmospheric water vapour and its relation to humidity at ground level, and
on the infra-red emission of the centre of the solar disc.[363] A relatively
small part of their work was devoted to the spectrum of sky and global
radiation between 400 nm and 1 μm, for which they used a prism
monochromator and caesium photomultiplier, with automatic scanning and recording, and a tungsten filament lamp standard. The curves
published were very small and their main value is in the comparison of
sky radiation at 3700 m altitude and $h=53°$, with values calculated for
60° by Hinzpeter. The agreement was poor, owing to much less radiation below 400 nm than in the calculated curve. Above about 450 nm
the experimental curve diverged more and more above the theoretical.
Other work on total flux and atmospheric transmission between 350
nm and 13 μm used balloons at a much greater height: the result was
a new determination of the solar constant.

Stair and Ellis made spectroradiometric observations for the purpose of solar constant determination.[161] The absolute values of irradiance were notably lower than in Johnson's standard curve, hence their lower value of S. Stair and Ellis worked below $m=3$ and therefore used $m=\sec z$ without corrections in extrapolation to $m=0$. Their values of H_λ are given in Table 3.

THE NASA PROJECT

In their programme with airborne instruments the Goddard Space Flight Center had two separate objectives. The first was the determination of the solar constant, the second the measurement of extra-terrestrial solar irradiance, in which five instruments contributed to the recorded results.[88] The first was a Perkin–Elmer monochromator. A beam of light, reflected into the instrument by a diffuse mirror of aluminium evaporated on to ground glass, passed four times through the single lithium fluoride prism. Detection by photomultiplier or thermocouple covered the range 300 nm to 4 μm. The second instrument was a Leiss double monochromator with fused silica prisms, and an integrating sphere and magnesium oxide screen to illuminate the entrance slit. A photomultiplier was used, but in this case the infra-red detector in the 300 nm to 1·6 μm range was a cooled lead sulphide cell. Calibration also was somewhat different from that in the first instrument where a tungsten–iodine working standard was used in flight, whereas secondary tungsten standards served the Leiss instrument. The data from the latter had to be normalized by previously known spectral irradiance data and the total flux measurements made in the solar constant project. This was due to unexplained drifting in the calibration. The third instrument was a filter radiometer used with an RCA 935 photocell detector for twenty-two narrow-band filters covering the range 310 to 600 nm, and an RCA 917 photocell for eleven more filters in the 600 nm to 1·1 μm region.

The last two instruments were unusual in this sort of work. One was a polarizing interferometer covering the spectrum from 300 nm to 2·5 μm. Radiation entering by an integrating sphere passed through polarizers with an oscillating Soleil prism between them to produce path retardation. The beam was then split between a photomultiplier and a lead sulphide detector: in the results reported only the long wavelength detector was used. Here a useful transmittance curve from 720 nm to 2·2 μm was obtained for the atmosphere above the working level (11·6 km). Solar irradiance at one hundred and twenty wavelengths in the same range was measured, but calibration by a standard lamp was relative only, and the spectral data were therefore normalized from the other instruments. The interferograms, recorded on magnetic tape, needed computer assistance for the Fourier analysis. Therefore the equipment could not be adjusted while in use.

The last instrument was a Michelson interferometer, operating only between 2·6 and 15 μm. The detector was a thermistor bolometer corrected for its own radiation. Results from this instrument were a spectrum showing absorptions by water vapour, carbon dioxide, methane, nitrous oxide, and ozone (9·6 μm), and of a general shape falling more steeply with increasing wavelength than the full radiator distribution for 6000 K.

Interpolated values for atmospheric attenuation, taken from accepted standards,[364] were found insufficiently accurate for the present work. They were corrected by straight line plots of the logarithm of photocurrent versus air mass for seventy-seven wavelengths, using data from the flights. The final table for $m=0$ was a weighted mean, extended to the 140 to 300 nm range by previous rocket measurements, from 300 nm to 4 μm on the monochromators and filter radiometer, thence to 15 μm by the Michelson interferometer, and from 15 to 20 μm by extrapolation of the observed decrease in irradiation below the 6000 K full radiator distribution. Intervals in the table were 5 or 10 nm to 750 nm and wider above this. As mentioned before, a −0·1 per cent correction was applied throughout to achieve exact agreement with the solar constant measurements. Figs. 55 and 56 show the resolution attained by the Perkin–Elmer monochromator in this work; the vertical scales have been adjusted to equality.[89]

The spectral power distribution was slightly modified once more by Thekaekara and Drummond to take account of filter radiometer measurements,[162] and was finally adopted almost unchanged by a NASA committee as a standard extraterrestrial spectrum.[167, 171, 365] The spectral range was extended at both ends by additions from other authors, and recorded in tables and log–log plots of wavelength versus irradiance from 5 nm to 5 mm, at intervals of 5 nm from 220 to 610 nm, and 10 nm to 1 μm. The values are averaged over 10 nm bands from 300 to 750 nm. The 10 nm values appear in Table 3. In a later publication tables were derived from the Perkin–Elmer data[89] for 300 to 610 nm at a 0·1 nm interval, with details of the normalization process making the spectrum conform to the standard table by giving equal

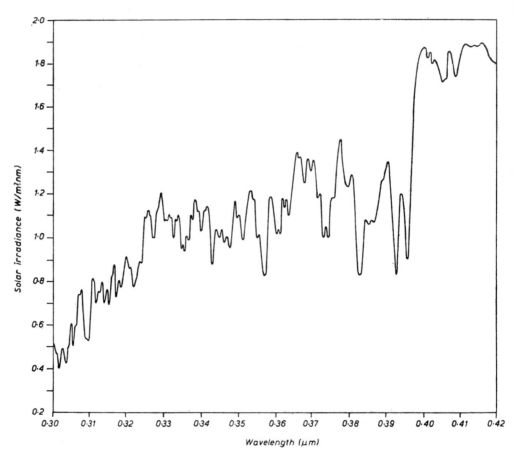

FIG. 55. Extraterrestrial solar spectral irradiance (ultra-violet) by Perkin–Elmer monochromator. (Thekaekara *et al.*)

values for integration over 5 nm intervals and by preserving the points of inflection.[366]

Another group in NASA recalculated the standard spectrum to air mass 2 for a water content of 20 mm and 3·4 mm ozone.[367] This distribution, in Table 5, shows some differences from Moon's values but the general trend is similar except in the water absorption band at

850 to 1000 nm. Between 300 and 800 nm the total irradiance is 442·2 W/m²; Moon's table gives 439·8 W/m² for the same region, but this is increased if the solar constant used by Moon is increased to that used by Brandhorst *et al.* Comparisons of chromaticity and ultra-violet content in these spectra for $m=2$ are shown in Tables 4 and 5.

Apart from these well documented data there remain some other determinations from aircraft which differ somewhat from the standardized ones. In measurements made by Webb *et al.* with a Leiss double prism monochromator during the same GSFC flights, the distribution originally extended from 300 to 1600 nm at 10 nm intervals.[168]

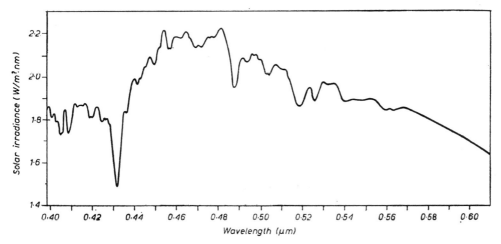

FIG. 56. Extraterrestrial solar spectral irradiance (visible) by Perkin–Elmer monochromator. (Thckackara *et al.*)

This spectrum was normalized so that its irradiance should become 1·351 kW/m² when supplemented by extensions at the ultra-violet and infra-red ends taken from Gast's tables,[360] 0·0246 and 0·142 kW/m² respectively. $S=1·351$ kW/m² was then accepted for the whole project before the final adjustments to 1353 kW/m². The Leiss measurements agreed with those made by the same authors at the same time with a filter radiometer, but were almost consistently a few per cent lower than the NASA means between 370 and 700 nm (Table 3).

During the same series of flights another group from NASA (Arvesen, Griffin and Pearson) made measurements with a Cary ratio-recording spectrophotometer having a first dispersion by grating and a second by prism.[169] An unusual feature was the admission of light from a 50 mm integrating sphere which viewed alternately, at a frequency of 30 Hz, the sun and the 1 kW tungsten–iodine irradiance standard. Extrapolation by Bouguer lines gave extraterrestrial values which were considered to be of an absolute accuracy of ±3 per cent over most of the range, 300 to 2500 nm. Addition of ultra-violet measurements from rocket flights, and of infra-red from the 5800 K grey body distribution, completed the smoothed values in a table intended for engineering use (Table 3). The values are consistently higher than the NASA ones between 400 and 830 nm. The full table for 300 to 2495 nm at much smaller intervals shows many Fraunhofer lines, and records the total irradiance from the shortest wavelengths as 1·15 per cent at 300 nm and 96·65 per cent at 2495 nm, whereas the NASA table gives 1·21 per cent and 96·27 per cent.[167] A high value of S resulted from these measurements. The engineering table corresponds to a chromaticity on the purple side of the Planckian locus, but the detailed one gives a markedly different point (Fig. 75).

RADIANT EXITANCE OF THE SUN

The problem here is to measure the radiance of the centre of the sun's disc, and to obtain a continuum uninterrupted by Fraunhofer absorptions by using estimates of the line absorption intensities. The multitude of lines, especially among the shorter visible wavelengths, makes this a difficult measurement, having a primary interest for astrophysics. Its relation to extraterrestrial irradiance, either total or spectral, is easily calculated after restoring the Fraunhofer absorptions to obtain the 'integrated values' of solar radiance, together with a (continuum) correction for the uneven luminance of the sun's disc. In this way solar radiance $I_{0\lambda}$ is related, somewhat indirectly, to terrestrial daylight.

Labs and Neckel have surveyed in detail a number of investigations of this kind (see below). Apart from the two recent determinations by Peyturaux, and by Labs and Neckel, the work of Sitnik may be mentioned. Two sites were used during 1952 to 1959, one of them near Alma Ata at 3000 m elevation.[368] Both used similar grating spectrographs with photomultiplier detectors and calibration by a tungsten ribbon lamp. Smoothed values for $I_{0\lambda}$ and $H_{0\lambda}$ were found over the working range of 328 nm to 1·24 μm, and extended by extrapolation with measurements by Peyturaux and by Pierce.[369] The results are included in Tables 3 and 5: the $H_{0\lambda}$ values and derived colorimetric values differ considerably from the others in the table. Sitnik discussed the change in atmospheric turbidity usually occurring near midday and its possible invalidation of Bouguer line extrapolation unless times for measurement were carefully chosen.[370]

Peyturaux conducted work on this subject during 1960 to 1965 at Mont Louis in the eastern Pyrenees, at 1600 m altitude.[371, 372] His equipment included two prism monochromators, the first to select a wide spectral band, the second to isolate the narrow band for measurement. The detectors were a photomultiplier and a lead sulphide cell, and calibration was by a full radiator held at 2600 K and checked by an optical pyrometer. At each of the fifteen wavelengths selected for measurement, the profile of the radiation was measured at the exit slit to provide corrections for line absorption. These corrections varied irregularly from 9·4 per cent at 447·7 nm to 3·3 per cent at 863·8 nm, which were the wavelength limits; the lowest correction was 0·5 per cent at 607·1 nm. Other corrections concerned mirror reflectances and polarization.

Atmospheric extinction measurements involved a spectrometer, thermopile and galvanometer providing an independent, continuous record of one wavelength daily. The recordings were used for the usual straight line Bouguer plots. The final results, estimated to be accurate within 3 per cent, appear in Fig. 57 but not in Table 5 owing to the lack of interpolated values. They agreed reasonably well with Sitnik's work but not with that of Labs and Neckel[373] especially at wavelengths less than 520 nm. The comparison here does not include the 1968 information from Labs and Neckel.[50]

Measurements like those of Peyturaux were made by Labs and Neckel over the same period of time. They worked at the Jungfraujoch (3600 m) and made observations only on days of apparently perfect sky conditions. This was an extension of earlier work at the Pic du Midi (2860 m)[373] and in Tenerife (2400 m).[374] In this main project the instrument was a grating double monochromator calibrated by a standardized tungsten lamp instead of the carbon arc used previously.

Measurements were made at 2 nm intervals between 328·8 nm and 1·248 μm, with suitable photomultipliers for different spectral regions. The final paper of the series gave tables of observational material, and calibrations of standard photometric atlases of the spectrum arising from the purely spectroscopic aspects of the work.[50] The chief table is of the continuous spectrum of the solar centre from 260 nm to 100 μm, the part beyond the observational limit being derived from a model of the photosphere continuum, similar to other models[375, 376] but normalized on Labs and Neckel's values for 500 nm. The continuous spectrum table had 20 nm intervals in the visible region, 10 nm in the ultra-violet and increasing intervals in the infra-red. It was considered accurate to ±2 per cent except in the extreme ultra-violet

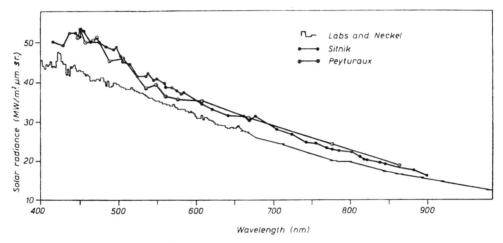

Fig. 57. Extraterrestrial solar spectral radiance (centre of disc) from Mont Louis, compared with other determinations. (Peyturaux.)

wavelengths. Small corrections had been applied on account of the revised gold point temperature but not for the present value of c_2. Table 5 quotes part of this $I_{0\lambda}$ distribution. The main unexplained disagreement was with the higher values found by Peyturaux, attributed by Labs and Neckel to possible inaccuracy in optical pyrometry and corrections for polarization; but Peyturaux himself, considering only the comparison with earlier work of Labs and Neckel, suggested that the extrapolation to $m=0$ was the most likely cause of error in his own work. Compare Sitnik's views on this technique.[370]

Comparison of the model solar continuum with sets of experimental values showed those of Labs and Neckel to agree closely from 400 to 850 nm but to be too low in the ultra-violet region. Other recent authors were much less consistent. Plaskett's values (Chapter 7) agreed

well between 400 and 670 nm in spite of internal evidence of uncertainty in that early work.

Labs and Neckel calculated solar irradiance at the earth's mean
distance, H_λ, by the relation

$$H_\lambda = I_\lambda \, \frac{\pi r^2}{D^2} \frac{F}{I} \, (1-\eta)$$

where r is the sun's radius, D its distance, F/I the ratio of mean to
central intensity of the solar radiation (about 0·8 at 550 nm), and η the
fraction of the radiation absorbed by Fraunhofer lines (the 'line
blanketing' coefficient). Values of F/I and η were obtained by weighting
earlier published material. F/I varied from 0·52 at 260 nm, 0·713 at
400 nm, 0·862 at 800 nm, 0·942 at 2 μm, to near unity at 50 μm. The
value of η varied with wavelength and waveband used, and for the
whole spectrum was assessed at 14·5 per cent, and 10·1 per cent for
the spectrum above 330 nm. Values of $H_{0\lambda}$ were given by Labs and
Neckel at 10 nm intervals from 205 nm to 1·05 μm, and at wider

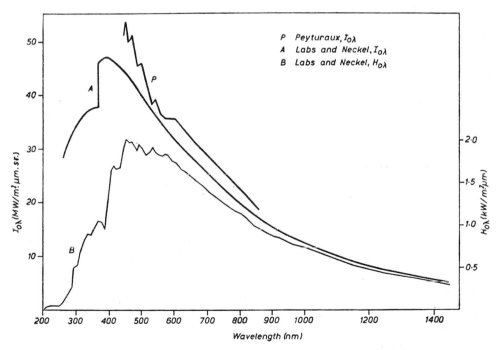

FIG. 58. Extraterrestrial solar spectral radiance (centre of disc) from
Jungfraujoch in 1961 (Labs and Neckel). Comparison with Peyturaux,
Fig. 57.

intervals and wavebands beyond up to 95 μm. In Table 3 interpolated values corrected to constant bandwidth are included, but these are from a slightly modified table with further corrections for line blanketing and the use of $c_2 = 1\cdot4388 \times 10^{-2}$ m K in the calculation of S.[173] This reduced the $H_{0\lambda}$ values by 1 to 2 per cent and altered the CCT by -20 K to 5915 K. This 1970 revision did not alter the $I_{0\lambda}$ curve (Table 5) but a new table was provided of 156 peak intensities between 329 and 658 nm, which agreed roughly with Houtgast's ultra-violet measurements at Mount Wilson[377, 378] and Holweger's calculations for a model continuum corrected for line absorption.[379]

The relative difference between solar exitance and solar irradiance appears in Fig. 58. It is due to the F/I and η factors. Corresponding chromaticity differences are appreciable, with CCT of 7205 and 5915 K respectively. Labs and Neckel made plots of the ratios of other irradiance data to their own, with the results that the closest agreement was achieved by Moon's 1940 values between 540 and 800 nm. The figure gives another comparison with the higher values found by Peyturaux (compare Fig. 57). Curve B in Fig. 58, drawn from the original data[50] is indistinguishable on this scale from the 1970 revision. Fig. 59 includes a section of curve A to show the agreement with the measurements of Houtgast (selected near 10 nm intervals), and with Holweger's calculated values. Differences from the model curve A are due to the many faint Fraunhofer lines having a large aggregate effect, including the masking of the discontinuity at the band head of the Balmer H lines (364·7 nm) and uncertainty as to its quantitative value.

Fig. 59 also shows some of Labs and Neckel's peak intensities, and the trend of a later model for $I_{0\lambda}$, due to Gingerich *et al.*, the 'Harvard Smithsonian Reference Atmosphere'[380] which is another example of correction for 'veiled line haze'.[381] Table 5 includes a few points from the model which corresponds to a solar temperature of 5768 K.

Labs and Neckel treated their project most extensively, particularly in discussions of the problems involved, the techniques used, and the findings of other workers. The solar constant aspect has been discussed in Chapter 5. The 1968 paper[50] gave the best general survey, but a diagram from the 1962 paper[374] (Fig. 60) may be compared with that of Dunkelman and Scolnik (Fig. 19) as a record of Rayleigh scattering and other attenuation factors. At one time in the course of their work

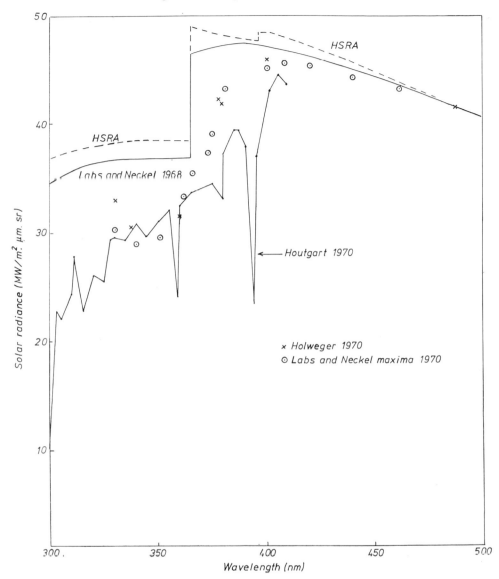

FIG. 59. Extraterrestrial solar radiance. (After Labs and Neckel, Houtgast, Holweger, Gingerich *et al.*)

Labs and Neckel found wider divergences from the ideal extinction curve.[172] In the review paper this was attributed to volcanic dust in the atmosphere from the 1963 eruptions of Mount Agung in Bali, and of Surtsey off the coast of Iceland. The former produced aerosols on a scale smaller, but of the same order, as those from Krakatoa, and of much more significance to atmospheric transmission than the material from atomic explosions. The Bali eruption was also blamed for the reduced total irradiation at Mauna Loa in 1963, reported by Drummond and Ångström.[273] Different levels of aerosols up to 41 km were shown to arise from the 1974 eruption of Fuego in Guatemala.[382]

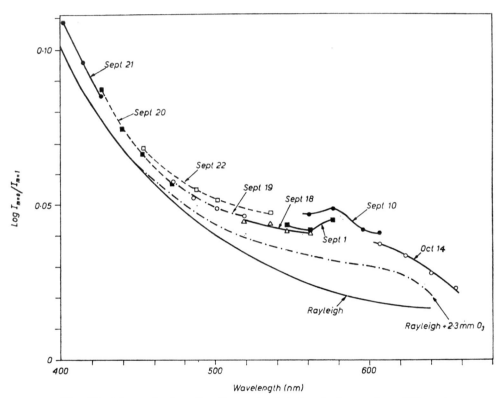

FIG. 60. Atmospheric extinction, calculated and observed on different days at Jungfraujoch; expressed as logarithm of ratio between irradiances at air mass 0 and 1. (Labs and Neckel.)

Atmospheric extinction in urban, desert and oceanic areas was studied by Guttman, who obtained several sets of Bouguer lines.[383] He also showed experimental curves similar to Fig. 60 with more realistic nominal curves for spectral extinction, including Rayleigh and aerosol scattering, and ozone absorption. These nominal curves were derived from Elterman's atmospheric attenuation model.[133] The experimental curves are of interest here though not from high altitude measurements. Fig. 61 is an example for a mid-Pacific sea-level site

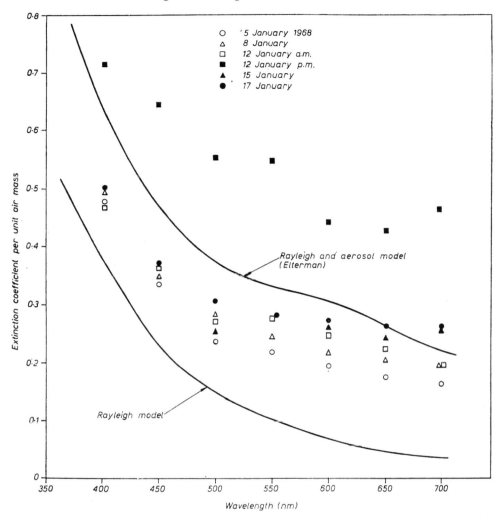

FIG. 61. Extinction coefficient of clear atmosphere in the Pacific
Ocean, with variable high transparent cirrus clouds. (Guttmann.)
The Elterman curve includes ozone absorption.

which shows how the atmospheric extinction varied from day to day,
mainly by the presence of transparent layers of cirrus cloud. An earlier
case of detection of diffusing clouds, not visible except when seen edge-
wise at heights of 6 to 12 km, was reported by Packer and Lock.[384]

The discrepancies between models and experimental measurements
emphasize again the disturbing effects of unpredictable atmospheric
conditions on all solar measurements, however carefully executed, and

suggest that the future of exact measurement must lie in extending the use of rockets or spacecraft, of which some early examples have been mentioned. Oertel and Epstein have reviewed the subject and other topics in solar spectroscopy.[710] The new NASA projects (Chapter 5) should eventually help to solve outstanding problems.

A reassessment of earlier work was made by Makarova and Kharitonov in 1969.[176] They were only partly concerned with high altitude measurements and used data from the SAO (1902–22), from Wilsing, Pettit, Stair and more recent investigators including Sitnik, Peyturaux, Labs and Makarova. The results did nothing to clarify the position since the weighted mean for $H_{0\lambda}$ differed seriously from most other values in Table 3. $I_{0\lambda}$ was calculated for both continuum and integrated spectra of the centre of the solar disc, and again for the whole disc. For the centre disc continuum (Table 5) there were similar differences from the Labs and Neckel work, while the derived value of S was also high (Chapter 5). Makarova and Kharitonov gave a separate table based on six investigations at higher spectral resolution, as evidence to support their opinion that the solar continuum has small waves near 480 and 710 nm in an otherwise smooth spectral curve. Later they asserted the greater accuracy of averaged Russian data, both old and new, and from both high level and surface measurements, in comparison with other European and American work which was criticized adversely, though with the admission that still more accurate measurements were required.[711]

The addition of these spectral power distributions to the list of earlier measurements must cause disappointment that, after so much careful work over the last fifty years, the values still show differences of the order of 10 per cent. By integration into the solar constant the differences are reduced since they occur in different parts of the spectrum. A similar integration of spectral power distributions to chromaticities also conceals some differences but may accentuate them if arbitrary wavelength intervals are chosen for the calculation (Chapter 13).

12

Colorimetric and other Measurements
on Daylight
from 1960

Light is something that cannot be reproduced. It must be represented
by something else, by colour for instance.
PAUL CÉZANNE, in conversation with E. BERNARD and K. X. ROUSSEL.

Chapter 10 showed that colorimetric values for daylight are useful in
themselves, apart from spectroradiometric data needed to calculate
them. Three main investigations based on direct colorimetry occurred
in the period under review. Their evidence sometimes confirmed,
sometimes differed from that already discussed.

The first to appear was that of Chamberlin, Lawrence and Belbin.[385]
This work was inspired by the lack of local measurements on daylight
that had encouraged the investigation by Henderson and Hodgkiss
(Chapter 10). The authors measured the chromaticity of the north sky
at Salisbury, Wiltshire, at an elevation of 45°, and at noon during a
year from March 1961. The method was unique in this field of work
by its use of a Lovibond–Schofield tintometer. This required a match
to be made between a magnesium oxide screen exposed to daylight,
and a similar screen in the instrument under Standard Illuminant C,
viewed through a selection of closely graded red, yellow and blue glass
filters; that is, by subtractive colorimetry. Records were made of
weather conditions and the illuminance of the screen exposed to the
sky. The latter varied between 200 and 7500 lux.

The chromaticities were close to the Planckian locus except for CCT
more than 135 mireds (less than 7500 K), when several points were on
the upper side of the locus. The range of CCT was from near zero to
147 mireds with a most unusual distribution of 20 per cent of the
values in the 120 to 135 mired interval, 40 per cent between 130 and
150 mireds, and none at all at higher ratings. Fig. 62 shows the

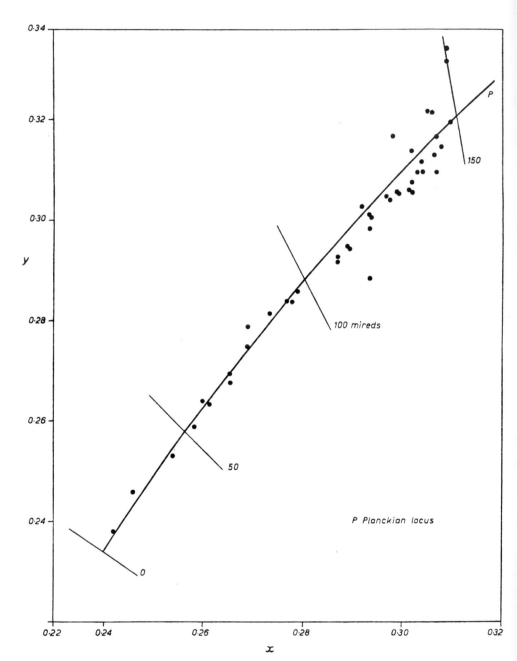

FIG. 62. Distribution of chromaticities of north skies at Salisbury, Wilts, 1961 to 1962. (Chamberlin, Lawrence and Belbin.)

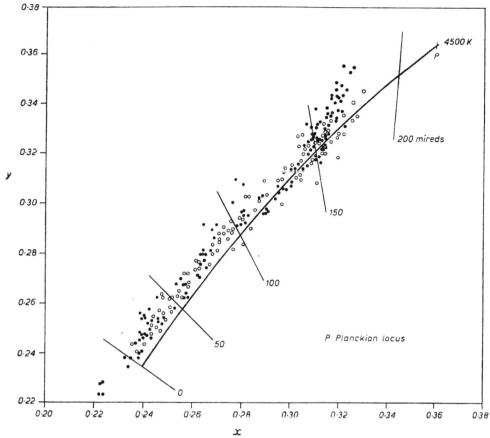

FIG. 63. Distribution of chromaticities of north skies at Ottawa, 1962
(Nayatani and Wyszecki). Isotemperature lines to Kelly's formulation have been added. The solid points are for determinations made in
the first third of the period.

chromaticity plots with isotemperature lines according to Kelly. The
observations were made through slightly green window glass, which
can hardly explain the unusual features of the results.

Another colorimetric study of north skylight occupied about a
month (March 1962) and produced upwards of two hundred chromaticities. In this case the light from about 45° elevations was reflected
into a Donaldson six-stimulus colorimeter by a barium sulphate screen.
The observations were made by Nayatani and Wyszecki at Ottawa,
Ontario,[334] leading to the distribution of chromaticities seen in Fig.
63, which covered the area found in most previous spectrophotometry.
It was very similar to that reported by Henderson and Hodgkiss, much

wider than that found by Budde, and less extensive than that given by Sastri and Das. Possibly these different areas depended on the extent of sampling in the experiments. The scattering of points in this visual colorimetry was marked above the full radiator locus for CCT at 7000 K or less. A new correlation was that of mired value with illuminance on the barium sulphate plate, both increasing over the range observed for clear skies up to 100 mireds. Above 100 mireds there was a wide scattering of illuminance associated with the lower colour temperatures and variability of overcast skies.

A two-year series of measurements of sky CCT was made by Collins at the NPL, Teddington, as a contribution to the revision of the artificial daylight standard (Chapter 10).[386] The Harding visual colour temperature meter was used to view four quadrants of the sky from 13° elevation to the zenith, with exclusion of direct sunlight. Other observations took a vertical view of the whole sky from about 32° elevation upwards, including sunlight when present. Numerous histograms presented the mired values for 09.00, 12.00 and 15.00 hours, for north-, east-, south- and west-facing measurements, and for the four seasons in two successive years. The morning change to lower mired values noted by Tarrant was contrary to Collins' mean results except for the east-facing ones, towards London, where mired decreases smaller than Tarrant's were often observed. Similar histograms for the whole sky covered a smaller range at higher mean mired values. For autumn and winter the maximum occurred at noon, for spring and summer the mean value was nearly constant from 09.00 to 15.00 hours.

With no indications of greenness compared with the Planckian locus, the data cannot be effectively compared with chromaticity or spectrophotometric measurements. The overall means were 150·3 mireds for quadrant measurements (6650 K), 142·9 mireds for north sky (7000 K), and 167·0 mireds for total sky (5990 K). Fig. 64 gives Collins' histogram for a year of total sky measurements. The proposed 6500 K daylight standard is well bracketed by these results, though the process of converting CCT values to mireds, averaging them and converting back to CCT is somewhat questionable in view of the very skew distributions which were found.

A survey of daylight by means of colour temperature meters was made by Ando, Ikemori and Kawakami. Values recorded were between 4500 and 7500 K. The paper has an abstract in English.[387]

A summary of Japanese work on daylight colorimetry and spectrophotometry was given by Okada *et al.*, who also developed a correlation between north sky chromaticity and amount of cloud cover (0 to 2,
3 to 7, 8 to 10 tenths for the three categories of weather condition
used). From a full year's data obtained at Nagaoka the mean annual
mired value of sky colour could be calculated for any other site
according to the proportion of fine, partly cloudy and cloudy days
recorded. From records made in Japan, U.S.A., U.K. and several
European countries over about 90 years the yearly averages of sky
colour temperature vary little though the data apply to varied periods
of time and number of recording stations. The extremes were 7650 K
for Japan and 8530 K for Italy, with the U.K. at 8160 K.[388]

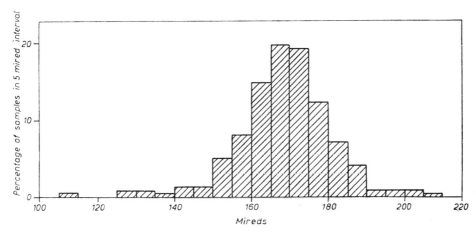

FIG. 64. Distribution of CCT (mireds) of total sky, 282 sets of
measurements over a year from September 1962 at 9.00, 12.00 and
15.00 hours. Mean value 167.0 mireds. (Collins.)

Twilight colours have been measured both on earth and from
satellites and are discussed in Chapter 15.

Measurements with pyranometers and Schott OG1, RG2 and RG8
filters gave approximate spectral components of the total radiation at
Mauna Loa in Hawaii (3380 m) under cloudless conditions, and in the
same project direct solar radiation was found by a normal incidence
pyrheliometer. Over a period of sixteen months the monthly mean of
solar radiation was nearly constant if corrected for sun to earth
distance. The maximum for 1959 to 1962 was 1·68 L/min, a value very

close to those found at other high level sites, namely Tenerife in 1896, Mount Whitney in 1908 to 1913, and at 5710 m near Mount Everest in 1961. The effects of clouds, aerosols and water vapour were discussed in this work by Drummond and A. K. Ångström.[273]

Lopukhin reported some percentages of ultra-violet (wavelengths not stated) in the total solar radiation for different values of solar altitude, and in a number of tables proposed six lowland and seven mountain types of spectral composition for direct sunlight.[389] Neither the method of deriving the tables nor their intended application is clear.

A year-long investigation by Elnesr and Hegazy at Giza, near Cairo, obtained values of direct solar irradiation by an actinometer and the usual Schott filters OG1 and RG2.[390] The tables gave the information in the bands below 525, 525 to 630, and above 630 nm, with air mass as a parameter. They showed the decrease of percentage flux in the first band and increase in the third as m increased from 1·5 to 4·0; also the remarkable constancy of the proportion of each band throughout the year at fixed m. For example, at $m=2$, the first band varied only from 23 to 25 per cent, the second 10 to 11 per cent, the third 63 to 66 per cent.

STANDARD SPECTRAL POWER DISTRIBUTIONS

A revision of Moon's 1940 spectral curves was completed by Gates after calculations made with more recent quantitative information.[391] His main assumptions were the extraterrestrial solar distribution due to Johnson, a solar constant of 2·00 L/min, and an ozone content of 3·5 mm. The other parameters selected for groups of distribution curves were air mass 1 to 8, precipitable water 5 to 30 mm, aerosol concentration 2×10^8 to 8×10^8 particles per m³, and elevation above sea level 0 to 5 km. For global radiation on a horizontal plane Gates used the calculated values for Rayleigh scattering derived by Deirmendjian and Sekera,[299] whose curves are included in the paper. A novel feature was the representation of most of the spectral distributions on wavenumber abscissae (cm⁻¹), with ordinates given in watts/cm² (wave number) of solar power incident normally. Total power was also given in each of the thirty-seven curves presented. In those reproduced in Fig. 65, precipitable water is 10 mm, aerosol

2×10^8 particles/m³. The resemblance of these curves to Langley's prismatic spectrum for daylight is interesting (Fig. 4). In Fig. 66 another set of curves, on a wavelength scale, demonstrates the difference between solar and global irradiation on a horizontal surface, for $m=1$ to 4.

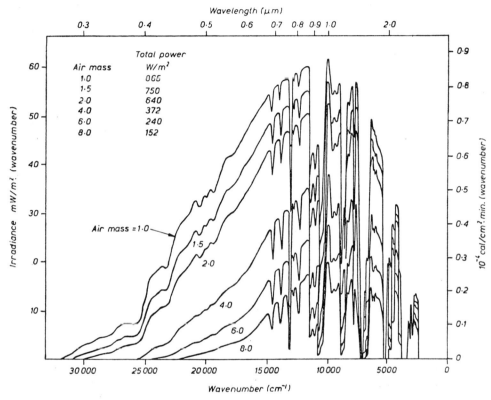

FIG. 65. Spectral power distribution of direct solar radiation incident normally at sea level; dependence on air mass. S 2·00 L/min, O_3 3·5 mm, H_2O 10 mm, aerosol 200 particles/cm³. (Gates.)

The scale of the published curves was small, and no tabulated numerical data were included. Later determinations of the extra-terrestrial irradiation spectrum and the solar constant have modified these fundamental quantities somewhat from the values used by Gates for $H_{0\lambda}$ and S. If these recent data are confirmed, Gates' extensive collection of material would need modification in the scale of power received from the sun, and to a smaller extent in the spectral distributions. It should be noted that in some of the published curves the units for the ordinate scales were given as 10^7 watt ... and 10^4 cal ... instead of 10^{-7} and 10^{-4} respectively.

Another set of spectral distributions was prepared by Dogniaux for use in architecture and building where it is necessary to consider

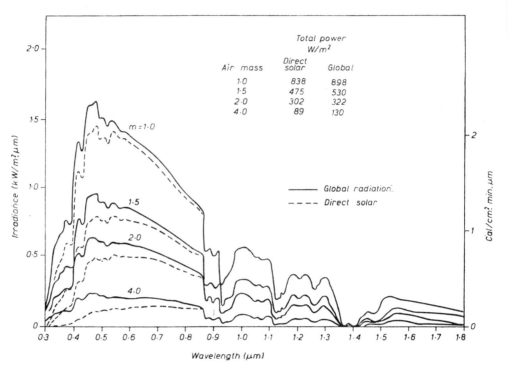

FIG. 66. Spectral power distribution of direct solar and global radiation incident on horizontal surface at sea level; dependence on air mass. Earth albedo zero, other conditions as Fig. 65 except H_2O 20 mm (?). (Gates.)

solar radiation entering buildings and affecting their heat balance as well as their illumination.[392] Previous standardized daylight spectral distributions, including the RD series, often ended in the visible range and neglected the infra-red spectrum which is important in the total power input to a building.

The proposals were of four types of distribution extending from 300 nm to $2 \cdot 15$ μm. They were tabulated in absolute units and in relative proportion to each other determined in earlier work by Nicolet and Dogniaux.[393] The chosen solar altitude was 30°, and the solar constant $2 \cdot 00$ L/min. The distributions were for normally incident sunlight, from a clear sky, for diffuse skylight from a clear sky, for global radiation comprising the two preceding distributions in appropriate proportions, and for overcast sky. They had chromaticities close to the locus of RD distributions at CCT of 5089, 14388, 6072 and 5547 K respectively, and were extended from 3 to 50 μm by values of the full radiator at 260 K to represent radiation of atmospheric origin.

Dogniaux made other calculations on scattering and absorption by atmospheric constituents.[394] The results were given as direct solar irradiance from a clear sky, in 40 nm bands from 380 to 800 nm and wider bands to 3 μm, and for different values of h, water vapour content W, Ångström coefficient β, Linke turbidity factor T, with $S=1\cdot95$ L/min ($1\cdot364$ kW/m²). Ozone content was not mentioned. Since diffuse irradiance was given only as a total for each combination of h, W, β and T, without spectral distributions, the calculation of global distributions or chromaticities is not possible.

Schulze prepared two versions of a spectral power distribution in 40 nm bands for global radiation at the earth's surface. The first[395] was based on Johnson's work with $S=1\cdot39$ kW/m², and included corrections to be applied to the values of total radiation for variations in h, W and T. The second[396] used a modified spectrum and $S=1\cdot36$ kW/m². Radiation at wavelengths beyond 3 μm was neglected since the 2 per cent of extraterrestrial radiation in this band is almost completely absorbed in the atmosphere.[105]

These and the Dogniaux proposals have lost most of their value by reason of later CIE recommendations on similar lines. The CIE formulated spectral distributions for the simulation of solar radiation for testing purposes, first relying on Hinzpeter's calculated values for sunlight and skylight, and the D_{65} distribution (Chapter 13);[397] then about 1969 using Schulze's data[396]; finally by adopting the NASA values for S and the extraterrestrial spectrum, from which irradiation at the earth's surface was calculated. The final CIE recommendations included tables of irradiation values in 40 nm bands from 280 to 800 nm and wider bands to 3 μm, with factors to be applied to the total for variations in h, W, and T. Table 5 gives the distribution for $h=90°$, $W=10$ mm, $T=2\cdot75$, $\beta=0\cdot05$, ozone$=2$ mm. Discrepancies between this distribution and other proposals are considered in Chapter 13. As in Schulze's papers, 1 kW/m² $\pm10\%$ was recommended as a standard for testing purposes.

STANDARD DISTRIBUTIONS OF SKY LUMINANCE AND GROUND ILLUMINANCE

The 1969 CIE proposals were used by Krochmann *et al.* to calculate the illuminance on a horizontal plane under a clear sky, for sun and sky separately, versus *h*, and to express these relations in empirical equations.[398] The same authors used an early version of the CIE report on relative luminance distribution on the clear sky to derive an equation for the luminance of any point relative to that of the zenith; and another equation for zenith luminance in terms of *h*. In this respect the final CIE report was not materially altered from the earlier version.[399] It supplemented a former proposal for the luminance distribution on an overcast sky.[400] This related the luminance at an angle θ above the horizon to the zenith luminance L_z by

$$L/L_z = (1+2 \sin \theta)/3,$$

a widely used formula.

The CIE clear sky distribution of luminance is given by a formula due to Kittler including the zenith angles of the sun and the point of measurement, and the angular distance between them.[401] The numerical constants may be replaced by another set to allow for higher atmospheric turbidity ($T \geqslant 5$). Other local conditions may be important in view of the 150 sets of measurements made in Madrid by Juan and Cruz, with *h* varied from 30° to 70°, where an alteration of most of the constants in the formula was necessary in order to represent the results well.[402] In the CIE report tables and stereographic projections of the sky allow the position of the 'dark point' where minimum luminance is found, to be related to *h*. In a vertical plane through the sun, the angle between the sun and the dark point changes from 96° at *h*=0° to 59° at *h*=90°, and is 90° as previously assumed[235] only when *h*=18°.

Many empirical formulae for luminance and illuminance, without spectral distributions, have been collected.[403] There is also a useful list of revised values of illuminance, luminance, efficiencies and temperatures of the sun based on the NASA 1971 solar constant. This gives, for example, the total radiation temperature of the sun as 5763 K, a value close to that calculated in Chapter 2.

The tables and formulae mentioned, together with standard data quoted elsewhere[115, 133, 364, 312, 404, 519] provide a varied choice of representative values for total radiation or daylight at the earth's surface. For a selected case of $h=90°$, $W=10$ mm, $\beta=0{\cdot}05$, $T=2{\cdot}75$, $O_3=2$ mm, the transmittance on to a horizontal surface of total solar radiation is about 83 per cent, but 91 per cent of light calculated as power: conversion of power to lumens makes little difference.[155] It is important to realise that the information available often refers to normal incidence of sunlight, or alternatively to radiation from the hemisphere received on a horizontal surface. Modifications are therefore necessary in applications to situations where solar power is not received under these ideal conditions. For calculations on its utilization for heating or conversion to other forms of energy the total received is important. In this country the amount is roughly 1 MWh/m² per annum.

Improvements in radiometry are important in this field of measurement. Drummond and Ångström compared sunlight or daylight illuminance measured photometrically with solar or global irradiance measured by a pyrheliometer with filters.[405] From experiments on land up to 3·4 km elevation, and in aircraft up to 13 km, radiometry was found to give illuminance values as accurate as those from direct photometry. For global illuminance they developed the relation

$$E=150\,(1+0{\cdot}102m)\times W$$

where E is in klx, m is air mass corrected to sea level pressure, and W is irradiance in L/min for the radiation at less than 700 nm determined by the difference between the readings with the Schott filters WG7 (clear glass) and RG8 (red).

13

The Processing of
Daylight Spectra

Knowledge, a rude unprofitable mass,
The mere materials with which wisdom builds,
Till smooth'd and squar'd, and fitted to its place
Does but encumber what it seems t'enrich.

WILLIAM COWPER, *The Task*, Book 6.

In this chapter we consider recent developments in calculating standardized distributions of spectral power, relative or absolute, based on comparatively large amounts of experimental material. The method of approach and presentation of results are different from those in the work of Taylor, Moon, Gibson, Middleton, Dogniaux and Gates, detailed in earlier chapters.

In 1951 the CIE Colorimetry Committee made a recommendation to its members 'to investigate the most suitable energy distribution to adopt for average natural daylight'. In 1955 the recommendation was merely 'to examine the suitability of light sources for reproduction of natural daylight', and in 1959 the subject was not mentioned in the recommendations. The provision of distributions by Schulze and Middleton during this period has been discussed in Chapter 9.

In 1963 it was recommended 'that definitions of standard sources of four correlated colour temperatures, 3900, 5500, 6500, 7500 K, be developed to supplement CIE standards A, B and C. With the exception of the source of correlated colour temperature 3900°K, these supplementary sources are intended to represent phases of daylight over the spectral range 300 to 830 nm.'[406] Of the new standards proposed, 7500 K was chosen to satisfy opinion in the U.S.A., 6500 K to be close to Standard Illuminant C, and 5500 K as a possible photographic source, as was 3900 K. In subsequent work an illuminant at 4800 K was developed instead of 3900 K, being close to Standard Illuminant B; and the 10 000 K illuminant was also included.

So far as work in this country was concerned, the simplest kind of data processing had been found adequate to obtain a typical spectral distribution for 6500 K, which had been selected here as a suitable standard. From the measurements made by Henderson and Hodgkiss (Chapter 10), the mean spectral response at each 10 nm interval was taken from forty-five curves normalized at 560 nm. These curves were all of equivalent CCT of 6500 K ± 10 mireds, and their mean resulted

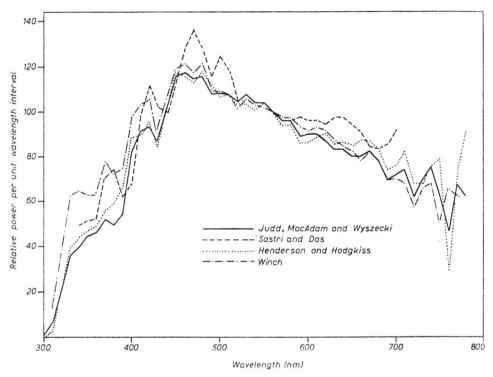

FIG. 67. Representative spectral distributions for daylight at 6500 K, normalized at 560 nm.

in an equivalent CCT of 6440 K, chromaticity $x=0.314$, $y=0.329$. It was proposed to the BSI as a standard daylight distribution in April 1963. Later by the adoption of Kelly's isotemperature lines, the choice of curves had to be altered slightly. There were now fifty-four curves available, with equivalent CCT of 6500 K ± 10 mireds, and a mean of 6498 K, chromaticity $x=0.3130$, $y=0.3275$ (Table 6 and Fig. 67). The mean was proposed to the BSI in August 1965, but was later superseded by the RD distribution described below.

Winch *et al.* took the mean of spectral distributions over short intervals of CCT and interpolated the values for 5500, 6500 and 7500 K. The results are summarized in Table 6, and the 6500 K values included in Fig. 67. Tarrant also grouped spectral curves in this way and derived mean spectral power distributions given in Table 6. His

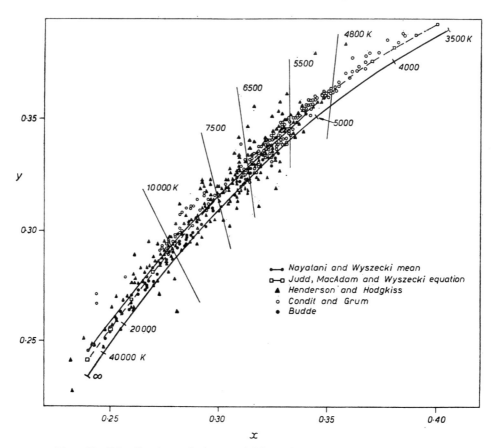

FIG. 68. Distribution of chromaticities of daylight used in reconstituted daylight formulation. (Judd, MacAdam and Wyszecki.) For Nayatani and Wyszecki mean see reference 334.

north- and south-facing groups for 150 to 160 mireds are plotted in Fig. 52: both correspond to CCT very close to 6500 K.

The CIE recommendations of 1963 resulted from a considerable amount of work which had been done previously. The technical colorimetry committees in the U.S.A. and Canada undertook about October 1962 a joint evaluation of the copious experimental material becoming available. Their task was to try to find typical daylight spectral distributions for a wide range of CCT. This was eventually limited to values recommended by the CIE, but the method of calculation allowed an extension to the whole range of possible daylight distributions, in so far as the original data were comprehensive. The events leading up to this project were described by Judd, MacAdam and Wyszecki in a paper which also included the essential results.[335]

The material for this investigation comprised 622 spectral distribution curves or sets of values of S_λ obtained by Henderson and Hodgkiss, Condit and Grum, and Budde, as described in Chapter 10. Chromaticities of the whole group are shown in Fig. 68. The required 10 nm spectral interval had been used in the first subset of measurements; for the second the values were determined by interpolation on the curves, and in Budde's curves, means over 10 nm intervals were used. For further uniformity, measurements below 330 nm and above 700 nm were discarded because all three subsets did not extend beyond this spectral range: to cover the wider range desired, Moon's 1940 values were adopted from 300 to 320 nm and 710 to 830 nm, and fitted empirically to the ends of the mean experimental curve. This was done after the characteristic vector method had been applied to the experimental material over the 330 to 700 nm range.

CHARACTERISTIC VECTOR ANALYSIS

This method is based on the assumption that all members of a set of curves to which it is applied can be represented as linear combinations of a number of fundamental curves. Each set of n response curves forms a matrix of r columns (wavelengths) and n rows (S_λ values). The mean is the mean of the n ordinates at each of the r wavelengths. By treating the matrix in a way described by Simmonds[407] a series of vectors is extracted, the first representing the largest amount of variability in the original set of curves, the second the next largest amount, and so on (see Appendix 1). Each vector is a set of r numbers, in this case thirty-eight. Multiples of the vectors, added to the mean curve, will reconstitute the original response curves more or less exactly, according to the number of vectors used and the underlying (but possibly undetected) similarities in the original set. Strictly, n vectors would be required for perfect reproduction of all the originals. Judd *et al.* calculated four vectors for each of the three subsets of curves and for the whole set. Their published results referred mainly to the whole set (622 curves), and showed that adequate representation of the originals was achieved by only two vectors.

The first vector corresponded to variations of chromaticity in the yellow–blue direction. The second vector showed some anomalies between the subsets, but for the whole set implied green–pink variations. It is not essential to the success of the method that the vectors should represent physical realities, though in this case they did so. Tables of the vectors were given, and some comparisons are shown in Figs. 69 and 70. Details of the vectors from the Condit and Grum

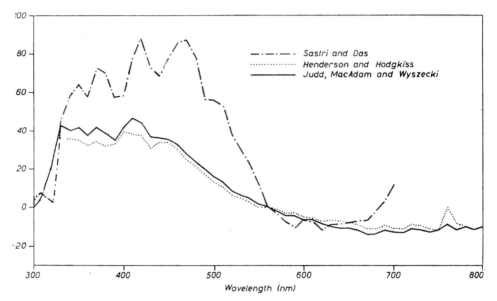

FIG. 69. First characteristic vector from groups of spectral power distributions of daylight.

measurements were given later by Condit.[408] One immediate anomaly is seen in the differences between the second vector for the Henderson and Hodgkiss results according to the wavelength range selected. Judd *et al.* gave several examples of experimental and reconstituted daylight distributions (RD) of similar CCT which confirmed the good agreement at some of the CIE-recommended colour temperatures. Fig. 71 includes the standard curves derived in the work. The power of the method lies in the fact that it can produce by interpolation any combinations where no experimental material existed at the outset. This facility was used by the authors, and subsequently adopted by the CIE.

Reconstitution of a spectral distribution depends on first choosing the chromaticity point. The 622 distributions were represented on the

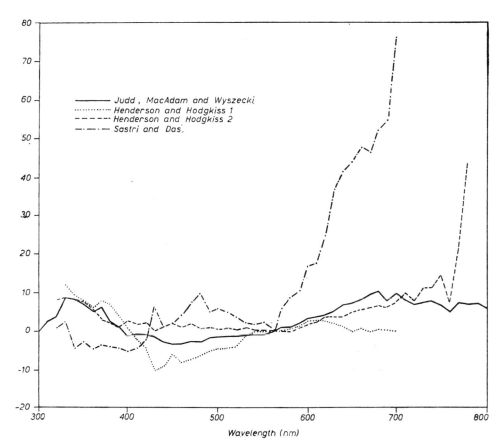

FIG. 70. Second characteristic vector from groups of spectral power distributions of daylight. Curve 1 refers to analysis of Henderson and Hodgkiss data to 700 nm only, curve 2 to 780 nm.

chromaticity chart by a line drawn through them by visual inspection, with some slight adjustment above 10 000 K to take account of the direct colorimetry results of Nayatani and Wyszecki, and Chamberlin *et al.*

The representative line was then found to be given by

$$y_D = -3 \cdot 000 x_D^2 + 2 \cdot 870 x_D - 0 \cdot 275 \qquad (1)$$

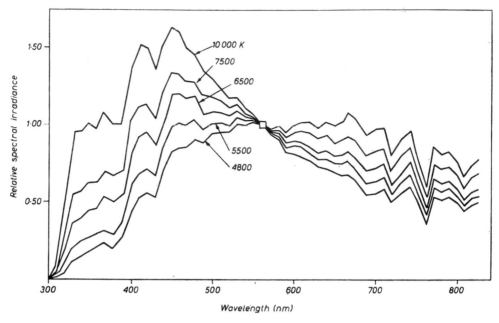

Fɪɢ. 71. Spectral distributions for typical daylight at selected CCT, reconstituted from first and second characteristic vectors of 622 measured distributions, normalized at 560 nm. (Judd, MacAdam and Wyszecki.)

where the suffix D refers to a daylight chromaticity (see Figs. 20, 68 and 85). If at a wavelength λ the spectral response is S_λ and the mean for the set or subset is $\bar{S}_\lambda$, then

$$S_\lambda = \bar{S} + \Sigma MV$$

where M is the scalar multiple of the vector V. The tristimulus values are calculated by

$$X = \Sigma S_\lambda \, \bar{x}(\lambda) \Delta \lambda$$

etc. and since for a single curve, M is independent of λ,

$$X = X_0 + \Sigma M_i X_i$$

where X_0 is the tristimulus value for the mean $\bar{S}$, and X_i is that for the vector V_i. Since $x = X/(X+Y+Z)$, etc. the chromaticity can be calculated from X_0, Y_0, Z_0 and values of MX, MY, MZ for each vector. Hence the required values of M follow in terms of x, y, X_0, Y_0, X_i, Y_i, etc. For the material analysed,

X_0	102434	Y_0	106769	Z_0	123630
X_1	1866	Y_1	1914	Z_1	34810
X_2	2133	Y_2	762	Z_2	-2355

The values of M_1 and M_2 became

$$M_1 = \frac{-1{\cdot}3515 - 1{\cdot}7703 x_D + 5{\cdot}9114 y_D}{0{\cdot}0241 + 0{\cdot}2562 x_D - 0{\cdot}7341 y_D} \tag{2}$$

$$M_2 = \frac{0{\cdot}0300 - 31{\cdot}4424 x_D + 30{\cdot}0717 y_D}{0{\cdot}0241 + 0{\cdot}2562 x_D - 0{\cdot}7341 y_D}. \tag{3}$$

APPLICATION TO A SELECTED CCT

It is often more convenient to use a colour temperature description of a daylight spectral distribution than a chromaticity specification. The CIE Colorimetry Committee proceeded to calculate equations for the conversion of daylight chromaticities to CCT, represented by T_c. For the range of T_c values between 4000 and 7000 K,

$$x_D = -4{\cdot}5993\frac{10^9}{T_c{}^3} + 2{\cdot}9645\frac{10^6}{T_c{}^2} + 0{\cdot}09905\frac{10^3}{T_c} + 0{\cdot}244063. \tag{4}$$

For T_c between 7000 and 25 000 K

$$x_D = -2{\cdot}0031\frac{10^9}{T_c{}^3} + 1{\cdot}8997\frac{10^6}{T_c{}^2} + 0{\cdot}24734\frac{10^3}{T} + 0{\cdot}237040. \tag{5}$$

Equations (1) to (5) were adopted by the CIE in August 1966.[409] Calculations may be largely avoided by the use of tables of x, y, u, v, M_1 and M_2 versus T_c. The standardized RD distributions were referred to as D6500 etc., according to their CCT. The symbols were later altered to D_{65} etc., and the distributions referred to as 'representing a phase of daylight' instead of RD. Though any value may be calculated, the CIE emphasized the need to maintain D_{65} as the predominant standard, with D_{75} and D_{55} as possible alternatives if necessary.

Further slight changes are caused by the modification to the International Practical Temperature Scale initiated in 1968 by the Commission Internationale des Poids et Mesures. This required Planck's second radiation constant c_2 to be increased from 1·4380 to 1·4388 × 10^{-2} m K and involves the alteration of the constants in the equations

TABLE 6. Daylight spectral power distributions for CCT 5500, 6500, 7500 K, normalized at 560 nm.

nm	D$_{55}$	W	SD	D$_{65}$	W	SD	HH	TN	TS	D$_{75}$	W	SD
300	0·02	0·3		0·03	1·3		0·9			0·04	2·0	
310	2·1	3·9		3·3	12·6		7·3			5·2	19·2	
320	11·2	14·5		20·2	37·2		21·8			29·8	54·6	
330	20·7	26·5		37·1	63·0		39·9	32·9	47·6	55·0	90·0	
340	24·0	30·6	35·4	40·0	64·4	49·4	43·3	39·9	42·2	57·3	90·6	61·0
350	27·9	35·4	36·1	45·0	63·1	51·4	46·5	46·3	44·8	62·7	89·2	64·4
360	30·7	38·7	38·4	46·7	63·0	52·1	49·1	57·2	47·1	63·0	85·1	63·8
370	34·4	48·6	54·0	52·2	77·6	70·6	56·0	77·1	67·9	70·3	98·6	85·5
380	32·6	47·4	57·7	50·0	72·8	74·6	59·0	71·8	73·7	66·8	91·5	89·2
390	38·2	51·0	49·6	54·7	74·2	63·3	65·9	67·7	60·3	70·0	91·3	75·1
400	61·0	69·6	54·0	82·8	98·0	68·1	88·4	94·7	78·3	101·9	119·0	80·2
410	68·6	75·0	80·8	91·6	103·2	99·6	90·6	104·4	95·3	111·9	123·8	115·6
420	71·6	79·0	90·8	93·5	105·0	111·9	95·6	102·6	97·4	112·8	124·0	130·0
430	67·9	69·3	84·9	86·8	89·2	102·6	84·9	93·5	93·0	103·3	104·1	117·7
440	85·6	84·6	84·3	104·9	105·3	101·0	104·0	103·8	100·2	121·1	120·6	115·1
450	98·1	100·4	95·4	117·1	119·4	114·2	118·0	118·8	115·8	133·0	135·6	130·3
460	100·4	102·9	107·7	117·8	119·8	128·4	116·3	119·6	116·8	132·3	133·8	146·1
470	99·9	103·2	115·0	114·9	116·8	136·0	113·8	119·0	117·0	127·2	129·5	153·8
480	102·6	108·7	108·8	115·9	120·8	127·3	117·8	120·6	118·7	126·9	131·7	143·0
490	98·0	101·0	102·0	108·8	110·3	115·4	112·8	113·5	110·2	117·7	121·8	126·9
500	100·7	104·6	111·5	109·4	110·0	124·8	108·4	115·1	111·9	116·5	117·8	136·2
510	100·8	104·0	106·1	107·8	108·4	118·7	108·2	111·4	110·1	113·7	114·5	129·5
520	100·0	98·0	95·3	104·9	101·8	103·9	104·2	102·9	101·7	108·6	105·7	111·3
530	104·2	102·4	100·3	107·7	105·6	107·4	104·2	107·6	107·1	110·5	108·6	113·5
540	102·1	101·6	99·2	104·4	103·0	104·5	101·2	105·2	104·4	106·3	105·8	109·1
550	103·0	101·8	98·7	104·0	102·6	101·4	103·2	106·4	105·9	104·9	103·5	103·7
560	100·0	100·0	100·0	100·0	100·0	100·0	100·0	100·0	100·0	100·0	100·0	100·0
570	97·3	99·5	96·7	96·4	97·7	96·1	94·6	100·5	101·1	95·6	96·9	95·5
580	97·7	99·7	96·6	95·7	97·2	95·0	94·8	97·8	97·7	94·2	95·3	93·6
590	91·4	99·5	99·5	88·6	93·2	97·1	85·9	84·5	84·6	87·0	92·0	94·5
600	94·4	98·5	97·6	90·0	91·5	96·2	86·4	93·8	98·7	87·3	89·2	94·8
610	95·1	97·3	98·1	89·6	92·5	96·2	88·6	92·3	89·2	86·2	88·5	94·4
620	94·2	95·5	97·3	87·6	89·8	94·5	89·6	90·3	87·5	83·6	85·7	91·9
630	90·4	92·8	99·1	83·3	86·5	96·8	85·4	84·7	83·7	78·7	82·0	94·6
640	92·3	92·5	99·3	83·7	85·2	97·2	86·4	87·0	87·7	78·5	80·0	95·2
650	88·9	91·8	95·5	80·0	82·2	93·6	84·9	82·4	83·0	74·8	77·7	91·7
660	90·3	87·8	92·5	80·2	78·8	90·8	87·7	76·5	77·4	74·5	73·5	89·0
670	94·0	90·6	87·1	82·2	81·9	85·6	86·4	79·7	81·6	75·5	76·0	83·9
680	90·0	87·3	83·6	78·3	78·6	83·1	83·0	82·0	84·7	71·7	72·8	82·3
690	79·7	79·6	84·2	69·7	69·4	85·2	74·1	54·1	54·9	64·0	64·0	85·9
700	82·9	80·8	89·1	71·6	70·4	91·9	75·6			65·2	65·0	93·9
710	84·9	78·9		74·3	68·1		82·5			68·1	62·6	
720	70·2	67·5		61·6	57·5		68·1			56·5	53·6	
730	79·3	72·8		69·9	67·5		69·1			64·3	59·3	
740	85·0	73·9		75·1	69·0		75·6			69·2	60·3	
750	71·9	54·4		63·6	50·6		79·3			58·7	43·5	
760	52·8	69·2		46·4	65·6		30·4			42·7	55·9	
770	75·9	67·1		66·8	62·6		76·1			61·4	53·7	
780	71·8			63·4			91·1			58·4		
790	72·9			64·3						59·2		
800	67·4			59·4						54·8		
810	58·7			51·9						48·0		
820	65·0			57·4						53·0		
830	68·3			60·3						55·6		
x	0·3324	0·3327	0·3315	0·3127	0·3134	0·3140	0·3130	0·3128	0·3140	0·2991	0·2999	0·3005
y	0·3475	0·3442	0·3370	0·3291	0·3247	0·3220	0·3275	0·3263	0·3284	0·3150	0·3116	0·3105
Δy	+·0065	+·003	−·033	+·006	+·001	−·0025	0		+·003	+·004	+·006	+·0025 0

D: Reconstituted daylight, from vectors (Judd et al. original values, ref. 335).
W: Winch et al. interpolated from mired groups.
SD: Sastri and Das, from vectors.
HH: Henderson and Hodgkiss: experimental, mean of fifty-four points.
TN: Tarrant, north-facing group T8: experimental, mean of forty-four points.[353]
TS: Tarrant, south-facing group T28: experimental, mean of fifty points.[353]
Δy Distance of chromaticity point above (+) or below (−) Planckian locus.

for x_D in terms of T_c (see Appendix 1). The change in c_2 has the further effect that D_{75}, D_{65} and D_{55} become the distributions for approximately 7504, 6504 and 5503 K respectively. The effect on Standard Illuminants A, B and C is given in Chapter 18. If the modification of the CIE 1960 UCS suggested in 1975 is adopted, the CCT values of the RD illuminants will become approximately 7375, 6380 and 5400 K, assuming their spectral power distributions to be unchanged (Appendix 1).

Table 6 is an abridged comparison of the RD standards and some experimental distributions. Fig. 67 shows some of these different versions of the 6500 K spectral power distribution plotted at 10 nm intervals. The values in Table 6 are calculated for $c_2 = 1 \cdot 4380$ m K but the later change has little effect: for D_{65} most of the spectral values are unaltered and none is more than $0 \cdot 1$ from the revised figures, while the chromaticities for D_{75} and D_{65} need to be increased by $0 \cdot 0001$ in both x and y.[605]

The RD distributions were based on measurements between 330 and 700 nm, with extensions to 300 and 830 nm by extrapolations which do not include data that were available in some (but not all) of the measurements considered. Similarly the interpolations to 1 nm intervals in the RD tables do not represent experimental results, and the accuracy quoted, like that in other tables of the publication, far exceeds that of the original measurements.

RD AND OTHER STANDARDIZED DISTRIBUTIONS

The RD distributions for terrestrial daylight were obtained as directly as possible from experimental observations. Other proposed distributions have involved calculations of scattering and absorption in the atmosphere applied to the extraterrestrial spectrum, itself found by extrapolation from measurements within more or less of the atmosphere. It is not surprising that the two types of spectrum do not agree exactly. Fig. 72 shows some other comparisons: the Moon values for $m = 2$ with the RD at the same CCT (5089 K), and the NASA and Labs and Neckel spectra with the full radiator for 5755 K, which is their equivalent calculated from the solar constant (Chapter 2). The Figure also compares the two $H_{0\lambda}$ spectra with parts of the full radiator spectra of equivalent chromaticities (see below). The agreement is worse than for the 5755 K curve. Fig. 73 shows the NASA $H_{0\lambda}$

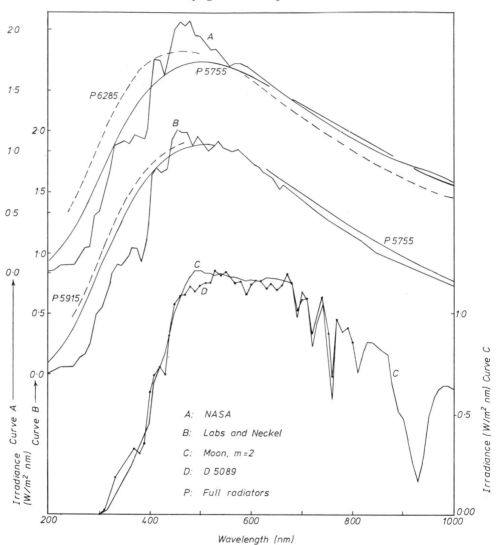

FIG. 72. Comparison of solar equivalent full radiator (5755 K) and parts of chromaticity equivalent full radiators (broken lines) with extraterrestrial spectra of NASA, and of Labs and Neckel; also of Moon spectrum for $m = 2$ with RD for 5089 K. Normalized at 560 nm.

spectrum[167] integrated to 40 nm bands, and its reduction to air mass 1, $h=0°$, clear skies and normal incidence, proposed in a CIE publication on the simulation of terrestrial solar radiation.[155] For comparison are included the D_{65} spectrum in the 10 nm form and in 40 nm

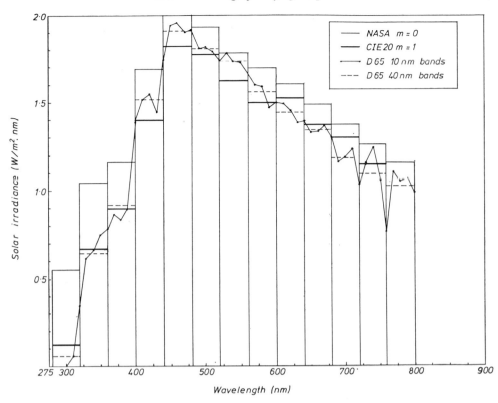

Fig. 73. Comparison of NASA extraterrestrial spectrum, the same reduced to sea level, and D_{65} normalized to equal areas (total W/m^2) between 280 and 800 nm. D_{65} shown in both 10 nm and 40 nm bands.

bands. The fit is far from satisfactory. Somewhat better matches, though still not exact, occur between D_{62} and the CIE proposal, and between D_{60} and the Labs and Neckel $H_{0\lambda}$ spectrum[174] reduced to $m=1$ in the same way. D_{65} is used in colorimetry, for which the wide band CIE distribution is unsuitable, though having advantages in its extension to 3 μm and a scale of power per unit area. In Fig. 73 the D_{65} curves are normalized to the same total power between 300 and 800 nm as the CIE proposal.

THE DAYLIGHT CHROMATICITY LOCUS

The line representing the 622 chromaticity points was chosen by visual inspection, and equation (1) connecting x_D with y_D calculated from it.

Daylight and its Spectrum

TABLE 7. Constants in $y_D = a + bx_D + cx_D^2$.

Authors	Number of points	a	b	c
Condit and Grum	249	−0·0934	1·818	−1·484
Henderson and Hodgkiss	274	−0·2644	2·910	−3·242
Budde	99	−0·1847	2·225	−1·990
Nayatani and Wyszecki[334]	553	−0·1076	1·773	−1·241
Chamberlin et al.[385]	60	−0·1170	1·837	−1·402
Least square method*	622	−0·196	2·393	−2·287
	1235	−0·176	2·223	−1·964
Judd et al.[335]	622	−0·275	2·870	−3·000
Kawakami et al.[351]		−0·160	2·224	−2·102

* Least square method applied to 622 points used by Judd *et al.* or to all 1235 points available.

TABLE 8. Values of y_D at chosen values of x_D according to the equations of Table 7.

Authors		x_D		
	0·24	0·28	0·32	0·36
Condit and Grum	0·2573	0·2992	0·3362	0·3686
Henderson and Hodgkiss	0·2474	0·2963	0·3349	0·3632
Budde	0·2419	0·2908	0·3332	0·3696
Nayatani and Wyszecki[334]	0·2465	0·2917	0·3328	0·3700
Chamberlin et al.[385]	0·2430	0·2873	0·3271	0·3624
Least square method* 622	0·2466	0·2947	0·3356	0·3831
1235	0·2466	0·2947	0·3345	0·3700
Judd et al.	0·2410	0·2934	0·3362	0·3694
Kawakami et al.[351]	0·2527	0·2979	0·3364	0·3682

* Least square method applied to 622 points used by Judd *et al.* or to all 1235 points available.

A more satisfactory method would have been to determine least-square equations to fit the experimental points. This was subsequently done by Nimeroff,[410] with results appreciably different from those in the calculations by Judd *et al.* Table 7 gives the revised constants determined in this way for equation (1), with different groups of experimental points. The last line of the table quotes an equation which

Kawakami, Ando and Ikemori found to represent their chromaticity determinations on the north sky at Nagaoka (number of points and method of finding the equation unknown).[351] These authors also gave their equation in the CIE 1960 UCS:

$$v = -0.8316 + 10.45u - 23.61u^2$$

Table 8 shows some of Nimeroff's calculated values of y_D for selected values of x_D from the different equations. The differences are not obviously related to the values of the constants a, b and c, which perhaps gave an exaggerated impression of the variability of the subsets of data. Some of the chromaticity differences are certainly rather large, but they lie within the variations of the original experimental points. However, the subsets seem to be not entirely compatible with each other, as was suggested by the work of Judd *et al.* on the first three subsets (CG, HH and B), when the vectors for the subsets differed noticeably. Sampling methods may be responsible, or there may be real differences in the daylight itself.

FURTHER USE OF
THE CHARACTERISTIC VECTOR METHOD

This project for the processing of data from independent sources was obviously limited in its supplies of experimental material. The subsequent investigations in South Africa, India, Japan and London produced about 1500 more spectral distribution curves, which might be included in some future analysis by the vector method. If the small differences already noticed between subsets of data are significant, it would be more useful to analyse homogeneous sets of observations from single locations than to lose the small differences by combining a large, heterogeneous collection of curves.

An example has occurred which emphasizes the value of smaller scale analysis. Sastri and Das resorted to the characteristic vector method and found vectors very different from those of Judd *et al.* Figs. 69 and 70 include their first and second vectors.[343] Possibly because of the limited number of curves available for analysis, Sastri and Das found rather poor agreement between original and reconstituted distributions; nevertheless the authors' use of the method is useful in helping to demonstrate its possibilities of revealing differences between similar populations of curves. In another case Sastri and Manamohanan analysed their sixty curves to obtain vectors differing widely from those derived from the Delhi curves, and from the others in Figs. 63 and 64.[344] The method was also applied to some of the measurements made at Amagasaki in 1965 to 1967.[346] The first vector accounted for nearly 98 per cent of the spectral power distributions and agreed well in shape with the CIE one (Fig. 69), and the second vector was not very different from the second CIE one (Fig. 20).[348]

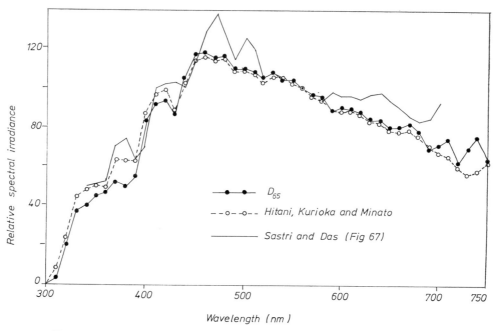

FIG. 74. Comparison of D_{65} with reconstituted distributions for 6500 K by vector analysis. Normalized at 560 nm.

The differences were mostly in the ultra-violet region (Fig. 74). To this Figure has been added the reconstituted 6500 K distribution calculated by Sastri and Das for their Delhi observations. These curves suggest real differences between the two types of daylight spectrum.

Vector analysis is useful in other problems where spectral curves are concerned. Nayatani and Takahama investigated the colour matching properties of fluorescent lamps[411]; and in the classification of the spectral reflectance curves of soils, Condit found identification of soil types and prediction of curves to be simplified by the method.[408]

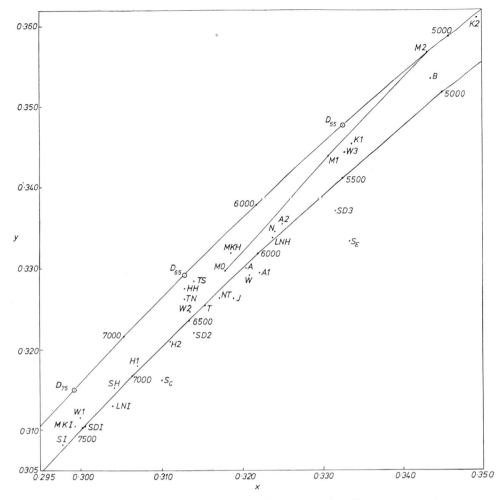

FIG. 75. Chromaticities of solar radiance and irradiance spectra in relation to Planckian locus (below) and RD locus (above).

SI, SH: Sitnik, $I_{0\lambda}$ and $H_{0\lambda}$

MKI, MKH: Makarova and Kharitonov, $I_{0\lambda}$ and $H_{0\lambda}$

†*W 1,2,3: Winch et al., 7500, 6500, 5500 K

†*SD 1,2,3: Sastri and Das, 7500, 6500, 5500 K by vectors

LNI, LNH: Labs and Neckel, $I_{0\lambda}$ and $H_{0\lambda}$

*H 1,2: Hisdal 1975, Spitzbergen

S$_c$: Standard Illuminant C

†*TN, TS: Tarrant, north and south facing

†*HH: Henderson and Hodgkiss, 6500 K mean

T: Thekaekara 1968, $H_{0\lambda}$

NT: NASA and Thekaekara 1971, $H_{0\lambda}$

M 0,1,2: Moon, sunlight at $m = 0,1,2$

J: Johnson 1954, $H_{0\lambda}$

A: Abbot 1920, $H_{0\lambda}$

W: Webb et al. 1970, $H_{0\lambda}$

A: 1,2: Arvesen et al. 1971, $H_{0\lambda}$ engineering (1) and extended (2) tables

N: Nicolet 1951, $H_{0\lambda}$

SE: Equi-energy spectrum

*K 1,2 Kok, sunlight at $m = 1,2$

B: Brandhorst et al. 1975, $m = 2$ calculated from NT

* Daylight observed at ground level   † As in Table 6

CHROMATICITIES FROM SPECTRAL DISTRIBUTIONS

It is not clear how the spectral distributions in Tables 3, 5 and 6 differ unless they are plotted. A more concise way of comparing them is by their chromaticity points (Fig. 75) though the spectral range used is 380 to 760 nm and therefore includes only half the total radiation (compare Fig. 72). Striking differences are revealed, some due in part to the selection of spectral power values at arbitrary intervals for computation (usually 5 or 10 nm) whereas the full table of experimental data may be recorded at smaller intervals (compare Fig. 76).

Spectra of different origins show remarkable consistency in that most of the points for I_0 and H_0 lie close to the Planckian locus. Variations along the locus for each group suggest that spectrally selective absorption and scattering have not been entirely eliminated in the extrapolations applied to the observations. The greenish hue of terrestrial daylight, by comparison with that of H_0, seen in many of the Figures of Chapters 10 and 12, and in some points included in Fig. 75, must be due to atmospheric effects and not to the Fraunhofer G absorption at 430·8 nm which was formerly suggested as a possible cause.

Figs. 20 and 75 use x, y coordinates since most of the published information is on this scale rather than the 1960 CIE UCS with u, v coordinates.

14

The Daylight Spectrum:
Achievements and Possibilities

All works are overcome by amplitude of reward, by soundness of
direction, and by conjunction of labours. The first multiplieth
endeavours, the second preventeth error and the third supplieth the
frailty of man.

FRANCIS BACON, *Of Proficience and the Advancement of Learning*.

Daylight is a comparatively simple natural phenomenon, scarcely
affected by man's activities, and not the mysterious product of his own
sophisticated experiments, like superconductivity or atomic fission. Its
study has therefore not been marked by serious controversies on
theoretical matters, and in viewing the efforts of the last hundred years
we see the pioneers struggling with limited equipment and technique
rather than with inadequate ideas on the nature of the phenomenon.

On the experimental side a number of well-organized explorations
of the field have been made, especially in the American foundations
and in the high altitude Swiss observatories. Besides this there has
been a great deal of uncoordinated effort, often seeming now to be
trivial, often difficult to correlate with present-day measurements or
requirements, sometimes lacking the circumstantial experimental
details which would have been valuable to later workers, and too often
presented without numerical data to supplement the published small
scale diagrams.

In assessing the experimental work on the daylight spectrum before
the present decade, the outstanding effort of the SAO on the extraterrestrial spectrum and the solar constant must be mentioned once
more. Abbot's weighted mean served its purpose for many years, and
there has recently been a remarkable return towards a value of S close
to that found by the SAO. It is however difficult to avoid some regret

that the observatory should have made so many hundred determinations of S, excellent examples of precision of measurement as they were, only to have their accuracy questioned because of uncertainty in the absolute scales of radiation, and further that so much work should have been done on a so-called constant when accuracy closer than 1 or 2 per cent is probably unattainable by measurements on the earth's surface. Similar considerations apply to projects like those of Dunkelman and Scolnik, and of Labs and Neckel, both admirably conducted, but not certain of having eliminated all atmospheric effects. However, Labs and Neckel provide the best wide spectrum of $I_{0\lambda}$[50] with the HSRA as a theoretical model.[380] The best representations of $H_{0\lambda}$, though slightly different, seem to be the NASA spectrum in extended or abridged versions,[366, 167] and the Labs and Neckel table.[174]

The measurement of the daylight spectrum at the earth's surface is simpler. The main problem is to deal with its extreme variability especially in regard to the effects of aerosols and cloud. In this respect daylight has some similarity with the weather, which indeed depends on some of the same variables. In both there are many factors which determine the final product, factors which appear in infinitely varied combinations. The daylight spectrum has the added complication of two different forms, for direct and scattered sunlight. The consequence is that not only do random experimental determinations of daylight spectra resist comparison or classification, but calculations of their structure from basic principles are complicated to perform, invariably depend on simplifying assumptions, and in the outcome have been somewhat unconvincing. Such moderate success that these predictions have had is of value because it confirms the general correctness of assumptions about atmospheric effects on incident light, their relative importance and magnitude. One of the more useful conclusions is that the spectral distribution of global radiation tends to be constant over a wide range of solar altitude, other things being equal, and this has been confirmed in some observations in favourable conditions.

The identification and measurement of atmospheric ozone has been the most considerable by-product of the study of the daylight spectrum. The information it provides on water vapour is also important though this depends more on the infra-red spectrum. The chemistry of atmospheric reactions has developed largely from theoretical ideas on the extraterrestrial spectrum of sunlight.

STANDARDIZED DATA ON DAYLIGHT

Measurements of the daylight spectrum on earth have been used repeatedly to provide standardized values for specified conditions. Kimball, Moon, Taylor and Kerr, Gates and Schulze have in turn attempted to supply this need. They have not all used the same method. There is a distinction between those who used experimental material directly, and those (Moon, Gates and Schulze) who adopted an extra-terrestrial solar spectrum and calculated the absorption which this suffers by the time the radiation reaches the earth. The scattered sunlight, or skylight, must be calculated separately from less certain premises. Perhaps the most useful tables are those of Moon for $m=2$, and the CIE abridged spectrum for $m=1$ in spite of its differences from the RD distributions.

The recent period has seen the development of the characteristic vector method of treating groups of experimental spectra. This can be applied without weighting, extrapolation, arbitrary corrections, or the grafting of extensions on to the ends of the spectrum, all of which have been practised before, provided that the material used in the vector method is homogeneous in wavelength range and preferably also in type of sky and place of origin.

Standardized daylight spectra at ground level, of wider application and more general acceptance, may result from new measurement projects planned in several countries (1975). These are intended to include the near ultra-violet and infra-red ranges.

The term 'average daylight' has been used frequently in the past, but with little justification if applied to experimental measurements of the spectrum. Sampling has rarely extended systematically over the whole of the daylight hours and never in this way for a long period. The matter is raised because daylight is used physiologically, with spectral discrimination, by land plants and many animals whenever available, much of it outside man's active hours. The total of received daylight is important, as it is also for life in the sea. Though there is plenty of evidence of the total power and total light received at different places and seasons, there is none of spectral power distribution. Average daylight must mean the relative spectrum of the total radiation received during a whole day or month or year, and such measurements of the spectrum might well be made with automatic recording, either continuously at selected wavelengths, or by fast repeated scans. So far determinations of the mean have scarcely progressed beyond the collection of many spectral curves on different days and at different times, usually with normalization of the curves at a selected wavelength. For total or a true average daylight, say in plant growth investigations, this procedure is inaccurate because light on a dull day should not be equated at an arbitrary wavelength with that from a bright day. The effectiveness of energy received is also dependent on threshold and saturation levels in photosynthesis.

THE ULTRA-VIOLET END OF THE SPECTRUM

The ultra-violet end of the daylight spectrum is variable and largely unpredictable. Its band structure due to Fraunhofer and other absorptions is well known, and good agreement on the wavelengths of the structural features can be quoted from numerous experimental curves. It is in the quantitative relations of the bands to each other and to the visible spectrum that aerosols produce their most noticeable effects. They raise doubts whether typical distributions are attainable at all, unless the user is required to insert factors for aerosol conditions, height above sea level, cloudiness, solar altitude, ozone content of the atmosphere, type of location and level of visible irradiance, some of which variables will not be generally known. Compare Green's contributions.[322] The extent of Bener's work necessary to standardize such distributions with most parameters controlled, shows the difficulty of the problem. The differences between the high ultra-violet values measured by Winch *et al.*, Boshoff and Kok, Tarrant, Sastri and Manamohanan, and those from other projects are not so far explained in any quantitative way and deserve further study.

The sensitivity of the ultra-violet region of daylight to conditions which seem to have little effect on the visible spectrum, is of course not sharply defined by the limit of the eye's vision. Atmospheric changes occur at all wavelengths, but increasingly with shorter wavelengths, and so are most noticeable at the ultra-violet end of the spectrum where they reach a climax at the complete cut-off near 290 nm.

THE INFRA-RED END OF THE SPECTRUM

Enough has been seen in earlier chapters to show the limited interest taken in the infra-red part of radiation on the earth, compared with what has been done in the visible and ultra-violet regions. Many applications are satisfied by measurements between 300 and 830 nm, but this is insufficient where total incident energy is concerned, in heat balances in buildings and the like. In most of these special cases the usual assumptions of a theoretical full radiator distribution in the infra-red region (of 6000 K or thereabouts), or of total irradiance from sun and sky, are probably adequate, though more accurate information is available in the CIE recommendations[155] and the earlier work of Dogniaux and Schulze (Chapter 12). Considerable quantities of power are emitted in the infra-red. Table 9 gives the percentage of extra-terrestrial radiation received at wavelengths *longer* than those quoted. The distributions for $H_{0\lambda}$ obtained by Webb *et al.*, Arvesen *et al.*, and Makarova and Kharitonov are similar to those in the Table, which includes other distributions to show the effects of the atmosphere in the Moon and CIE data, and of the artificial curtailment of D_{65} at both ends of the spectrum. The visible radiation of the sun between 400 and 830 nm is seen to be slightly higher than that for the full radiator.

TABLE 9. Percentage of extraterrestrial radiation and other spectral distributions at wavelengths longer than each wavelength quoted.

Wavelength	NASA $H_{0\lambda}$	L & N[174] $H_{0\lambda}$	Planck 5755K	CIE $m=1$	Moon $m=2$	D_{65}
250 nm	99·8	99·8	99·0			
300	98·8	98·9	96·9	99·7*	~100·0	~100·0
400	91·3	92·4	87·9	93·9	97·4	90·5
700	53·1	53·6	51·4	49 *	53·0	20·6
830	41·6	41·7	39·5	40 *	37·2	0·0
1 μm	30·5	30·6	28·4	28·2	24·0	
2	6·6	6·3	6·1	4·6	1·1	
3	2·2	2·0	2·1	0·0	0·0	
4	0·9	0·9	1·0			
5	0·5	0·5	0·5			
10	0·06	0·06	0·07			

* values integrated from wide bands

CONSIDERATIONS ON SPECTRAL MEASUREMENT

A few comments may be made on the physical details of measurement. Conditions that have not always been observed, though desirable, include:

(1) The exclusion of skylight when direct sunlight is being measured.
(2) The proper integration of light from the sky, which is very variable with direction of view, when this part of daylight is separated from sunlight: Budde has described how to improve on existing methods by a new integrating sphere.[361]
(3) The correction for polarization by reflection in the measuring instruments, usually a small but appreciable effect if there is no intermediate diffuser.
(4) The avoidance of the effects of uneven surface sensitivity in photocells and photomultipliers.
(5) The correction for or avoidance of rapid changes in the daylight spectrum during the measurement period: this is not the same as change of total irradiance, and seems to be amenable only to integration by photography or monitoring at several fixed wavelengths and perhaps discarding observations when certain limits of change are exceeded.

Spectral power distributions have been measured continuously with narrow slits by some authors, or by wider slits to integrate the Fraunhofer absorptions. The resolution, possible or attained, in many investigations is not easily discovered. Information may be taken from the curves at fixed wavelength intervals or at selected peaks and troughs and points of inflection.[357] If the readings are made at fixed intervals, or the curves not recorded but read only at these intervals, the points will lie at random on peaks and absorption bands in the spectrum, varying in position according to the wavelength accuracy of the spectroscope. Variation of slit width has a similar disturbing effect and may cause some, but not all, of the differences seen in the columns of $H_{0\lambda}$ values in Table 3.

All the absorptions should be included at their correct profiles for determinations of total irradiance or of chromaticity, while atmospheric bands need to be discounted where extraterrestrial spectra or the solar constant are concerned. In work on the photosphere radiance $I_{0\lambda}$, the envelope of the curve is one of the standard measurements, without Fraunhofer absorptions, as in Table 5.

The effects of 'instrumental blending' by increasing slit width were demonstrated by Wilson *et al.* for a spectrum of high resolution taken from the Utrecht atlas (see Fig. 76).[412] The daylight spectra obtained by Henderson and Hodgkiss, Winch, Sastri and Das and others may be compared with a solar radiance curve by Labs and Neckel.[173] Part of the latter is given in Fig. 77, where high resolution and selected wavelengths give more strongly marked absorptions; for example at the iron and magnesium lines near 517 nm, the F line of hydrogen at 486·1 nm, and particularly the G line of calcium and iron at 430·8 nm which falls exactly into the narrow measurement interval at this point of 429·6 to 431·6 nm.

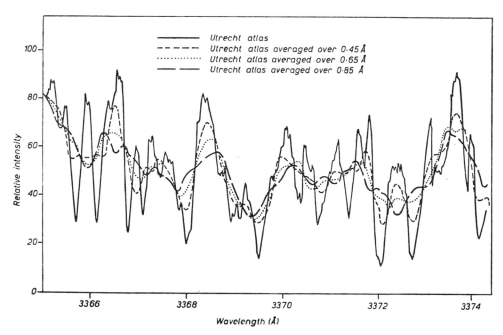

FIG. 76. The effect of 'instrumental blending' on detail of the solar spectrum. (Wilson, Tousey, Purcell, Johnson and Moore.)

References to filter measurements are frequent in this book, the method being widely employed where high resolution spectra are not essential. There are obvious limitations, particularly when the spectral power varies rapidly with wavelength as at the ultra-violet end of the terrestrial daylight spectrum. Difficulties in calibration and measurement arise.[413]

For short discussions of interferometric spectroscopy, mentioned in Chapter 11, see Jacquinot[414] and Gebbie.[415] A small instrument for the near infra-red (1·25 to 5 μm) suitable for use in aircraft or space-craft was described by Schindler.[416]

Fraunhofer line profiles are affected by atmospheric reactions to solar radiation. For example, in daylight spectra recorded with his own spectroradiometer, MacAdam found that absorption bands due to water vapour and oxygen, three of each between 550 and 720 nm, disappeared from the spectrum of a clear sky to an extent which could

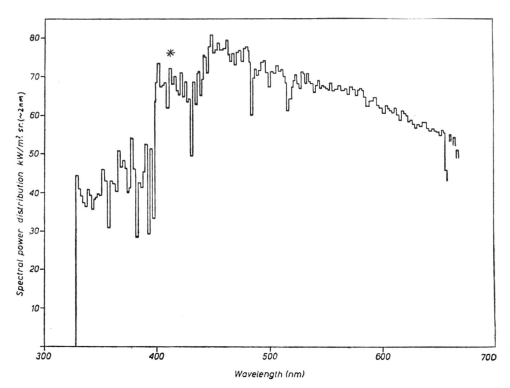

FIG. 77. Spectral power distribution of centre of solar disc, including Fraunhofer absorptions. Measured waveband is 2·05 nm at * (∼ 411–413 nm) and shorter wavelengths, but 2·00 nm at longer wavelengths. (After Labs and Neckel.)

not be judged exactly owing to the rather low dispersion of the instrument.[417] One suggested explanation was the filling up of absorption bands by fluorescence of the molecules concerned. Grainger and Ring, examining the slight ultra-violet fluorescence of the moon, found the profile of the Ca II line at 396·8 nm to be partially filled up when the source of radiation was changed from moonlight to blue skylight, and similarly for the H line at 486·1 nm by the change from sunlight to blue skylight. They concluded that there is a source of light in the daytime sky other than scattered sunlight.[418, 419] The most probable reason for the extra radiation now seems to be the rotational Raman scattering proposed by Brinkman[420] and fairly well confirmed by calculation.[421] This effect may be partly responsible for the comparative greenness of daylight discussed in Chapters 10 and 13.

REFERENCE SOURCES

Standards for calibration of spectral irradiance measurements have for forty or fifty years been almost entirely dependent on tungsten filament or ribbon lamps in envelopes of fused silica or glass of known spectral transmittance. Before this period carbon arcs were often used, for example by Fabry and Buisson[55] and by Plaskett[219] but only occasionally in recent times.[368, 372] Both sources have their disadvantages, and for the tungsten lamp the first is the low emission at shorter wavelengths, including the blue end of the visible spectrum. The second is the uncertainty of the emissivity at the shorter wavelengths if the lamp has been calibrated to a colour temperature standard, for example Standard Illuminant A. This makes it inadvisable to extrapolate the full radiator curve for 2855·5 K into the ultra-violet, and corrections are required by emissivity data,[422, 423] or by Preston's method.[424] Even in the visible spectrum there is a chance of slight errors in a colour-matched standard since the standard CIE 2° colorimetric data are somewhat suspect in the violet and blue parts of the spectrum. A full discussion of the radiometric properties of the tungsten ribbon and the carbon arc has been given by Schurer.[425] From the de Vos emissivity data[422] Jones derived tables of absolute spectral radiant intensity per candela for tungsten and for the full radiator, covering temperatures between 1700 and 3300 K and wavelengths between 240 and 2600 nm.[426] Tungsten emitters may be used for calibration in spectrophotometry with particular advantage in the infra-red, and lamps with sapphire windows extend the useful range to 6 μm.

The tungsten–halogen lamp has been developed since about 1959. In it the fused silica bulb is kept clear by operating at 250 °C or higher while the halogen included with the argon filling prevents the deposition of evaporated tungsten on the envelope, but instead deposits it on the filament. Since the filament may now operate at higher temperatures than usual (up to 3400 K) because the damaging effects of tungsten evaporation have been eliminated, the lamp has improved emission in the short wavelength region. Lamps calibrated for spectral and luminous emission under controlled power input have become available.[337, 427] Presumably they should be used for a minimum of time because of changes in the tungsten surface, and therefore of its emissivity, during the halogen–metal cycle in the lamp. In lamps of this type containing iodine it may be desirable to consider the superimposed vibration spectrum of the halogen which modifies the emission of the plain tungsten filament to a measurable degree,[428] and in those containing bromine the presence of lines due to impurity metals in the tungsten filament.[429]

Another source of ultra-violet emission is the deuterium lamp with its continuous spectrum from about 180 to 400 nm, which may be calibrated by synchrotron radiation for use as a standard of irradiance.

Radiation scales and measuring instruments still require continued scrutiny to avoid difficulties which occurred in the past.

THE VISIBLE SPECTRUM

More information could be extracted from the existing material on the daylight spectrum, for example the best-fitting locus to the chromaticity points in the form $y = a + bx + cx^2$. Variations in these loci between different groups of observations might be instructive, since it is evident by inspection that the chromaticity patterns are not all alike. If spectral curves of previous work were available, the extraction of characteristic vectors might be even more revealing.

The above two analytical methods should be applied to any future measurements on the visible spectrum and this should include separation into groups of spectral power distributions taken, for example, from the north sky, the zenith sky, the whole sky, the sun alone, and so on. Different seasons could be investigated, with variable h, and different surface albedos where uniformity prevails over large distances as on a Pacific island or on the Antarctic continent. The distribution of observing sites mentioned in this book is very uneven. The great majority are within the latitudes of 24° and 52° N with a few at higher latitudes but very few in the southern hemisphere. The only strictly tropical sites

($23\frac{1}{2}°$ N or S) are those in Hawaii, Bombay and the early SAO observatories in Chile. Arctic and Antarctic observations are very limited except for those of meteorological interest. It would be desirable to have large groups of measurements at some more scattered sites, when the results might lead to better identification of the processes affecting daylight spectra and chromaticities, and to evidence on the possibility of consistent variations from place to place as some sets of measurements suggest.

Another outstanding problem is the predominant greenness compared with the full radiator locus, and the increased greenness often appearing at higher CCT values. This locus is not an entirely artificial standard of comparison, for Table 3 and Fig. 75 show that where the chromaticity of extraterrestrial sunlight is calculated from proposed spectral distributions, it lies on or very close to the full radiator locus, even slightly below.

The full extent of daylight chromaticities is a related problem, seen in different aspects in the preceding figures. The strongly biased distributions of chromaticity in some cases need explanation, for example the lack of low mired values in Figs. 41, 44 and 50, and their frequency in Figs. 37, 43, and 63. Other sections of the colour gamut in daylight deserve more attention. Very little has been done concerning the chromaticity of sunrise and sunset skies and its possible correlation with climatic conditions. The chromaticity and spectral power distribution of the zenith sky at twilight, and of the whole sky in total eclipses of the sun, have not been frequently or very accurately recorded, and offer interesting problems both theoretical and practical. Information on the chromaticity of blue skies is scattered and not very extensive (Fig. 20).

OTHER POSSIBLE STANDARDIZING MEASUREMENTS

The ultra-violet part of the daylight spectrum suffers most disturbance from aerosols, and for this reason any new programme of work limited to the ultra-violet wavelengths would need to be long continued if designed as a basis of prediction. Its value would probably be strictly local, or applicable to very similar sites elsewhere, but the more urban and industrial the site, the more difficult the task and the more restricted the application of the results.

Standardized information on the visible spectrum at earth level would not be greatly changed by recalculation from recent new measurements. Methods might be explored to allow simple corrections to the daylight curves of Moon, Gates or Brandhorst, in order to include the many known variables. New measurements have been planned recently, and by covering a wider spectrum and using present knowledge of the variable factors these may eventually yield more useful standardized data for the expanding practical applications. Only as long-term average predictions, with little relation to daily variations, are the calculated spectra likely to be valid.

15

Less Common Forms
of Daylight

He had been eight years upon a project for extracting sun-beams out
of cucumbers, which were to be put in vials hermetically sealed,
and let out to warm the air in raw, inclement summers.
                    JONATHAN SWIFT, *A Voyage to Laputa*, Chap. 5.

This chapter collects together a number of less important ways in
which the sun's light reaches the earth. Most of them provide only
small amounts of light. Some are of astronomical interest, but not in
the main stream of the developments followed in this book.

ECLIPSES

Total eclipses of the sun, which have contributed much to the theory
of the sun's structure and radiation, have been studied for their
temporary effects on daylight. The interest has been limited, for the
main attraction of a brief total eclipse must always lie in the sun itself.
Measurements have frequently been made on sky luminance and earth
illuminance, but there are not many records of sky colour or spectral
power distribution.

Wright described a portable three-filter visual colorimeter capable
of moderate accuracy.[430] He used this during the eclipse of June 30,
1954 at Grebbestad, Sweden, to obtain information on sky luminance
and chromaticity. One area of sky at 40° elevation and an azimuth of
90° from the sun was chosen. The main result was a large change of
chromaticity towards the blue and back again after the eclipse,
covering the approximate range of 4500 K to a very purplish 11 000 K
(Fig. 20), the latter not usually seen in daylight. The relation of sun
to atmosphere, and radiation from the zenith are different during an
eclipse from those found at normal twilight, and more information on
this would be useful.

The purplish blue colour of the sky is evident from Wright's chromaticity measurements. It had been noticed long before, when the astronomer Halley observed the eclipse of April 22, 1715 from the Royal Society's house off Fleet Street in London.[431] He wrote: 'By 9 of the clock . . . when the Face and Colour of the Sky began to change from perfect serene azure blew, to a more dusky livid Colour having an eye of Purple intermixt, and grew darker and darker till the total Immersion of the sun, which happened at 9 h 9′ 17″ by the Clock, or 9 h 9′ 3″ true time.'

Spectral distribution information is of more value than colorimetry, and some was supplied by observations made by Sharp, Lloyd and Silverman at Hermon, Maine, during the eclipse of July 20, 1963.[432] The luminance of the zenith sky decreased by a factor of about 3500 during totality, of the same order as in Wright's measurements. Regarding spectral effects, they found that the distribution, though measured only over the range of 520 to 640 nm, did not change till the sun was obscured to the extent of 99·8 per cent. Then the shape of the zenith daylight curve changed in the direction of less red to green ratio, which was not inconsistent with Wright's results but difficult to compare without measurements on blue wavelengths. The change at wavelengths of 600 and 520 nm was very marked, from a ratio of 0·7 to 0·18 in totality and back to 0·65. The authors compared this changed colour to that of twilight with the sun 5·1° below the horizon. In another analysis of eclipse phenomena sky luminance was found to be proportional to percentage obscuration of the sun up to about 99·7 per cent, after which multiple scattering from outside the shadow zone made a noticeable contribution.[433]

Deehr and Rees observed the same 1963 eclipse from an aircraft at 9·6 km altitude in 61°N latitude over Canada.[434] They used a birefringent filter photometer with polaroids to scan small wavelength intervals preselected by interference filters centred on 427·8, 486·1, 557·7 and 630·0 nm. Apart from work on the daytime airglow, they measured the luminance of the zenith sky in the four wavelength bands during totality, establishing a relative decrease of the yellow and red bands (see Fig. 78). Comparisons with morning twilight measured at the zenith (at 1100 m at El Paso, Texas) showed approximate agreement in the relative band levels, and an absolute level equivalent to solar altitude $h = -7·8°$.

Like Sharp *et al.*, Dandekar reviewed earlier work on sky luminance and colour in eclipses.[435] In his own observations made on November 12, 1966, at Quehua, Bolivia (4210 m), he used a scanning spectrometer and photomultiplier to cover the spectrum from 530 to 660 nm, with control observations on the zenith sky in daylight and at twilight. The fall in sky luminance, measured in kilorayleighs per ångström, was of

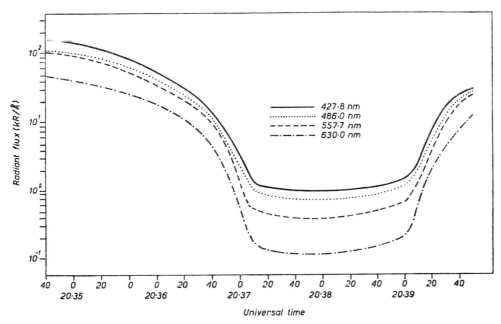

FIG. 78. Interference filter photometry of zenith sky during total eclipse of 20 July, 1963, from aircraft at 9·6 km height. (Deehr and Rees.)

the order of 2500 to 1 during the eclipse, but the elevated site reduced all the values by a factor of 2 or 3 compared with sea level observations. Other results often showed inconsistencies in respect of luminance levels. The sky luminance during the eclipse corresponded in the present case to twilight for $h = -4·1°$, whereas Deehr and Rees from their higher observing station had found still lower zenith luminance and an equivalent twilight of $h = -7·8°$. Moore and Lamb made measurements from a height of 12·3 km during the same eclipse, at seven wavelengths and various distances above the horizon. They found an equivalent twilight of about $-4°$.[436] The spectral curves given by Dandekar for normal zenith day sky, sky at mid-totality and morning twilight for $h= -4\cdot1°$ showed most difference below 550 nm and above 630 nm (thirteen wavelengths were recorded for each), and their corresponding colour temperatures reflected different scattering conditions when compared with the equivalence in luminance between eclipse and twilight conditions. The three curves gave 9100, 11 100 and 7800 K respectively. Spectral observations on this eclipse were made in South America by Velasquez,[437] and by Lloyd and Silverman who also reported measurements from an aircraft at $11\cdot6$ km over the south Atlantic.[438] Dandekar quoted earlier observations in the 15 500 to 16 400 K range, but found 11 000 K again during the eclipse of March 7, 1970.[439] Polarization values were consistent with the predominance of multiple scattering during the eclipse, as required by the luminance changes discussed by Lloyd and Silverman.

Shaw made polarization measurements at an eclipse in Kenya on June 30, 1973, when a fall of zenith luminance by a factor of 10^4 and a change to a more blue colour occurred. The horizon sky was seen to become more red.[440] Spectral changes in the zenith light during several eclipses were correlated by Hall.[441]

From the above it is clear that little effect on sky colour is likely during partial eclipses of moderate extent. Sastri observed the partial eclipse at Delhi on February 15, 1961, and measuring the sky chromaticity at 60° from the sun by spectroradiometry (Chapter 10), found no significant change during the eclipse compared with changes after the eclipse.[442] He quoted Jagannathan *et al.* who, unlike other authors, found the sky to become less blue during a partial eclipse (December 12, 1955). They used blue, yellow and red filters for measurement.[443]

THE MOON

The moon reflects a small amount of sunlight from its dark surface rocks, an amount which would be increased by fourteen times if the surface were a perfect diffuser. This albedo of $0\cdot07$ is the mean value for a surface showing marked directional reflection which increases the observed luminance of the full moon above that calculated from the data of Chapter 3, namely 3×10^3 cd/m². Under clear atmospheric conditions the luminance may reach 4×10^3 cd/m², or more than half that of an average 40 W fluorescent tube, sufficient for its use in spectrographic determination of atmospheric ozone, and in the mapping of its own infra-red emission spectrum which began with Langley in 1884,[444] and continued with Adel and others.[445] The infra-red emission varies with the moon's phase since the surface temperature varies from about 90 to 390 K during the cycle.

As a source of light the interest of the moon lies in its effects on the extraterrestrial solar spectrum. In Chapter 7 some of the earliest spectrophotometry by Vogel was described, including his work on moonlight.[196] His measurements at eight wavelengths 426, 444, 464, 486, 517, 555, 600 and 633 nm may be expressed as ratios of moonlight to sunlight, normalized at 555 nm. Vogel also measured sunlight reflected from different rocks. The ratios derived from a yellow-grey sandstone are compared with the other ratios.

Moon:sun   0·56, 0·50, 0·62, 0·68, 0·84, 1·00, 1·07, 1·06
Sandstone:sun   —, 0·58, 0·75, 0·73, 0·87, 1·00, 1·09, 1·11

Thus moonlight was shown to be somewhat yellowish compared with sunlight, and its surface might well resemble the yellow-grey sandstone.

At the same period of time Pickering found a much more marked tilting of the solar spectrum after reflection from the moon.[39] It is possible that his sunlight was mixed with skylight though this would hardly account for the large decrease in the violet band. Pickering's wavelengths were roughly 450, 520, 590 and 650 nm. The relative values for sunlight and moonlight respectively were

2971:250:100:45   and   363:155:100:87.

Seventy years after these pioneers, Stair and Johnston made an accurate spectral survey using the same equipment as in the solar work in New Mexico (Chapter 9).[276] The intention was primarily to measure the ultra-violet radiation, and the long wavelength limit was 530 nm. No complete record of the absorptive effects of the lunar surface was obtained, but a selective absorption at 380 to 390 nm was established, suggesting a yellowish glassy surface material. The samples recovered by the Apollo 11 Mission in July 1969 have confirmed the earlier deductions from spectrophotometry. Many other investigations show the ratio between the spectra of sunlight and moonlight when normalized at 500 nm: there is a steadily increasing advantage to

moonlight in the wavelength range of 500 nm to 2·4 μm, and to sunlight from 500 nm to 300 nm, hence the slightly more yellow colour of
moonlight.[446]

When the moon is just past the new phase and is visible soon after
sunset, most of its disc is illuminated by light from the earth, the bluish
white light mentioned in Chapter 4. Owing to the earth's higher albedo
it is much more bright, seen from the moon, than is the moon seen
from the earth, and this reflected earthlight illuminates the dark part of
the moon to a noticeable extent. It is called the 'ash-grey' or 'ashen'
light. This term is also applied to the darker part of the planet Venus
when it presents an illuminated crescent, though the cause is disputed
in this case. For the moon the ashen light is not the same as the dull
reddish appearance in total eclipse, when sunlight is refracted through
the earth's atmosphere on to the disc but with loss of the shorter
wavelengths by scattering in the long air path. This was explained by
Kepler in about 1600, and the ashen light by Leonardo da Vinci a
century before.

In contrast to these early discerning views there are others where
primitive ideas persisted among reputable astronomers of much later
date. For example, though it was not unreasonable for the eighteenth
century William Herschel to believe the moon to be inhabited, it is
astonishing to find the astronomer Pickering asserting, in 1924, that
apparent changes in the lunar crater Eratosthenes could be explained
by temporary snow fields and the movements of large groups of small
animals.[447]

THE ZODIACAL LIGHT AND OTHER SUNLIGHT EFFECTS

A minute proportion of the sun's light reaches the earth in the zodiacal
light, always present, but not well known because of the difficulty of
seeing its feeble luminance which is of the order of 10^{-2} cd/m².[448] No
appreciable illumination of the earth can be seen. On winter evenings
in the western sky, and in autumn before dawn in the east, this faint
pyramid of light extends along the plane of the ecliptic. It is probably
sunlight scattered by dust or electrons in this plane. The Gegenschein
is an even fainter glow of similar origin. Minnaert gives an attractive
account of these and other atmospheric phenomena.[449] More colour-
ful ones are due to atmospheric refraction and aerosol scattering, for

example the green flash from the setting sun,[450, 451] and the many appearances of blue or green sun and moon. One record is due to Edward Whymper who, when climbing Chimborazo in Ecuador, in 1880, saw an eruption of Cotopaxi, 60 miles away. The clouds of volcanic dust drifted between the sun and the mountaineering party, causing the sun to appear green, with other brilliant colour effects in the sky.[452] Soon afterwards the volcanic explosion of Krakatoa in the Sunda strait produced a few years of spectacular sunrises and sunsets by the vast amounts of aerosols released by the explosion, and appearances of blue or green sun and moon were frequent.[453] The 1963 eruption of Mount Agung in Bali had similar results.

Volcanic dust is considered to be the most effective of the aerosols causing loss of solar radiation to the earth, with the result that the sea level temperature is lowered by an estimated 1·3 K.[454] The quantity of naturally produced aerosol in the atmosphere has been put at $2·3 \times 10^{12}$ kg per annum, or about eight times that from man-made sources.[455]

The spectral distribution of the zodiacal light[456] and the Gegenschein[457, 458] have been examined to a limited extent. For the green flash, Jacobsen measured diffraction grating spectra on colour cine film, taken at sunset over the Pacific ocean at Oahu (Hawaii). The resulting curves of visual intensity showed a maximum near 525 nm for the flash, whereas the low sun gave a flatter curve with its maximum near 505 nm, confirming the excess of green light in the flash.[459] Soon afterwards, in Edinburgh, Wilson photographed the spectrum of a 'deep indigo blue' sun and attributed the lack of red wavelengths to a layer of fine particles lying between 9·5 and 13 km above the earth. The particles were probably oil globules from a forest fire three days earlier in Alberta, Canada. The cloud remained long enough to produce a blue moon the same night.[460]

Sunrise, twilight and sunset offer a great variety of colour effects which have scarcely been considered except qualitively.[449] In Chapter 10 some spectroradiometry of the horizon sky at these times was described, but there seems to be no systematic account of the range and variations of chromaticity found in the sky when scattering effects are most prominent, possibly because of the apparently infinite variety to be seen.

To return to the night sky: other light production comes from fluorescence of excited atoms or recombination of ions liberated in daylight by ultra-violet photons. This is the airglow, which includes lines and bands of O, O_2, N_2, Na, OH. The forbidden linc of atomic oxygen at 557·7 nm is prominent, as it is in some kinds of aurora, where excitation and ionization of oxygen and nitrogen atoms by fast charged particles, accelerated in the earth's magnetic field in the upper atmosphere, produce comparatively large and spectacular displays of light. These phcnomena have long been studied actively, their scope extending to the day airglow, twilight glows, and artificial glows made by contaminating the upper atmosphere by rockets.[461] Their energy is provided by the sun. Their spectra have no relation to daylight. Spectrophotometry of the night sky reveals a background of approximately daylight spectrum due to starlight and the zodiacal light (and moonlight if present), on which are superimposed peaks due to airglow lines, chiefly OI at 557·7 nm.[462] The maximum value recorded by Höhn and Büchtemann for this line was about 20 $\mu W/m^2 \mu m sr$, which is of the order of 10^{-6} times the maximum spectral radiance of the clear sky in daylight. The latter is itself roughly 10^{-6} times the maximum spectral radiance of the sun's disc, (Table 5), and since the disc occupies about 10^{-5} of the area of the sky hemisphere, the sun's direct contribution to daylight is roughly 10 times that of the sky under the clearest conditions of atmosphere and blue sky. This approximation agrees with published data.[394] Höhn and Büchtemann also gave a value for the photopic luminance of the clear moonless night sky (at 1350 m in the Austrian Alps) of about 5×10^{-4} cd/m^2.

THE ATMOSPHERE

Rayleigh scattering in a clear atmosphere produces the blue sky in full daylight, and this is observed from the earth. It has recently been seen as a bluish halo from outside the atmosphere, but more modest attempts have been made in the past to observe, from above, the bluish light which is scattered. It is scattered all through the air and not merely at a high level. The blue sky is, in fact, all around us. Schimpf and Aschenbrenner were interested in the photographic effects of this light.[463] They first photographed, as a reference, the spectrum of a titania-coated screen on the ground, followed by spectra of meadow and forest areas from aircraft at 100, 1000 and 2000 m. A step wedge in the prism spectrograph gave seven densities on each exposure.

Differentials between the 100 and 2000 m spectrograms showed the contribution of the intervening air. The dominant wavelength of the light from the air proved to be near 480 nm with 15 to 30 per cent saturation. These values agree with later work on sky colour by Lenoble and Bullrich (Chapter 9).

Others followed this line of investigation in measurements of the zenith sky spectrum from various heights. All the measurements available in 1952 were collated by Wenzel[464] in terms of dominant wavelength, purity, luminance related to horizontal luminance, and chromaticity on the MacAdam system (now the CIE 1960 UCS). Some of these chromaticities, shown on Fig. 20, are derived from the measurements of Schimpf and Aschenbrenner. The dominant wavelength was almost always between 480 and 490 nm, with a concentration of values near 482 nm. Saturation varied between 3 and 32 per cent. The effects of this scattered light have received much attention in respect to the change of colour in photographic reproductions of distant objects or the sky itself. Two accounts were given by MacAdam[465] and Masaki.[466] Both made chromaticity plots for the sky and different types of terrain, and collected useful material from other writers. The identification of natural features on the earth has been extended by measurement of their reflection spectra from spacecraft, for example in the 1969 Soyuz flight.[467]

DAYLIGHT UNDER WATER

A different approach by Stamm and Langel was to measure light coming from below, in this case from heights up to 610 m over the sea, which acts as the light source.[468] The contribution of the air was considered negligible. The prism spectrograph had a paper diffuser fixed in front of the slit to depolarize the light, and to act as an integrating surface for measuring irradiance from the hemisphere below. Spectral curves showed little structure, mostly falling smoothly from short to long wavelengths, though in coastal waters the green region of the spectrum might be increased and sometimes became dominant. It was not possible to separate completely the atmospheric effects from surface reflections and light returned from inside the water. Daylight penetrates the sea and, as in the atmosphere, suffers scattering, reflection and absorption though the relative importance of these processes is greatly changed. The visible results are more striking than in the atmosphere, and less easy to measure.

Halley, who made the comment on the sky colour during an eclipse, was an early observer of light in the sea. Newton recorded: '. . . an Experiment lately related to me by Mr *Halley*, who, in diving deep into the Sea, found in a clear Sun-shine day, that when he was sunk many Fathoms deep into the Water, the upper part of his Hand in which the Sun shone directly through the Water looked of a red Colour, and the under part of his Hand illuminated by Light reflected from the Water below looked green.'[469] From this Newton concluded, wrongly, that in the depths of the sea the light is an intense red because it is transmitted more freely than the other spectral colours, which are largely reflected back to 'compound a green'.

In Chapters 8 and 9 Lenoble's work on the ultra-violet content of daylight has been described. She also made contributions to the penetration of short wavelength radiation into the sea, both in Mediterranean waters and off the coast of Brittany.[470] A spectrograph was lowered to depths of 50 m to photograph the spectrum between 320 and 410 nm on 35 mm film.[471] For calibration a special lamp was used, devised by Chalonge and Servigne, in which a phosphor mixture emitting largely in the ultra-violet region was held outside the low pressure mercury discharge tube.[472] Visible radiation was removed by nickel–cobalt sulphate filter solutions in the underwater exposures, and in the use of the calibrating lamp.

The distribution of light in the sea by non-isotropic scattering was the main subject of this work. Extinction measurements were also made, resembling those in the atmosphere by the straight line plots of logarithm of flux versus depth. As an example, for radiation at 320 nm, the downward flux was reduced to one-tenth in a vertical depth of 10 m.[473] Applying the method usual in atmospheric calculations, namely Rayleigh–Chandrasekhar scattering in a non-selectively absorbing medium, but assuming an infinite extent of water instead of air, Lenoble showed that the colour would be blue to greenish-blue according to absorption coefficient and solar altitude selected.[474] Chromaticities varied from $x=0.157$, $y=0.209$ to $x=0.234$, $y=0.287$ with the dominant wavelength between 478 and 486 nm and saturation between 31 and 79 per cent increasing with h: these are shown in Fig. 20, as are also chromaticities calculated for distilled water.

Lenoble used a uniform 'white sky' for calculation, possibly Nicolet's spectral distribution though this is not stated. She concluded that selective absorption in the water explained the observed colour without recourse to assuming selective diffusion as well.

Attention to this subject has increased in recent years. Shuleikin tried different exponents of the wavelength, much as in earlier atmospheric work by Ångström, to explain scattering by aquasols of various particle sizes.[475] He constructed standard spectra for values of the exponent between 1·5 and 4·0. These would need modification for selective absorption in addition to the scattering.

Measurements by Sullivan on the very similar absorption spectra of distilled water and artificial sea water suggested that real sea water had some other absorbing components in the spectral range 580 to 790 nm.[476] A wide programme was carried out by Duntley who also reviewed related topics, especially light distribution in the sea.[477] He gave a few spectral curves for light arriving at the surface from below, and some of the same origin spectrographed at 1300 m above the surface. The important factors recognized were selective absorption in water, strong scattering by large suspended particles in a manner almost independent of wavelength, and a small amount of Rayleigh scattering by water molecules. The transmission maximum was given as 480 nm, and light of this wavelength has been detected at a depth of 600 m. Changes in the incident light are of importance only in the surface layers, and in most waters dissolved substances exercise the predominant spectral effects. At depths below about 50 m, in clear water without dissolved coloured organic matter, plankton containing chlorophyll, or suspended particles, the light is almost constant in its bluish green colour and in its polar distribution. The transmission maximum wavelength varies somewhat in different reports.

Among others who studied light transmission in the sea, Jerlov concluded that with increasing depth upward and downward light converged towards a peak wavelength of 462 nm.[478] In a series of measurements to depths of 600 m in locations of different transparencies off Madeira and the Canary Islands, Kampa[479] found the ultimate transmission peaks to be all close to 475 nm in the deep waters, though spectral resolution was low owing to the use of seven narrow band interference filters. One unusual measurement was of the penetration of light from the full moon, with a transmission curve similar to that of sunlight and a peak near 470 nm at 125 m depth. The irradiance at this wavelength was 5×10^{-8} W/m² nm, or about 2×10^{-6} of that due to sunlight at the same location and depth, in close agreement with the relative luminances of the full moon and the sun (4×10^3 and 2×10^9 cd/m²).

Tyler advanced the technique of spectroscopy in water by a submersible double grating monochromator, and measured the colours of water under daylight illumination.[480–482] Working with exceptionally clear water at Crater Lake, Oregon, he found a dominant wavelength of 483 nm, and a chromaticity approximately $x=0.20$ $y=0.25$ (Fig. 20).[483] A later study at the same lake gave different results. The upwelling light had a maximum at 400 to 420 nm, with the downward peak near 420 nm.[484] This lake probably contains the purest known natural water ('the natural analogue of distilled water'), and its deep blue colour is primarily due to scattering by water molecules.

A similar case of very clear water in the sea was described by Kullenberg who measured scattering to a depth of 400 m in the Sargasso Sea by means of a laser beam. Wavelength-dependent backward scattering was largely due to the water molecules.[485] Scattering in sea ice in Alaska, measured by Roulet *et al.* with a submersible spectrophotometer, resulted in transmitted daylight having a sharp spectral peak near 465 nm. Thus a 1·2 m thickness of ice produced a similar effect to that found below much greater depths of clear sea or fresh water.[712]

Among the more theoretical studies are calculations on the transfer of radiation between atmosphere and water, including multiple scattering by aerosols and hydrosols and the effect of a rough ocean surface.[486] The spectrum of light rising from the sea was calculated with moderate success by McCluney with the intention of eliminating reflected and atmospherically scattered light which is unavoidably included in measurements made high above the surface.[487] The observed spectrum in this case, at 305 m above the Sargasso Sea, had its maximum near 415 nm.

Visibility of colours under water is restricted by the decreasing spectrum of the light, and varies with the type of water. Fluorescent colours were found to be best,[488] and various metal vapour arcs have been proposed for lighting in the depths of the sea.[489, 490] A review by Lythgoe and Northmore discussed colour, hue discrimination and other topics of the underwater scene.[491]

THE COLOUR OF THE SKY

Measurements and calculations relating to the blue sky colour have appeared in earlier chapters. The explanation of the typical blue colour, so different from sunlight, has been effected by the theory of molecular scattering in the atmosphere, together with the absorption by ozone in the visible spectrum. The subject appears to offer nothing of great interest for future research. Apart from the quantitative work on the blue sky, there is, however, some interesting material of a qualitative kind. Some has been briefly reviewed by Middleton;[492] more extensive, mainly subjective accounts have been given by Minnaert;[449] and a discussion by Volz, with coloured illustrations, deals with sky and shadow colours resulting from different aerosol conditions.[493] Some further quotations may be added to these, to show a little of the growth of ideas in a qualitative way until they led to the analytical solution.

One notable opinion was that the blue of the sky was due to the combination of dark sky and sunlight scattered by dust and vapour in the air, probably the view of the Arab philosopher Al-Kindi who lived in the ninth century.[494] Leonardo da Vinci was much concerned with colour problems from the painter's viewpoint. He discussed the appearance of the sky at some length in his *Notebooks*: 'I say that the blue which is seen in the atmosphere is not its own colour but is caused by the heated moisture having evaporated into the most minute imperceptible particles which the beams of the solar rays attract and cause to seem luminous against the deep intense darkness of the region of fire that forms a covering above them. . . . The same is true of the atmosphere which excessive moisture renders white, while little moisture acted upon by heat causes it to be dark and of a dark blue colour. . . . If the atmosphere had this transparent blue as its natural colour it would follow that wherever a greater quantity of atmosphere intervened between the eye and the fiery element it would appear of a deeper shade of blue as is seen with blue glass and with sapphires, which appear darker in proportion as they are thicker. The atmosphere under these conditions acts in exactly the opposite way, since where a greater quantity of it comes between the eye and the sphere of fire, there it appears much whiter. . . . It follows therefore, from what I say, that the atmosphere assumes its blueness from the particles of

moisture which catch the luminous rays of the sun.'[495] The 'sphere
of fire' means the region beyond the atmosphere.

Newton, who had produced the analytical solution after centuries
of speculation about spectrum colours, was less helpful in the present
case. He was thinking of the relation between the colours of natural
objects and the colours seen in thin plates (much later called interference colours), and attributed the sky colour to minute particles of
condensed vapour: 'The *blue* of the first order, though very faint and
little, may possibly be the Colour of some substances; and particularly
the azure Colour of the Skys seems to be of this order. For all vapours
when they begin to condense and coalesce into small parcels, become
first of that bigness, whereby such an Azure must be reflected before
they can constitute Clouds of other Colours. And so this being the first
Colour which vapors begin to reflect, it ought to be the Colour of the
finest and most transparent Skys, in which vapours are not arrived at
the grossness requisite to reflect other Colours, as we find it is by
experience.'[496]

The discovery of the law of attenuation of light in the atmosphere
was only one of Bouguer's achievements. Middleton has referred to
his views on visibility through the atmosphere;[493] Bouguer also
realized how scattering in the atmosphere might leave a residual blue
light: '. . . transparency of the atmosphere, and from the fact that it
produces a real separation in the light, the red rays penetrating much
farther, and the blue rays having, on the contrary, less strength to pass
through in a straight line, are reflected much more easily. . . . The
shadows of morning and evening take on a very blue tint. . . . This
phenomenon is caused by the aerial colour of the atmosphere which
illuminates these shadows, and in which blue rays predominate. They
are reflected obliquely in quantity, while the red rays, which lose
themselves farther on in following their original direction, cannot
modify the shadow because they are not reflected, or are reflected
much less.'[497]

In the middle years of the nineteenth century the problem seems
to have been dominated by the authority of Clausius, who adopted a
hypothesis of scattering by minute water bubbles, not water drops
which would not lead to the observed effects. About the same time
Brücke made experiments with sols of gum in water, observing that
blue light was scattered in these primitive artificial skies.

Tyndall, the skilful experimenter, took this further and used the decomposition of organic vapours by powerful beams of light. For example, a low pressure mixture of butyl nitrite and hydrochloric acid, in a long tube traversed by a narrow beam of light, produced brilliant blue clouds polarized at right angles to the beam, and as the particles increased in size so the saturation of the colour diminished, together with change of the polarizing angle.[141] The idea of aerial particles persisted, but already Brewster and the younger Herschel, as Tyndall related, were thinking of sky polarization and its angular relationships as being indeed due to the reflection of sunlight, but since the polarizing angle was 90°, it could not be reflection by water or ice. Therefore, 'the reflection would require to be made *in* air *upon* air'.[498] This seemed inexplicable if the air was of constant density. Later Rayleigh examined the possibility of scattering by spontaneous fluctuations of air density, as well as the more useful hypothesis of scattering by suspended particles. Rayleigh's assumptions on particle scattering also imply fluctuations of density, and according to Kocinski the theory of critical opalescence based on such fluctuations among the molecules of air can explain the blue colour of the sky.[499]

The inherent blue colours of oxygen, or water, or hydrogen peroxide have often been regarded as a contributory cause of the sky colour. Chappuis and Hartley thought ozone might act in this way, and some thought its fluorescence a possible cause. Pernter was convinced that the natural colour of the air contributed nothing, but that turbidity was the real cause. The matter was largely settled by Rayleigh in 1871, though he did not then arrive at the conclusion that his minute scattering particles were air molecules, which now seems to be strongly suggested by Herschel's remarks on polarization of skylight. Rayleigh knew that his theory was not a complete explanation of sky colour. In a later paper he wrote, 'the residual blue appears to be capricious in its appearance and to depend on conditions not fully known.'[500]

Rayleigh demolished the water bubble theory of Clausius by several arguments, one being that the bubbles would act as thin plates and give reflection according to the inverse square, not the fourth power of the wavelength. In confirmation he pointed out, quoting Brücke, 'the blue of the sky is a much better colour than the blue of the first order' which would arise from reflection by thin plates. Thus Rayleigh also disproved Newton's suggestion of a blue sky due to interference colours.

Nichols summarized the nineteenth century history of this matter,[501] including an ingenuous subjective theory of his own, to the effect that the red and green vision processes in the eye increased in sensitivity with increasing stimulus:[502] in this way sunlight would give an impression of yellowness, while moonlight, and presumably skylight, would raise the relative importance of the blue process and give the appearance of a bluer light.

The Tyndall type of experiment was revived at the end of the century by Bock, whose steam jet into which hydrochloric acid was injected produced similar effects.[503] He measured scattered light intensities at different wavelengths, compared them with red skies and with calculations by the Rayleigh inverse fourth power law. Some of the clouds he produced were bluer than the best sky he could measure.

Attempts to record the colour of the sky began in the eighteenth century with pigment scales used by de Saussure (a scale of fifty-three tints of blue), and by Parrot (a rotating disc mixture of white, black and blue sectors). These were superseded by measuring instruments based on polarimetry by Arago, Bernard, Wild and others. Crova related this in a paper describing skylight measurements at Montpellier and on Mont Ventoux in Provence.[504] Crova did not mention Michel Paccard, who made observations on the summit of Mont Blanc during the first ascent in 1786.[505] De Saussure did the same in 1787, but his more notable work on blue skies took place in 1788 during seventeen days spent on the nearby Col du Géant at 3370 m.[506] A later instrument of the polarizing kind was devised by Priest.[240] With it he classified light sources according to colour temperature and spectral centroid values, including some blue skies in his list.

A return to pigment scales was convenient for Hendley and Hecht in determining approximate chromaticity areas occupied by natural objects, where Munsell chips were used.[507] They found the mean blue sky colour to be 2·7 PB/6·2/3·6, and their extreme range in the blue–yellow direction is shown in Fig. 20. Another scale was based on the Ostwald colour system and was included in a discussion of blue scales by Linke.[508]

The accurate measurement of sky chromaticities has now become a normal feature of work on the daylight spectrum. It refers sometimes to selected areas of sky, sometimes to half the hemisphere or to the whole of it. It is normally accomplished by spectrophotometry and calculation. The familiar blue colour is often lost, and it must be admitted that after all there is no 'typical' blue colour for the sky, except perhaps in the zenith sky at twilight.

THE RAINBOW

The rainbow is a concentrated form of daylight and at the same time is dispersed into an imperfect spectrum. It was discussed in Chapter 1 to show its influence on the growth of knowledge on the spectrum. Apart from this it offered optical problems long after the elementary theory had been established, showing the first two bows with radii of 42° and 52° and 'Alexander's dark band' between them to be explained by the laws of refraction and total internal reflection of sunlight in raindrops. Young supplemented the simple ray tracings of Descartes, Newton and Halley by an attempted explanation of the faint supernumerary bows, seen inside the first and outside the second bow, in terms of interference between rays of differing path length. This idea was superseded when, about 200 years after Descartes' work, the treatment by wave fronts and diffraction was begun by Airy and continued by other mathematicians, with some modifications to the angles of the bows and calculation of their dependence on raindrop size, shape, temperature and refractive index. By continuing the process of internal reflection the optical properties of bows up to the twentieth have been calculated;[509] of these the third and fourth, produced at angles of about 40° and 45° but round the sun instead of in the opposite direction, were predicted by Halley and possibly have never been seen in the sky. Bows up to the nineteenth have been identified in laboratory experiments,[510] and a photograph of the first, second and supernumerary bows in the infra-red has been made by the use of filtered sunlight.[511]

Polarization is high in the bows, particularly in the two familiar ones. This and many other topics are included in the review by Rösch.[509]

THE ZENITH SKY AT TWILIGHT

This part of the sky is normally of a deep blue colour if free from clouds and haze. The reasons were established by Hulburt, following earlier contributions with Tousey to the theory of atmospheric scattering and polarization.[122] Hulburt found by experiment that the zenith luminance, while variable from place to place, was much less than that calculated by primary Rayleigh scattering.[512] This discrepancy was removed by adding to the calculation a correction for absorption in the Chappuis bands of ozone. It was, however, removed only for $h=0°$ to $-6°$, and with increasing depression of the sun below the horizon calculated luminance became too small owing to neglect of secondary scattering. Hulburt gave small scale spectral curves which showed how the ozone absorption was responsible for changing the zenith colour from what would be expected in the absence of ozone, namely pale greenish blue at sunset and yellowish later. As the sun declines the ozone absorption becomes more important than scattering in the appearance of the deep blue colour.

Le Grand and Lenoble came to similar conclusions shortly afterwards.[513] The dominant wavelength of the blue zenith sky was said to vary little from 480 nm while the sun set from $h=4°$ to $-4°$, but the saturation was minimal at $h=0$.

The portable colorimeter used by Wright in the eclipse provided Gadsden with chromaticity observations on the zenith twilight sky.[514] As h declined from $-1°$, saturation increased by movement towards the z corner of the chromaticity diagram, roughly from $x=0·26$, $y=0·26$ to $x=0·21$, $y=0·205$, or colour temperatures from 10 000 K to beyond infinity (Fig. 20). Calculation based on the Abbot extra-terrestrial distribution (Table 3), an assumed ozone concentration and distribution of molecular density with height confirmed these colours approximately. Gadsden observed the maximum saturation at $h=-7°$ on several occasions, in agreement with Hulburt's assumption of secondary scattering when h exceeded $-6·5°$. Siedentopf also found a maximum colour temperature at $h=-10°$.[515]

There is a considerable literature on many aspects of twilight. Rozenberg's thorough survey included a number of small scale spectra curves, but higher resolution and the colorimetric aspects are somewhat lacking.[516] Ashburn quoted a number of authors on the colorimetric determination.[517]

The sequence of twilight colours surrounding the earth was recorded by photographs and spectrophotometry from about 250 km altitude in the Soyuz flight of January 1969. The aim was to compare the vertical distribution of spectral luminance and the chromaticities with calculated values from an atmospheric model, assuming molecular scattering alone, or with aerosol present, or with aerosol and ozone.[518] The maximum luminance in the 'twilight aureole' was at 480 nm, and the chromaticity sequence agreed roughly with calculations based on Elterman's molecular atmospheric model,[519] but only in the blue to white colours, not in the further change to purple shades.[518, 467] The limiting blue colour was close to the points VH in Fig. 20. Kondratiev *et al.* also discussed observations made from Gemini 4.

Twilight has a more mundane importance in the planning of street lighting. Recommendations for the illuminance levels below which supplementary artificial lighting is required specify 90 lux in the evening and 45 lux at dawn.[520]

16

Whiteness

It comes to pass, that Whiteness is the usual colour of Light; for,
Light is a confused aggregate of Rays indued with all sorts of Colors,
as they are promiscuously darted from the various parts of luminous
bodies. And of such a confused aggregate, as I said, is generated
Whiteness, if there be a due proportion of the Ingredients.

ISAAC NEWTON, *Phil.Trans.Roy.Soc.*, 1672, **6**, 3083.

The concept of whiteness, of present interest in several fields of theory
and practice, has a direct connection with daylight because the phases
of daylight have long been accepted as the standard of 'white light' in
spite of their variations. The special visual effects produced by whiteness in lights or illuminated surfaces may be the result of long adaptation of the human eye to daylight, but apart from a general agreement
on the importance of daylight as a reference, and on some colorimetric
limitations to whiteness, there is little else that is universally agreed.
This is essentially a subjective matter in which chromatic adaptation
of the eye to light sources alters the perception of colours (white
included), while personal preferences play some part and may be
decisive in the commercial applications of white materials. The variety
of materials, industries and artificial light sources increases the complexity and prevents any common standard from being acceptable. In
many cases difficulty occurs in the correlation of subjective observations with objective measurements and standards. This practical aspect
of whiteness is concerned more with the appearance of surface colours
than with lights; it has produced an extensive literature which can be
surveyed only by a few examples from the recent period, and in any
case natural daylight is of little importance in the many grading and
measuring methods now in use.

WHITE LIGHT: EXPERIMENTAL EVIDENCE

This is a less controversial area of the subject and it has a longer history than the study of whiteness in materials which are not self-luminous. Some early workers on the daylight spectrum, notably Jones, Nichols, Ives and Priest, considered the lack of white light standards and the possible use of natural daylight for this purpose (Chapters 7 and 8). Their investigations are mentioned in the later treatment of artificial daylight sources. Some of Priest's contributions on whiteness perception are interesting in the present discussion.

The acceptance of variable daylight, except perhaps in some highly coloured sunrise or sunset, as a standard of white light depends largely on the process of chromatic adaptation in the eye, which is conditioned by the ambient light and accepts it within wide limits of chromaticity as the reference for other coloured stimuli. It was well said by Ayrton, the chairman of a discussion on artificial daylight sources in 1892, that 'People talk about white light as though it were some definite thing. . . . White light is what you see most of'.[521] When there is no physical standard of comparison the effects may depend on memory. About the time when the tungsten lamp was becoming predominant over all other artificial light sources used in photometry, Priest made tests to determine the neutral stimulus or 'hueless sensation'.[522] His observers chose from a variable, near-Planckian distribution produced by the rotary dispersion method, with no standard for comparison. Their judgments fell in the range of 4800 to 5400 K, in rough agreement with the views of Ives (Chapter 7). Curves were shown for each observer to record the probability that a stimulus would be recognized as blue, white or yellow. In other experiments 'artificial noon sunlight' was compared with a light described as close to equal energy distribution in the visible spectrum, but without details of its design. Two-thirds of the observers judged the latter to be less saturated or nearer white than the artificial sunlight, and Priest decided that it was a suitable neutral stimulus for colorimetry.[523] If it had an equal energy distribution its CCT would be near 5500 K.

Among the best known later experiments are those of Hurvich and Jameson. Their three observers viewed a white screen against a dark background and illuminated at 10 to 10^4 cd/m² by approximately full radiator spectra at intervals in the range 2482 to 10 000 K. Threshold contours at which the screen appeared white were plotted as colour temperature of the source versus luminance of the screen, and these showed throughout flat minima varying in position with field size,

time of exposure to viewing, and the observer. The minimum luminance to evoke the sensation of whiteness was for colour temperatures
of 5500 to 7500 K. Similar curves were found with an illuminated
surround, when the threshold luminance was lowest for approximately
equal colour temperature of test field and surround. The colour
temperatures seen as white over the widest range of conditions were
6500 and 7500 K. These experiments emphasized the fact that with
increased luminance a source appears white over a wider range of
chromaticity, with no upper limit to whiteness within the colour
temperatures and luminance levels used. Hurvich and Jameson suggested the existence of a white perception mechanism in the eye additional to the chromatic ones, to explain the increase of whiteness area
at high luminance as well as the disappearance of the hue sensation at
very low luminance.[524] These effects have no doubt influenced the
wide chromaticity specifications for white light sources and black
surfaces, discussed later, and the minimum luminance for perception
of white in the region of 5500 to 7500 K has significance in the television application.

About the same time Sproson attempted to find the white point as a
reference standard in colour television. Observers selected chromaticities by a colour naming technique for an area with an adapting surround. The mean white selected increased from about 3500 to 4500 K
as the surround changed from 2700 K of tungsten light to 6000 K of
simulated north skylight.[525] The luminance used was up to 10 cd/m².
The area recommended for a standard white light is shown in Figure
79: it has a slight bias towards the green side of the Planckian locus.

Honjyo and Nonaka devised two methods for finding the limits of
white areas chosen by observers who viewed an opal screen subtending
10° and illuminated at less than 10 cd/m², with no surrounding field.
In the first method they altered a mixture of red, green and blue lights
to compound an acceptable white. In the second they judged stimuli
presented to them as white or not white. The chromaticity areas
chosen in the first method differed widely among the observers, with
CCT variations in the 50 per cent probability ellipses from 5000 K to
∞ and almost entirely below the Planckian locus; that is, on the purplish
side. In the other trials the resulting ellipse extended from 5500 to
8500 K, centred near the chromaticity of Standard Illuminant C,
resembling in size and position that marked W in Fig. 79.[526]

Similar experiments with a smaller area at 30 cd/m² with a dark surround, and a more complicated colorimetric procedure, were made by Valberg.[527] He found 'achromatic areas' to vary with the observer, from 5600 to 7500 K but on the greenish side of the Planckian locus. One much older observer chose an area near 11 000 K and further on the green side.

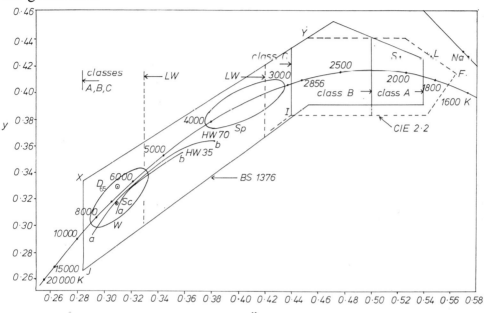

FIG. 79. Chromaticities of white standards and experimental results in relation to the Planckian locus.

A,B,C: maximum x limits in BS 1376 for different restrictions.
CIE 2·2 area: broken lines YLFI, thence to JXY.
LW: Lunar white, part of BS 1376 class C, $0·330 \leqslant x \leqslant 0·420$.
S: High pressure sodium lamp (SON).
Na: Sodium D lines.
LF: Junction between CIE areas for white and yellow.
W: Approximate area of preferred white surface colours.
Sp: Area for television white and adaptation $\sim$ 2800 K. (Sproson.)
HW 35, 70: Neutral points with adaptation at 35 and 70 cd/m², and of CCT 17240 K (at a) to 2985 K (at b). (After Hunt and Winter.)

A final example of white selection by colour naming is that used by Hunt and Winter. A light source varied in colour round the neutral point was viewed in a mirror against a background illuminated by light of full radiator chromaticities. Neutral points selected in this way were all on the purple side of the Planckian locus: their general trend, omitting minor variations, is shown in Fig. 79. For a surround field of 35 cd/m² the mean neutral CCT increased from about 4500 to 7000 K for adaptation between 3000 and 17 000 K; for 70 cd/m² adaptation field the changes from the adaptation colour were less, from 4000 to 9000 K. The smallest changes between adaptation and chosen neutral chromaticities, and in distance of the latter from the Planckian locus, occurred near 6000 to 7000 K.[528]

The above items selected from the literature show no decisive trend in favour of white lights on one side of the Planckian locus rather than the other, at any rate under the conditions of viewing described. This is not so in everyday applications of artificial lighting. Most subsequent experiments have compared subjective judgments of whiteness with colorimetric data, but on surface colours where the mode of viewing is different and standards are more critical.

Among more theoretical studies, Howett produced formulae for the chromaticity of a test spot which appears achromatic against an extended chromatic background, in terms of the background chromaticity and luminance ratio between spot and background. This work was based on experimental data of earlier investigators. Howett stressed the difference between luminance and lightness (colloquially 'brightness'), since for equal luminance a chromatic field looks brighter than a white one.[529]

APPLICATIONS OF WHITE LIGHT

The main applications where light is to be viewed directly are in signalling systems for all forms of transport, and in television reception. In signalling, the light may be seen as a point source of varied intensity, from high to near threshold brilliance, (which is measured by illuminance at the observer's eye), and apart from visibility the main requirement is that there should be no confusion with other colours. The practical limits of recognizable whiteness are therefore very wide, a result of long experience tempered by the arrival of new light sources. Fig 79 shows the area of the 1931 CIE chromaticity chart adopted as the British standard of 'signal white', which extends from bluish hues near 10 000 K to yellowish ones below 2000 K. The latter limit is well below the range of frequent daylight colour temperatures and is a concession to the survival of oil lamps; it is unfortunate that this area (class A) also includes the chromaticity of the high pressure sodium lamp (S). Classes B and C, restricted in the x-coordinate direction, have closer control over the colour, though even class C still extends to 3000 K. A long-standing British railway colour, 'lunar white' (LW), is specially mentioned in the specification as a section of class C.[530]

The CIE recommendations for the colours of white light signals are almost identical with the BSI standards for the limits B and C, but the most yellow area extends even more towards the spectrum locus near the sodium D lines, and at LF in the Figure adjoins the area for yellow light signals.[531]

If a signalling system uses four colours including yellow, the white area, called 'coded white', is restricted to $x \leqslant 0.500$ as in BS class B. Class C corresponds to CIE 'recognisable white', and to achieve 'distinguishable white' when both white and yellow are visible the latter must have a value of x at least 0.05 greater than that for white.

Those whose occupations depend on the use of light signals are often given simple tests for the recognition and distinguishing of such signals in portable lanterns, usually with the most important colours, red, green and white. The white signal, from a low power incandescent bulb, will be in class B or even A.

There are white light standards in television. Black and white pictures have varied considerably in the past, but the chromaticity now widely accepted is given as 9300 K plus 27 MPCD or minimum perceptible colour differences, the term commonly but inaccurately used for units of standard deviation in MacAdam's colour matching experiments.[532] This is a chromaticity of $x=0.278$, $y=0.321$.

In colour television the chromaticity of D_{65} is the standard white, produced by unit quantities of the red, green and blue signals from the camera. Change of illuminant in the scene is corrected by altering the relative gain in the three channels. D_{65} is much more blue than the chromaticity area found by Sproson; the change seems to reflect the more general acceptance of bluish whites, seen in much of the later work reported in this chapter. A more positive reason suggested by Bartleson[533] is that a television white in the middle range of colour temperatures conforms with the effects demonstrated by Hurvich and

Jameson, namely that in this hue region the sensation of white is produced by the lowest values of luminance. The use of the well-defined D_{65} point rather than some full radiator chromaticity acknowledges the importance of daylight in this application.

However, the usual adaptation in domestic viewing of television is to tungsten incandescent light, and from the experiments quoted[524, 525, 528] a picture white located near 4000 to 4500 K appears to be more suitable visually though less convenient in practice.

WHITE LIGHT FOR ILLUMINATION

A much more limited interpretation of white light than for signal purposes is that of the reconstituted daylight distributions (Chapter 13), where for lack of experimental evidence the series is restricted to a range of 4000 to 25 000 K. Prior to recent work on the chromaticity of daylight it was thought, for example by Priest, that full radiators would be acceptable white light sources if only they could be reproduced physically above the low temperature limit set by tungsten filaments. When it was established that terrestrial daylight, unlike extra-terrestrial, was slightly green in colour compared with full radiators, a few light sources of appropriate colour were produced for commercial use, chiefly near 6500 K since this seemed to be the most popular version of daylight and is in the region of most frequently observed colour temperatures. Development was hampered by the impossibility of close imitation of the daylight spectrum except by complicated filter systems (Chapter 18). There is some slight evidence that a somewhat greenish light is preferable in colour assessment[598] and some experiments by Sproson and Valberg had a similar trend. This was not the case for the other experiments quoted above, and it will be seen that illuminated white surface colours of greenish tints are usually not preferred.

The choice of lighting for domestic, commercial and industrial purposes has been determined in the past, first by the availability of light sources, with the tungsten filament lamp of colour temperature below 3000 K dominant for many years; since then the selection of fluorescent tube colours for non-domestic use has depended partly on cost or the efficiency of light production, and partly on the relations of the chosen illuminance level to acceptable colour and colour rendering

properties. These matters are discussed further in Chapter 18. Whiteness of these sources scarcely arises as a criterion, but instead there is
a strong preference for the warmer colours, mostly outside the normal
colours of natural daylight, and with chromaticities below the Planckian
locus or approaching that of the tungsten filament.

Whiteness in a light source does not require the presence of all
wavelengths in the visible spectrum as Newton suggested, but it can be
produced by incomplete spectra such as three bands,[627, 648] or even
by two bands in the combination of blue and yellow light. Deficiencies
in the spectrum will be revealed by poor colour rendering though not
by colour appearance since the eye cannot analyse light in the way that
the ear can analyse complex sound. A familiar example is the black
and white television screen which if used as a light source has poor
colour rendering properties due to the lack of parts of the visible
spectrum.

WHITE SURFACES

Here we are concerned with materials having smooth reflectance curves
throughout the visible spectrum. Those needing most careful control
are reference standards used in the measurement of the spectral
reflectance of coloured objects for quality control, identification and
similar purposes. The primary standard is a theoretical one, a perfect
reflecting diffuser with 100 per cent reflectance over the whole spectrum
range involved. Realizable working standards are needed. Magnesium
oxide smoked on to a surface from the burning metal, or as a compressed powder, has been widely used, and was assumed to be a
perfect diffuser until the introduction of the theoretical standard reduced its spectral reflectance to approximately 98 per cent, but its
instability and fragility led to the substitution of compressed barium
sulphate powder which has an equally high uniform spectral reflectance
and great stability to the atmosphere and light. These materials need
calibration for absolute measurements since their reflectance falls
gradually with decreasing wavelength in the visible spectrum.[534]
Their luminance factors are very close to unity. A microcrystalline
organic fluorinated polymer with a slightly higher reflectance than
magnesium oxide has been proposed as a hard-wearing substitute.[535]

Opal glasses with white diffusing particles dispersed through the mass
or applied as a surface layer are used for 'neutral' transmission or

reflection in optical work. They also need calibration to correct for selective spectral effects, and reflection from a glossy surface must be considered.

NON-FLUORESCENT MATERIALS

These include paints, papers, plastics, and textiles and their natural constitutents (wool, cotton). Apart from visual inspection they are normally graded by chromaticity and luminance factors under a standard illuminant, formerly Source C but now generally D_{65}. The connection with natural daylight therefore exists but is often no more than formal because the white products are as likely to be viewed under incandescent or other 'warm' lighting as in daylight. In special cases chromaticities may be measured by calculation from spectral luminance factors, but this process is too lengthy and difficult with samples of very close chromaticities, and direct measurements with filter colorimeters are mostly acceptable. Attention to several factors is essential: (1) the selection of a light source: this was no problem with Source C but there is no exact equivalent of D_{65} (Chapter 18); (2) the correlation with visual appearances of the chromaticity values, or their differences by one of the many existing formulae; (3) the selection of an area of chromaticity for the particular sort of product to contain the preferred whites, or perhaps the acceptable whites; and (4) the selection of an adaptation illumination in the visual appraisal. An incidental difficulty is that measurements made on different types of instrument often lead to divergent results.

FLUORESCENT WHITE MATERIALS

Materials in the class just discussed have become less common during the last thirty years because of the introduction of fluorescent brighteners to improve the colour of many near-white organic materials, which are usually slightly yellowish or become so on ageing. The additives are colourless dyestuffs excited to fluorescence by the ultra-violet and the short wavelength end of the visible spectrum, and they emit light in the violet and blue regions, occasionally in the green or red regions. The widest application is to textiles, where laundering maintains the effect by the presence of the dyes in most detergent powders. Here natural daylight becomes important due to its ultra-violet content and the striking changes in appearance of the treated fabrics.

The prevalence of bluish whites has resulted in a widespread preference for this type of colour mainly because of its enhanced 'brightness' rather than its hue, and there is a change in the preferred area of 'white' chromaticity as well as in the everyday meaning of 'white'. Much more serious are the problems of measuring the chromaticity and providing the same (or equivalent) light sources for viewing and measurement. The correct measurement method, separating reflectance from fluorescence, was described by Donaldson,[536] followed by many attempts to simplify the process and obtain approximately correct tristimulus and chromaticity values for dyed white materials under a light source of known spectral power distribution.[537–539] Conversely, colorimetry of fluorescent whites is used in some methods of checking the ultra-violet emission of light sources (Chapter 18).

White materials, whether fluorescent or not, are all subjectively of low saturation and high lightness or Munsell Value; in terms of objective measurement they are of high purity and high luminance factor (Appendix 1). A report by Berger defined an area centred near the D_{65} chromaticity;[540] Mori quoted twelve white points below the Planckian locus preferred by observers in four independent experiments in ranking paper and paint samples;[541] and other proposals for a visual whiteness reference point were below the locus but nearer to it and close to the S_C point.[526, 542] In Fig. 79 the area W covers most of these examples.

Standardization of surface colours for signalling has been under study by the CIE for some time past, and it is probable that white surfaces will be specified in the same general area, while black surfaces, being difficult to identify as having any hue, will cover a much wider range of chromaticities, expanding the white area.

WHITENESS SCALES

Industrial whiteness tests often determine the ranking of samples by inspection, or by their numerical values on some scale of whiteness, and there may be pass–fail limits. Numerical scales are most widely used, generally based on photometric or colorimetric determinations and normalized to a maximum value of 100. Well over 100 have been proposed, showing the varied requirements of different industries and the difficulty of defining whiteness numerically in a way which allows for the subjective factors. Selection of a scale depends largely on convenience in a particular application, and not by the theoretical perfection of the scale. In some cases assessment by inspection is adequate, and for this purpose permanent sets of reference samples are available, similar to the Munsell atlas, graded usually in the blue–white–yellow direction: for example, the Ciba–Geigy white scales consisting of (1) eighteen graded cotton samples with increasing amounts of a fluorescent brightener; and (2) twelve graded plastic samples, four with varied yellow dye content and eight with a brightener.[543] Some types of scale are briefly considered in the following. The references quoted provide many detailed examples.[541, 542, 544–550]

In many materials a departure from ideal whiteness is shown by increasing yellowness towards a dominant wavelength near 570 nm. Indices of yellowness have been used, particularly in plastics, to give a simple measure of the difference from a white standard. Such indices are derived from a combination of reflectance measurements (or transmittance for transparent or translucent samples) at blue, red, and sometimes green wavelengths. Whiteness may be measured as the inverse of yellowness or, as in the case of paper, where the property is called brightness, by the reflectance at 457 nm. Tristimulus values X, Y, Z determined by a colorimeter are more useful since they provide the essential lightness factor. Another simple case is that of scoured wool, which can be graded by the Z value, or a combination of X, Y and Z, for example $\frac{1}{3}(X+Y+Z)$.[544]

A type of formula used in many forms, first applied to paper technology by Judd, produces a scale of whiteness (maximum 1) calculated by 1 minus (the colour difference between the white standard and the sample, divided by the colour difference between the white standard and black).[545] Most other formulae emphasize the lightness factor, often including the distance from the chosen white point on a scale of chromaticity or some related parameter. One such, related to the Munsell scale of Value, is the Adams–Nickerson–Stulz system, AN*Lab*, which has a lightness coordinate L and two chromatic values a and b for the distances along red–green and blue–yellow axes respectively (Appendix 1). In this case whiteness is calculated as the sum of L and different multiples of a and b.

At the present time attention is concentrated on formulae based on measurements of Y, x and y; or ANLab; or measurements through blue, green and amber filters of the Zeiss Elrepho photometer with a xenon lamp as light source where the value of whiteness is given by the sum of multiples of these filter readings. Relations between these and their interpretation as lines and surfaces of equal whiteness ('isoleukai') are discussed in a report by Ganz,[546] who has also proposed a universal formula for whiteness based on a preferred dominant wavelength (say 470 nm), a perfect white diffuser as a reference point, planes of constant whiteness, a unit of whiteness and, to allow for variable hue preferences, the 'hue preference angle' between the isoleuke and the line of constant dominant wavelength at the chromaticity of the sample. By choosing D_{65} as illuminant, the 2° colour mixture data, and suitable values of the variables to give approximate agreement with visual results, a simplified version for calculation is

$$W = 2Y - 2283x - 3284y + 1694 \cdot 8. \text{[547]}$$

An example of the *Lab* system, based on the Glasser cube root formulation instead of Munsell Values (Appendix 1), is the formula applied to paper samples by McConnell. Observer preferences under D_{65} illumination were found to be on the nominal neutral axis of the system ($a=b=0$) for papers with a blue reflectance B of 80 per cent, the usual value for many white materials before the introduction of fluorescent brighteners, which however increased B, and stronger blue tints were then preferred. At lower values of B more yellow tints were preferred, and these differences could be expressed by a 'neutral' axis with $b=0 \cdot 36$ ($80-B$), where negative b indicates a bluish tint and positive b a yellowish one. The axis could not continue in the same direction when B is less than 50 per cent, otherwise deep greys would be classed as yellow, and therefore an arbitrary change of slope had to be included, the line reaching the neutral axis at $B=0$.[548]

CIE spectral tristimulus values for both 2° and 10° fields of view have been used by different workers. It has long been suspected that the blue end of the 2° curve for $\bar{y}$ is incorrect, and in some cases the 10° curve gives better agreement of calculated values of chromaticity with visual judgments. This happened in the ranking of near-white samples of the rutile and anatase varieties of titanium oxide pigment, though correlation with visual appearances under north sky illumination was also improved in these experiments by a change from Source C to D_{65} or D_{250} in the colorimetry.[549]

A comparatively simple whiteness formula due to Grum *et al.* is based on the 10° field Z tristimulus value because of its good correlation with visual assessments of whiteness. The best agreement with ranking of twenty paper samples by twenty-one observers was achieved by adding a term representing the colour difference between the sample and the D_{65} illuminant, namely the excitation purity $P_{\xi,\eta}$ derived in the usual way but from the MacAdam geodesic chromaticity diagram (Appendix 1). The formula for whiteness W is then

$$W = -270 \cdot 1 + 3 \cdot 80 Z - 3 \cdot 647\, P_{\xi,\eta}.$$

Good correlation was found between W and judgments by other observers on other sets of paper and textile samples.[550]

ATTRIBUTES OF WHITENESS

Whiteness is determined by a chromaticity area, variable according to the observer, the conditions of viewing, and often by the material concerned, but generally located near the chromaticity of full radiators. Another and perhaps the dominant factor in whiteness is a minimum value of luminance or lightness, below which surface colours do not appear white. As brilliance of a light source, or lightness of a surface increase, whiteness persists and widens its chromaticity gamut. White is therefore to be regarded as a colour with more than one dimension, not merely as a hue.

A different view taken by Wiltshire and Savage, who excluded the hue factor, was to regard whiteness as one aspect only in the dual nature of perception of achromatic colours. Whiteness was considered to be the ordering of achromatic colours from black to white, and the other necessary attribute was 'brightness', or ordering from dim to bright on a scale shown to be independent of the black to white scale. They pointed out one difficulty in attempting to use a single criterion instead of two: that of distinguishing a grey surface under high illumination from a white one under low illumination, and they therefore deprecated the use of single number scales for achromatic colours.[551] In practice surface colours such as these authors used are normally judged under conditions of illuminance and background arbitrarily chosen to avoid difficulties of the kind mentioned. The importance of background was stressed by Minnaert.[449] He thought that the only quality in which black, grey and white differed was in their 'brightness' with the background providing a standard of comparison. He suggested that a snow-covered landscape may appear brighter than the overcast sky because of contrast between the snow and dark objects in the landscape, which physically cannot be brighter than the sky. Similarly Ganz remarked on the unpleasant effect such a snow background may have on the appearance of fabrics containing fluorescent brighteners.[547] This failure of brighteners to produce whiteness when apparent lightness is reduced by a high background luminance is an effect closely related to that studied by Stenius.[552] He found that reflectance of bluish whites could be increased above the optimum reflectance for non-fluorescent materials of the same chromaticities, the increase being a measure of the power of the brighteners to hide bluish tints.

These and other effects described above are all affected by adaptation of the eye, whether to luminance or to hue. They may suggest that the visual process is biased towards perceiving white, but under other conditions the results are quite different. Examples are the spectacular Land demonstrations of projecting in register two transparencies of the same scene, for example one in black and white from a photograph through a green filter, and one from a normal photograph projected through a red filter; together these produce the appearance of a wide range of colours.[553] A much older demonstration is Benham's top, where a rotating disc with black and white designs produces strong sensations of colour, known to depend on the presence of flicker in the scene. These are more complicated effects than those experienced in the usual viewing and assessment of whiteness, but they show that the visual process cannot be reproduced by photometry and colorimetry except in broad outline, since the process may attribute chromatic hues to stimuli perceived as white under normal circumstances, as well as discounting the hues produced by fluorescent brighteners.

In the complications resulting from the widening of the acceptable area of 'white' chromaticities, and by the many formulae for scales of whiteness, it is difficult to find any important role for natural daylight except its visible effects on brightened materials. The D_{65} distribution retains a somewhat insecure position in a changing technology.

17

Uses of Daylight

I see the bright sky through the window panes:
It is a garish, broad and peering day;
Loud, light, suspicious, full of eyes and ears,
And every little corner, nook and hole
Is penetrated with the insolent light.

PERCY B. SHELLEY, *The Cenci*, Act 2, scene 1.

The beholding of the light is itself a more excellent and a fairer thing than all the uses of it.

FRANCIS BACON, *Novum Organon*, Book I.

The nature of daylight has been explored thoroughly by many investigators, and their diverse results have led eventually to approximate agreement on typical spectra. It remains to consider how this daylight can be reproduced by artificial means, primarily to extend man's activities into the hours of darkness and to other environments. In this chapter we review some of the uses of daylight and where it needs to be replaced or modified or supplemented. In Chapter 18 the artificial daylight sources and their applications are discussed, while Chapter 19 gives some special cases.

HUMAN VISION

Man's ability to see is one of his most essential functions, and daylight makes its possible during the greater part of his activities. Human sight has evidently evolved to take advantage of the main part of the spectrum transmitted by the atmosphere. The match between the eye's spectral sensitivity and the sun's spectral emission is not close, and less so with the addition of scattered light from the sky which generally produces a maximum in global irradiance towards the blue end of the spectrum. The scotopic visibility curve of the eye, valid at luminance values below 10^{-2} cd/m², when the eye is 'dark-adapted' with a threshold of about 10^{-6} cd/m² under favourable conditions, has a maximum at 507 nm. The photopic curve with its maximum at 555 nm is less favourably placed with respect to the daylight spectrum; it applies to the more usual conditions of light-adaptation and luminances above 10 cd/m². The light-adapted eye is much less sensitive to blue or red light than to green or yellow, and very much less sensitive to violet light. Other animal eyes which have been examined have sensitivity curves differently placed in the spectrum and sometimes extending into the ultra-violet region.

The position of the peak sensitivity is of less concern here than the overall sensation of 'whiteness' which is experienced in daylight, and the eye's power of adaptation, not only to light intensity, but to colour (Chapter 16).

This adaptation to the colour of ambient light has made it easy for gas lighting, and then incandescent filament lighting, to be widely accepted, and without much realization of how they were altering the appearance of objects compared with their appearance in daylight. The needs of industry and commerce using coloured materials resulted in a critical assessment of the effects of these artificial sources[590] and eventually means were found for the measurement of 'colour rendering.'

LIGHT AND PLANTS

Agriculture and horticulture depend vitally on light, on the quantity and quality of light falling on the growing plant, and on other factors not directly concerned with light such as ambient temperature and climatic conditions. The range of spectral wavelengths promoting photosynthesis from atmospheric carbon dioxide is very wide, with a minimum in the green region where chlorophyll reflects part of the incident radiation. To extend or supplement daylight, greenhouse culture has often made use of convenient lamps regardless of their spectra, for example, medium pressure mercury discharge lamps. The growth of stem, leaf and root, the production of flowers and the germination of seeds, the reaction to light in phototropism, are now all distinguished as processes which may be separately stimulated or retarded, either by the spectrum chosen for the irradiating light, or its distribution in time. There are many different types of plants showing different reactions, but successful means of interfering with the normal life cycle have been found in a number of applications. These are particularly concerned with the night-length which controls flower production in many species. Extension of the day, or converting a long night into the equivalent of a short one by short periods of illumination at various wavelengths, these have produced remarkable effects. Some depend on the reactions of a protein pigment, phytochrome, which has two forms with different spectral absorptances, forms interconvertible by red (<700 nm) or far-red light (>700 nm) respectively.

Plant cultivation is likely to need convenient light sources of special spectral distributions for these control methods, while artificial daylight does not need very accurate specification in application to photosynthesis. A recent review covers principles and practice in lighting technique,[554] and there are several books on the subject.[555-557]

LIGHT AND ANIMALS

Animal husbandry is less convenient to relate to illumination levels and spectral quality of light. The advantages of extending the day length are known in some cases, and it is widely believed that the ultra-violet part of the daylight spectrum is of importance in maintaining animal health.

Ott has produced results suggesting the importance of extending the spectrum for various animals living under artificial light sources.[558, 559] The ultra-violet content was believed to be advantageous to health, and this view is widely but not universally supported in respect of the human animal.

The sea is an important source of animal life. Here daylight has a very limited range in depth, with rapid attenuation in radiance and spectrum (Chapter 15). Fish life depends on this for seeing, and is possibly affected by the diurnal variations. The visual pigments of deep sea fishes certainly have their absorption maxima close to the dominant wavelength of the residual light, 480 nm.

Man reacts psychologically to daylight, and is also stimulated physiologically.[560, 561] The periodicity of many processes is controlled by the alternation of day and night,[562] or by artificial changes in illumination.[563, 564] It has been claimed that present levels of artificial lighting are inadequate for optimum human well-being because they are so far below natural daylight levels.[565] On the other hand Hollwich *et al.* quoted many experiments on man and other animals to support their objections to present methods of using artificial light, on the grounds of spectral differences from daylight, excessive illuminance values, and the monotony of fixed installations, these being possible causes of physiological disturbances after long periods.[566] Short term experiments do not support these claims. The more obvious effects occur in the ultra-violet and extreme violet regions of the spectrum, and here the acceleration of daylight effects by exposure of the skin to levels of irradiation higher than normal, at mountain or snow or seashore sites, or by lamps with suitable output, is a well-known pastime as well as an accepted medical practice. Physiologically it has the effect of converting a cholesterol derivative in the skin to vitamin D, though most sun-bathers do not lack this vitamin. The visible effects are first, an immediate temporary tanning (depending on the power absorbed), which occurs for wavelengths between 300 nm and the blue or green part of the visible spectrum, with a maximum near 350 nm; second, reddening of the skin or erythema caused by the spectral region 280 to 315 nm, with a sharply marked peak near 297 nm. The spectral dependence of the effect causes it to increase when the incident solar radiation is even slightly extended towards the shorter wavelengths, as in clear atmospheres or at high altitudes, when the flux may also increase. Erythema has an onset normally delayed for some hours, and is followed in many persons by even more slowly developing tanning caused by the production of the dark pigment melanin. The wavelengths involved may be harmful to the eyes in large doses, though not nearly so seriously as the shorter wavelengths found in some sunlamps. Radiation of 253·7 nm wavelength predominates in low pressure mercury discharges in silica envelopes. It has bactericidal properties, and is deleterious to animal and plant life. The atmosphere protects us from the natural solar supply of this and other dangerous radiation in the UV-C wavelength band.

Excess of normal sunlight can cause ageing of the skin, and is responsible for a form of skin cancer. Sunlight effects are serious in porphyrin diseases where the skin is abnormally photosensitive. Magnus has surveyed all these skin reactions.[567] Working in Australia and New Guinea, Robertson measured ultra-violet irradiation by an instrument with a peak response at 300 nm, to record units of erythemally effective radiation.[568] He concluded that skin cancer is promoted by repeated high doses of ultra-violet radiation over long periods because a cumulative effect predominates over the recession which occurs with occasional exposure or low doses (unpublished work). Green's tables (Chapter 10) were intended to provide a map of ultra-violet climatology which would help in this problem.[322]

The infra-red content of solar radiation has important functions in the heat balance of earth and atmosphere, as previously discussed, but also specific effects which are used medically. At certain wavelengths it can penetrate the skin, providing valuable heating in depth for the relief of muscular pain and stimulation of the circulation. Such radiation is easily supplied by thermal radiators at comparatively low temperatures.

LIGHT IN BUILDINGS

In architecture, whether imposing or humble, the provision of daylight for those living or working indoors has, till recently, been an essential part of design. Planning of this facility has often been very poor, often very careful, sometimes with more enthusiasm for the use of a free and copious light source than the results have justified. The last category contains examples of buildings where excessively large window areas have caused discomfort or even disability glare to the occupants. Not only is the light admitted by windows converted into heat, but so is the infra-red radiation. Thus free light may become expensive if it means admitting too much light and too much heat, and the interiors have to be made comfortable by air conditioning, by infra-red absorbing glasses, by external or internal louvres or blinds,[569] perhaps in the future by windows of transparent photochromic materials which darken reversibly in light.[570]

The trend towards windowless buildings, particularly factories, raises psychological questions, but also more immediately soluble ones. It is necessary to ensure correct lighting and heating levels, as with natural daylight. It is also desirable to imitate a daylight ambience. The problem of what spectral quality and chromaticity of light to provide is less severe if there are no windows, because of the eye's powers of chromatic adaptation. In conventional buildings daylight may need to be supplemented early or late in the day, also in rooms too deep to receive enough light from the windows. In the latter case the supplementary lighting may be needed during the whole period of occupation, and it might appear reasonable that the added light should be close to daylight in colour appearance at least, selecting some 'average' or 'typical' daylight. It is often convenient to use lamps which do not match daylight colours, and there is a case for the deliberate introduction of colour contrasts between natural and artificial lighting: this can hardly be avoided to some extent with daylight varying as it does. Much more experimental evidence is needed before acceptable general principles can be established in this field. The problem has been discussed, among many others arising in the lighting of buildings, in a book by Hopkinson and Kay.[571] The influence of architectural styles was discussed by Bell in a review of development and practice in daylighting,[572] while Bellchambers gave a concise account of the data necessary to the planning of interior daylighting, with a description of the geometry and parameters required in the calculations.[573] An authoritative guide is available to the integration of daylight with artificial lighting.[574] There are many older books of continuing value.[575, 99]

SUNLIGHT

Direct sunlight, though often a major part of daylight, is considered separately in some cases since it has its special uses and produces special problems. In the daylighting of buildings it can be more easily controlled than diffuse skylight in order to avoid unwanted power input and glare. However, its psychological value seems to be important and has been the subject of many surveys, though without very firm conclusions.[576, 577]

The use of daylight irradiation for sun-bathing does not depend on direct sunlight for its tanning effects, which diffuse radiation also produces, but direct sunlight is appreciated more for its rapid action and the accompanying sensation of warmth. Simulation of sunlight for this purpose is described later.

Sunlight is optically controlled for use in solar furnaces, where large mirrors concentrate the radiation on to small areas for the high temperature treatment of materials. At the French Solar Energy Laboratory at Odeillo in the eastern Pyrenees a power input of about 1 MW has been achieved, and working temperatures of 3000 K or higher.[578] On a smaller scale stills and cookers powered by sunlight have been found effective in regions of dependable sunshine. In a still lower range of temperature the absorption of sunlight (as well as diffuse light) to heat water for domestic buildings has proved a valuable source of energy in several countries, requiring absorbing surfaces of the order of 10 m² for a dwelling house. There are many other plans for the direct use of solar energy to replace combustible fuels.

Extraterrestrial sunlight is a reliable constant source of power, used to provide electricity in spacecraft by means of large arrays of silicon photovoltaic cells.

PHOTOCHEMICAL ACTION OF LIGHT

The most extensive application is in photography. The usual materials made from silver halides are very sensitive to a limited part of the daylight spectrum, and so to a small part only of the power distribution. Sensitizing dyes extend the spectral sensitivity throughout the visible and near infra-red regions to supplement the initial blue, violet and ultra-violet response. To obtain pictures compatible with what is seen by the eye it is profitable to remove ultra-violet radiation by filters, especially in colour work. In transparency making, further manipulation of sensitivities and exposures of the film layers is necessary, for the results are usually viewed in the light of a much lower colour temperature than the daylight used for the original exposure. Problems arise in making printed copies from such transparencies. 'Daylight' colour film can be used without daylight if an equivalent light source is available, and developments of this technique are described in Chapter 18. Standards of sensitivity are based on daylight spectral distributions because of the predominance of daylight in photography, but the extensive control of emulsion sensitivity now possible makes the use of almost any light source feasible.

Daylight has destructive effects on many natural and manufactured products in common use. The action consists in absorption followed by conversion of the energy to heat or, more seriously, by activation of molecules by the absorbed quanta and subsequent chemical reactions. The ultra-violet quanta are the largest available in daylight and are often responsible for decay, decomposition, fading or darkening, sometimes for bleaching,[579] but in coloured products radiation of any wavelength absorbed may be effective. Physical or structural changes must also be considered in objects which have to operate in light for long periods.

The best-known types of product concerned are the dyes and pigments in coloured products, their substrates which are generally complex organic polymers, and foodstuffs. The chemical changes are complex and sometimes difficult to identify or prevent, especially in foodstuffs where extremely small changes may produce objectionable effects in appearance or flavour or odour. They often produce them rapidly, from which it might be hoped that control methods would give some guidance in treating labile compounds of much longer life, like dyed fabrics or oil paintings. However, these methods are not very helpful, being chiefly the avoidance of exposure to light, or the exclusion of oxygen which often cooperates in the decomposition processes. For example, hermetic packaging in the absence of oxygen prevents the fading of some kinds of cooked meat. In the case of beer the exposure to light is much reduced by coloured bottles, but if sufficient light penetrates it will cause the unacceptable 'sun-struck' change of flavour and odour. Potatoes exposed to light develop chlorophyll and solanine, which cause a green coloration and affect the flavour respectively.

Compared with perishable foodstuffs, domestic objects including dyed fabrics and decorative paints seem to be very durable. Most are exposed to light without precautions, and the main application of artificial daylight in this case is to measure the probable changes by acceleration under an increased amount of irradiation, which can also be kept constant unlike natural daylight.

A class of comparatively durable objects which has received much attention is that comprising drawings and paintings in art collections. Oil paintings form the most popular displays and need to be exposed to light for long periods of time. Owing to their rarity and value, the more important ones are expected to last for hundreds of years. They may have done so already, probably under far less illumination than is now considered essential for inspecting them. In daylight there is therefore the dilemma of increasing illuminance for viewing and reducing it for conservation. There is evidence that the ultra-violet part of the spectrum is the most dangerous. The situation under artificial light

sources was at first made worse when the fluorescent tube was introduced (see Chapter 19).

COLOUR MATCHING

Colour matching is an industrial technique of very great importance
and precision. Its purpose is to ensure that objects conform to prior
standards of colour appearance. Though instrumental methods of
measurement may be helpful, the final judgment is nearly always made
by the eyes of the colourist, which are generally more sensitive than
colorimeters. This is the situation as it applies to paints, dyed textiles,
paper, colour printing, pigmented plastics and ceramics, and cosmetics,
among other articles where appearance must be closely controlled.
Some products with a wider tolerance in colour may be controlled by
colorimeter, even automatically sorted (for example, fruit, peas, etc.).
Here we consider the more critical cases where the degree of control is
such that no visible colour difference will be seen by a large proportion
of users of the products. These users will often be able to refer to
authentic standards to check the closeness of colour reproduction.

The users of daylight considered in the foregoing sections of this
chapter are not seriously concerned about the exact colour or spectral
distribution of daylight, and for these replacement by artificial sources
should not offer much difficulty. This is not so for colour matching.
The oldest craft among those listed, that of dyeing textiles, has
traditionally used daylight from a north-facing window as the source
of illumination when two patterns are compared with a view to exact
matching. Some degree of overcast seems to be preferred, and the
absence of direct sunshine and therefore less variable illuminance is the
chief attraction in this kind of lighting. The Roman architect Vitruvius,
after recommending the cooler northern aspect for summer dining
rooms, continued: 'Not less should the picture galleries, the weaving
rooms of the embroiderers, the studios of painters have a north aspect,
so that, in the steady light (*propter constantiam luminis*), the colours in
their work may remain of unimpaired quality.'[580] The constancy of
the light is not true in respect of spectrum and colour, but the changes
are not often large enough, or rapid enough, to be obvious to the eye
with its adaptive power, and consequently the north sky has gained an
undeserved reputation for constancy. In spite of this, practitioners in

the art of colour matching have been very difficult to satisfy when artificial sources were offered, and the next chapter relates some of the long history of attempts to do this.

A form of colour matching called colour appraisal is more common than the textile dyer's skilled operators. It is colour matching without a physical standard of comparison to use beside the object under scrutiny. The standard is a mental image, based on anything from an exact memory of a well-known object in the particular situation, to the rather indefinite ideas of someone seeking a dress material, or a wallpaper to 'match' or 'harmonize' with other objects not at hand. Many of the resulting matches are not very exact, others are of vital importance. Good examples of the latter arise in medical practice. One is the examination for hypoxaemia, when it is essential to make rapid judgments of the state of oxygenation of anaesthetized patients. The judgment depends on the observer's memory of the condition. Such medical problems are particularly important in surroundings where artificial lighting has largely replaced daylight. The subject is discussed again in Chapter 19.

One form of testing with medical implications is the use of pseudo-isochromatic plates, for example those of Ishihara, or other coloured samples designed to test colour vision and any defects revealed by confusion of differently coloured printed or dyed objects. Daylight is usually specified for the illumination, possibly with an illuminance value but not with a colour temperature. When routine tests need to be done on numbers of subjects it is an advantage to use an artificial source of constant output, if it can be shown that its spectral distribution is suitable for illuminating the test material; in other words, if it has the correct colour rendering properties.

COLOUR RENDERING AND METAMERISM

The concept of colour rendering is essential to discussion of artificial light sources, daylight or otherwise, and the properties required of them. Colour rendering is defined as the effect of an illuminant on the colour appearance of objects, in conscious or subconscious comparison with their appearance under a reference illuminant. Certain conditions of viewing are required, but need not be discussed here. The reference illuminant or light source is most often some phase of daylight, or it may have a spectral distribution of similar shape, for example that of a full radiator. In the latter case the reference cannot be physically constructed unless it happens to be in the range of possible tungsten filament temperatures; otherwise the reference is merely a standard spectral distribution for use in calculation.

The colour appearance of a source (or an illuminated object) is due to the integrated effects of all the wavelengths in the spectrum of the source (or reflected from the object) at their relative emitted power values, and can be calculated as a chromaticity by standard methods. The eye has no power to analyse the spectral distribution of the light producing this simple colour impression and cannot distinguish by inspection sources having different spectra which happen to produce the same integrated colour sensation. The colour appearance of the light from a low pressure sodium lamp can be closely imitated by an incandescent lamp with a selection of yellow and orange filters. An object of saturated blue colour will appear black under the sodium lamp since it has zero reflectance for light in the yellow part of the spectrum, which, to a first approximation, is all the lamp provides. The filtered incandescent lamp has some small amount of emission all through the spectrum, and the blue object will be seen to be more or less blue. The sources look alike; they even make white objects look alike, but they render coloured objects differently.

The pair of light sources may be called metameric, though the phenomenon just described is also referred to as dichroism of the sample. The most common use of the term metameric applies to a pair of objects which look alike in colour under one light source, but different under another light source of different spectral distribution and usually also of different colour appearance. The chromaticity of each sample can be calculated from its spectral reflectance curve and the spectral power distribution of the light source by the use of standard tables (see Appendix 1). Differences in the chromaticities give an indication of the extent of change in appearance or metamerism when the light source is changed. There are also numerous more or less complicated formulae for calculating colour differences on various scales. Metamerism often occurs when a mixture of dyes or pigments is used to reproduce a given coloration in a material, and the mixture chosen is different from that in the original. It is a constant source of concern in the dyeing industry and reinforces the need for careful selection of any substitute for daylight.

The colour appearance of an object depends on a third factor as well as its spectral reflectance and the characteristics of the light source. This is the eye which views the illuminated object, and it is responsible for a less common, and often less severe type of colour difference. One observer sees two samples matching exactly under a light source. Another with somewhat different colour vision may see the samples as slightly mismatched: this also may be called metamerism. Variations in 'normal' colour vision are sufficient to show the effect in colour matching experiments, for example those conducted by Warburton.[581]

Colour differences in one sample produced by different light sources occur over a wide area of the chromaticity chart and in almost any direction and magnitude of shift. It is convenient to have a numerical value to represent the excellence of a source compared with a reference standard, or its degree of departure from the standard. Colour rendering indexes of this kind rely heavily on calculations from the spectral power distributions of the light sources, and make little if any allowance for the effects of adaptation in the eye, which tends to reduce the perceived differences compared with their calculated values. The chief methods used are:

(1) the NPL method developed by Crawford, based on spectral differences in the sources,[582] and
(2) the CIE method based on calculated colour shifts of a set of coloured samples, where daylight is largely used as a reference standard.[583] The index R_a is 100 if the source under test is identical to its reference illuminant, is 80 or more for sources of good colour rendering, but lower for many sources in extensive use both outdoors and indoors. Because of the increasing use of fluorescent materials in fabrics and other commercial products, and the consequent effects on their hue in daylight, it is now considered that colour rendering indices should take account of the ultra-violet content of light sources by the additional use of fluorescent samples, though the procedure is not yet decided. See also Chapters 16 and 18.

Metamerism is not easily reduced to a numerical index. The publication by Nimeroff and Yurow may be consulted.[347]

This subject has developed so much in recent years that the early efforts in this century seem quite primitive. It was the lack of means of measurement and of light sources rather than lack of ideas that delayed progress. Aristotle knew something of the eye's uncertainties which our modern formulae are still trying to reduce to order: 'There is an indescribable difference in the appearance of the colours in woven or embroidered materials when they are differently arranged; for instance purple is quite different on a white or black background, and variations of light can make a similar difference. So embroiderers say they often make mistakes in their colours when they work by lamplight, picking out one colour in mistake for another.'[584]

A few centuries later the poet Ovid recommended daylight for making important judgments: 'Consult daylight as to gems, and as to wool dyed in purple, and consult it as to the face and figure also.'[585]

As usual, Newton is worth quoting: 'If the Sun's Light consisted of but one sort of rays, there would be but one Colour in the whole World . . . the variety of Colours depends upon the composition of Light.'[3]

The uses of daylight have been discussed in this chapter with a view to considering alternative sources. In a different context there is one application of daylight which remains of permanent scientific value. This is the spectroscopy of daylight, the source of most of our knowledge of the sun, of much of the chemistry and physics of our atmosphere, and of the atomic and molecular states of matter. Though astrophysics is now mainly interested in the radio spectrum and the ultra-violet and X-ray region below 300 nm, the terrestrial solar spectrum is likely to maintain its importance for a long time to come.

18

Artificial Daylight Sources

A man who has candles may sit up late, which he would not do if he had not candles; but nobody will deny that the art of making candles, by which light is continued to us beyond the time that the sun gives us light, is a valuable art, and ought to be preserved.

SAMUEL JOHNSON, 1772, quoted by BOSWELL.

The thing to do is to supply light and not heat.

WOODROW WILSON, Speech at Pittsburgh, 1916.

ARTIFICIAL LIGHT

Earlier chapters have described the uses of daylight and the ways in which its spectral composition has been standardized in partly experimental, partly theoretical investigations. It remains now to show how the need for replacement of daylight by other sources has been met.

The extension of daylight to assist in man's activities, from survival upwards, has no known beginning but must have come with the discovery of fire, initially the burning of vegetable matter, chiefly wood. Vegetable and animal oils as fuels for lamps have the longest known history, extending over more than four thousand years. From the early centuries A.D., if not before then, tallow and beeswax candles added another source of light, and from then no significant advance occurred until the end of the eighteenth century. Coal gas was discovered about 1765, but came into general use only at the time of discovery of another light source, Davy's electric arc between carbon rods (1802). After another long delay this in turn became a usable light source, followed by an event which revived the ancient sources. This was the discovery and exploitation of the mineral oil deposits in the U.S.A. (1859), providing better oil for lamps and wax for candles. In 1879 the carbon filament lamp entered the field, its early triumphs outshone by the Welsbach mantle of 1893. Finally, as electric power supplies became more widespread and more reliable, and vacuum techniques improved, there were the first practical gas discharge sources, following the Geissler tubes of 1856. The landmarks in this phase were the Moore carbon dioxide tube of 1895, the Cooper–Hewitt low pressure mercury arc of 1900, and the Claude rare gas discharge tubes of about 1910; from 1930 a profusion of discharge lamps and phosphors for altering their spectral output became available, and the choice of light source colour became almost unlimited. So far as providing artificial light was concerned, most of man's problems seemed to have been solved. So far as a true substitute for daylight was concerned, much of this flood of discovery had been unproductive.

The position was not at all promising about 1890, when many technologies were finding a need for colour control, and demands for daylight substitutes were becoming urgent. Particularly in dyeing, but also in other fields of work requiring critical judgments on colour, attempts were made, with only partial success, to use the existing light sources, which mostly had unsuitable spectra, poor durability, instability in operation and limited light output. Future developments were foreshadowed by Trotter. In 1892 he compared the spectra of the positive crater of carbon arcs with sunlight, or 'daylight' as specified by Abney and Russell.[586] The arc was found to be deficient in the blue wavelength region, to have a large excess emission in the orange, and a smaller excess in the green. Filters or reflecting screens were proposed to make the necessary corrections in the light from the arc, though Trotter appreciated the fact that much of the output would have to be sacrificed, more than two-thirds if other sources like gas-light or 'glow lamps' were used. There was almost an anticipation of the Abbot–Gibson distributions in his opinion that the daylight spectrum could be represented by sunlight plus one-third of the difference between sunlight and blue sky. The Abbot–Gibson tables applied in this way show Trotter's estimate to be rather a poor one: the two possible alternative distributions give CCT of 9880 and 7840 K.

EARLY ATTEMPTS AT ARTIFICIAL DAYLIGHT

The 'white Moore light', first demonstrated in New York in 1896, might have made most subsequent work unnecessary if its inherent limitations could have been removed. It was a high voltage discharge tube containing about 0·1 torr of carbon dioxide; its light was white of CCT 6500 K produced by a crowded band spectrum, with good colour rendering properties.[587] Even at the time of Moore's discursive account of its past (and future) successes,[588] its failure to overcome the problem of gas clean-up, and its low efficacy of 2 lm/W, apart from the high voltage inconveniences of the time, meant that it was declining in importance compared with the improving incandescent lamp. There were a few commercial installations of Moore lamps in use up to the advent of the fluorescent tube, and one type of unit was manufactured in this country as 'Claudegen Daylight'.[589]

American work of a more fundamental nature seems to have begun in 1899 with Nichols and Franklin (Chapter 7). Their comparison of the spectra of daylight and artificial sources was continued by Nichols, who, later, turned his attention to variations of the daylight spectrum rather than to artificial source possibilities. Nichols believed the maximum energy of the solar spectrum to coincide with the maximum of luminosity for the normal eye (the spectral luminous efficiency).[195] Ives developed this idea further and with consequences of some utility. Disagreeing with Nichols' experimental results, he assumed that 'average daylight' was equivalent to clear noon summer sunlight and this, he thought, agreed closely with the spectral distribution of a full radiator at 5000 K. By a useful coincidence, this full radiator had its peak emission at 590 nm, the wavelength of maximum sensitivity of the eye, as then accepted. This spectrum was to be considered as 'white light', and it had the further property of being produced by a full radiator with nearly the most efficient visible light output, though admittedly the most efficient was at 6000 K according to his temperature scale. In later papers Ives examined the relation between colour of the illuminant and colour of the illuminated object.[590] He considered the mixture of light sources (mercury discharge and tungsten filament).[591] He described correction by coloured glasses and dyed gelatine filters in attempts to reach his ideal artificial source at 5000 K. His practical achievement was the colour matching booth of Plate 7, which included glass and gelatine filters and a gas mantle as the light source. Ives recounted the uses of this instrument in silk mills, cigar factories and department stores, and showed how surgeons, for example, could work under gas light wearing spectacles with the required filters. This development by Ives occurred in 1914. Fig. 80 shows his spectral curves for standard sunlight and its filtered gaslight equivalent where the mantle was of thoria with 0·7 per cent ceria. The typical blue sky spectrum was imitated by a special mantle having 0·25 per cent ceria and suitable filters, thus producing the light required by dyers and colour matchers. Fig. 81 compares a modern gas mantle emission with full radiator distributions, and includes some points from Rubens' thermopile measurements of 1905.[592] The mantle distribution corresponds to a chromaticity of $x=0·454$, $y=0·437$ (Fig. 20), and a CCT of 2982 K.

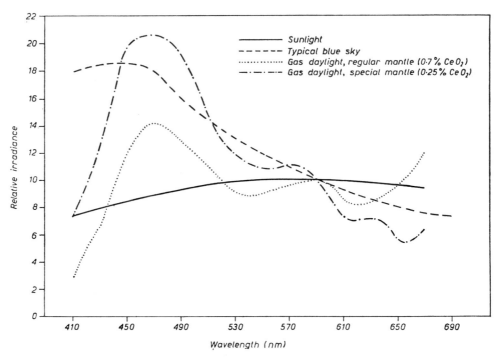

FIG. 80. Spectrophotometric curves of natural and artificial daylight, with Planckian distribution for 5000 K adopted for sunlight, normalized at 590 nm. (Ives.)

For some time the more academic interests were concerned with more accurate sources for colorimetry and spectrophotometry. The 1922 report of the Optical Society of America's Colorimetry Committee accepted the colour temperature of 5000 K for Abbot's spectral distribution, given for average noon sunlight in Washington, D.C. on the basis of twenty determinations.[593] For artificial sunlight two possibilities were considered. The first was a gas-filled tungsten lamp at 2848 K with a filter consisting of a quartz plate 0·5 nm thick cut perpendicular to the optic axis, and mounted between crossed Nicol prisms. This colour temperature changing filter was due to Priest, who had used a rotatory dispersion photometer with quartz plates to measure the colours of light sources.[240]

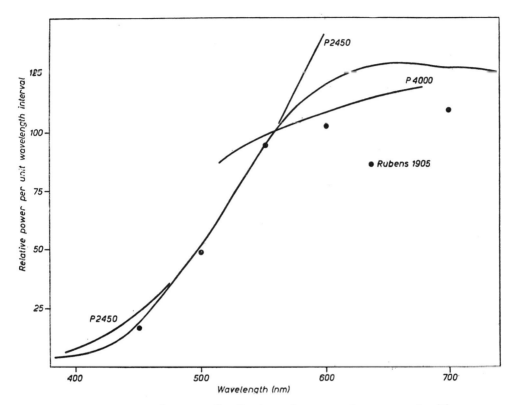

FIG. 81. Spectral power distribution of gas mantle compared with Planckian radiator and experimental data by Rubens, normalized at 560 nm.

The second proposition was to use an acetylene flame and a Wratten filter as a source of artificial sunlight. This was the type of acetylene flame which Jones advocated as a spectrophotometry standard (Chapter 7). The Kodak design was still in use when Cunliffe made his spectral measurements on daylight (Chapter 8).[594, 242]

Priest's work on subjective whiteness with near-Planckian spectral distributions has been described in Chapter 16. Another question which he took up, this time with inconclusive results, was the apparent preference, and occasionally increased comfort, of observers reading under light of daylight colour instead of the uncorrected incandescent light.[595]

CARBON ARC LAMPS

This high intensity source has been used as a spectroradiometric standard, as a photometric standard, for street lighting, and for testing the behaviour of photosensitive materials (Chapter 19). Few if any attempts have been made to adapt it to artificial daylight requirements since Trotter's measurements previously mentioned. Cunliffe and Lanigan discussed the suitability of arcs with impregnated carbon electrodes. They concluded that the excess of radiation below 400 nm, and the red colour compared with daylight, made them unsuitable. Many spectral power distributions were given for electrodes with metal additives.[594] Much more recently carbon arcs have become sources of high irradiance in testing materials for space research[596] and in a different context Schurer concluded that the anode could be used as a highly reproducible standard of spectral radiance.[425]

THE TUNGSTEN FILAMENT LAMP

Applications of tungsten lamps as artificial daylight sources began soon after the new filament had displaced carbon, osmium and tantalum. In 1913 Langmuir and Orange, describing some of the earliest gas-filled lamps (using nitrogen), included a reference to blue glasses which would increase the colour temperature of a vacuum lamp to that of the gas-filled type.[597] The commercial development was largely due to the Macbeth Corporation in the U.S.A. A brief history of this related how colour matchers had a choice of filter glasses, giving them resultant lights which were of variable degrees of greenness compared with the full radiator locus, as well as a number of nominal colour temperatures between 4000 and 7500 K.[598] Fig. 82 shows the green and pink (or least green) limits: the difference between them was narrowed in the course of time, always remaining on the upper side of the full radiator locus. Later measurements of daylight show the great majority of chromaticity points to lie on the same side: this may be taken as slender evidence of a subconscious preference among colour matchers for the natural daylight colour. Fig. 83 gives the spectral distribution of some filtered incandescent sources of this type in comparison with D₇₅. Similar distributions were obtained by glass filters made in the U.K., for example Chance OB8. The paper by Macbeth and Reese included a summary of the U.S.A. standards for various colour matching or assessment practices.

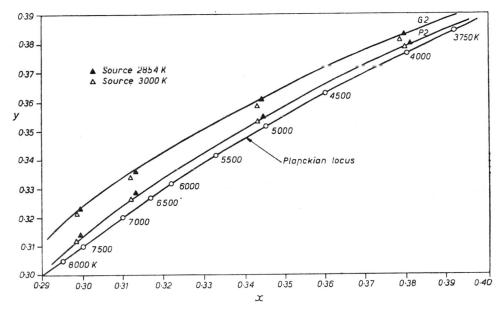

FIG. 82. Chromaticities produced by tungsten filament sources and Macbeth glass filters, showing range of tolerances acceptable in 1946 between the most green (G2) and the most pink (P2). (Macbeth and Reese.)

Removal of unwanted parts of the incandescent filament spectrum was most conveniently done by transmission through blue glass filters, which could also be used to make lamp envelopes. Another less efficient method was seen in at least one daylight lamp (the Shering-ham) in which the long wavelength parts of the spectrum were absorbed by reflection from a shade of a blue colour.[600]

An incandescent lamp and a blue glass filter were easy to assemble into a 'daylight lamp', but not all were properly designed. In 1923 Ord described experiments with a Nutting spectrophotometer in which light from the north sky was compared with that from one of these daylight lamps of unspecified make.[601] It proved to have a spectrum steadily diverging from skylight with increasing wavelength, with three

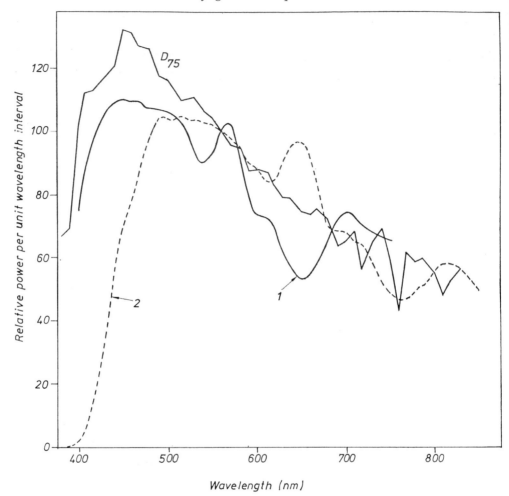

FIG. 83. Distribution of Macbeth filtered incandescent sources at 7500 K. Curve 1: Former combination with 3000 K source.[599] Curve 2: 'Super skylight'. Normalized at 560 nm.

or four times the correct amount of radiation in the red end of the spectrum. This may have been an exceptionally bad case, but the lack of standards made it possible. A short survey of these lamps was given by Cunliffe and Lanigan in 1928.[600]

CIE STANDARDS

A considerable advance was made by the CIE in 1931, when the present system of colorimetry was established, and three illuminants for use in colorimetry were standardized. They were called Standard Illuminants A, B and C, and are now preferably named 'sources' because they can be realized physically. The term 'illuminant' is now applied to spectral distributions which may, or may not, be capable of embodiment in a real lamp, for example D_{65} or a Planckian distribution.

Source A was originally defined as a gas-filled lamp operating at a colour temperature of 2848 K on the international temperature scale of 1927, and its spectral distribution was assumed to be that of a full radiator at 2848 K.[602] In the 1948 scale the Planck radiation formula replaced that of Wien, and the constant c_2 was changed from 1·435 to 1·438 × 10⁻² m K, resulting in an altered specification for source A. This became 2854 K in 1951, rounded from 2853·954 K, the values of colour temperature being proportional to c_2. By the 1968 change of c_2 to 1·4388 × 10⁻² m K a further change was required and the colour temperature became 2855·54 K. These values of colour temperature retain the same nominal spectral distribution for the illuminant, and similarly for Sources B and C. This condition of unchanged distribution is retained in colorimetric calculations because of its convenience; but if it is not required the complete change from the scale of 1948 (T_{48}) to that of 1968 (T_{68}), involving an additional change of the freezing point of gold from 1336·15 to 1337·58 K, means that a source formerly considered to be at 2854 K must now be described as 2858·7 K. A slight change in the spectral distribution formerly attributed to it is due to the use of new values of the Planckian function based on $c_2 = 1·4388 × 10⁻²$ m K. These consequences have been more fully explained by Preston[603] and Morren;[604] they include discarding the degree sign in the Kelvin scale.

The definition of Source A has been altered in a CIE restatement of its recommendations.[605] The source is now described as representing light from a full radiator at approximately 2856 K (T_{68}) and is to be realized by a gas-filled coiled-tungsten filament lamp operating at a CCT of 2856 K.

Source B was intended to represent direct sunlight or one of the more yellow phases of daylight. It consists of source A with two liquid filters formulated by Davis and Gibson. Its CCT is approximately 4871 K (T_{48}), which becomes 4874 K (T_{68}).

Source C was intended to represent one of the more blue phases of daylight and resembles source B. The liquid filters contain the same ingredients in different proportions. The CCT is approximately 6770 K (T_{48}), which becomes 6774 K (T_{68}). The CCT of Standard Illuminants B and C will be altered on the proposed $u'v'$ UCS (Appendix 1), increasing by reason of their chromaticity locations below the Planckian locus. Source B will become about 4900 K and Source C about 6860 K.

The chromaticity points for sources B and C are slightly on the lower (purple) side of the Planckian locus (see Appendix 1 and Fig. 20). Their spectral power distributions are given in Fig. 84, and details of the filters are in the CIE publication on Colorimetry.[605]

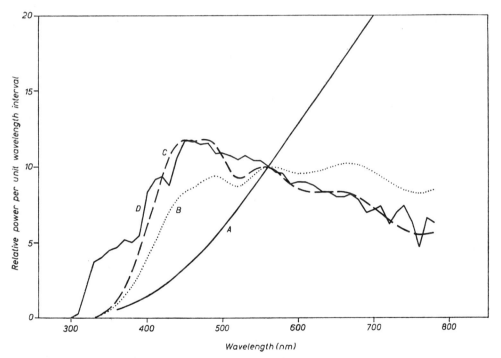

FIG. 84. Spectral power distribution of CIE Standard Illuminants A, B, C and D ($\equiv$ D$_{65}$) normalized at 560 nm.

These sources were intended for use in colorimetry, as calibration standards or for illuminating samples under measurement. Those with filters were unsuitable for providing high illuminance on large areas for colour matching work, and there has been uncertainty about the reproducibility and stability of the filter solutions. This was disturbing, especially because the sources are specified with no tolerances expressed or implied. The extension of the spectra of sources B and C to wavelengths less than 380 nm was not recorded till 1951. The ultra-violet radiation content then proved to be much less than was required by comparison with daylight.

Source A has been very extensively used in spectrophotometry, and still is. The other sources clearly have several drawbacks, in spite of which source C has been widely used in colorimetry, as a light source or an illuminant for calculation. The CIE Colorimetry Committee expects these filtered sources to pass out of use in favour of the RD illuminants. There are two problems caused by this change. The first is that many years of records based on source C will become obsolete. The second is that sources B and C do actually exist, however difficult to reproduce they may be, whereas no close approximation to an RD source is yet available in a form suitable for easy application.

The use of artificial daylight sources in colorimetry would improve on natural daylight whose normal variations make it unsuitable. Sunlight was used for a standard at one time (Chapter 7), but as a calibrating source for a colorimeter it can evidently not compete with sources A, B and C in reproducibility.

The history of the CIE 1931 sources may be briefly extended. The CIE has made no changes apart from an attempt to replace liquid filters by glasses. Several reasonably good corrections were obtained, but the matter has not proceeded further. To introduce a glass filter as an international standard would be a formidable task in view of the large quantities and high precision and uniformity required in manufacture. Another proposal at this time was to define sources A, B and C by full radiator spectral distributions at 2854, 4800 and 6500 K.[606]

In the use of the CIE sources, many workers have replaced the liquid filters by suitable glasses like Chance OB8, and made the appropriate corrections to the equivalent values for source B or C by means of the known spectral transmittance of the glass filters. An increase of ultra-violet content, necessary in the colorimetry of fluorescent materials, was proposed by Harrison in a combination of two source A lamps. One was converted to a near-source C distribution by a filter of Chance OB8 glass, the other with a Chance OX1 filter to add mainly ultra-violet radiation; or, to avoid adding any visible radiation by the second source, this could be a 3000 K source with combined OX1 and OB10 filter glasses.[607]

Similar approximations were made by Hisdal, using a normal source C and other incandescent lamps of higher power with glass and liquid filters. His aim was to reproduce the spectral distribution of Middleton's 1955 (CIE) data. The best fit in the ultra-violet region required two auxiliary sources, each with filters of Schott glass and nickel sulphate solutions.[608] Hisdal discussed variations possible in source A due to differences in the emissivity of tungsten.[609]

OTHER INCANDESCENT LAMPS WITH FILTERS

Attempts at filtering the output of incandescent lamps went on independently of the CIE standards. Gage discussed the colours of artificial sources in terms of colour temperature or Priest's mired scale, and calculated possible changes for some filter glasses.[610] Estey considered numerous filters for altering colour temperature and was concerned about energy matches as well as chromaticity values.[611] A theoretical investigation of the possible changes in visible light emitted by full radiators, when the spectrum was subjected to a Fabry–Perot interference filter, was made by Blottiau *et al.*[612] Calculated values were given for chromaticity, luminance and some spectral distributions, the mirrors in the filter being assumed to have selected reflection factors and to be separated by distances between 0·24 and 0·19 μm. It proved possible to arrive at a near-full radiator spectrum at 5500 K starting from a full radiator at 2800 K, and chromaticities up to 7000 K if the initial radiation was provided by source A.

The use of glass filters with incandescent lamps is still a useful technique despite the disadvantage that, if an upward movement in colour temperature is required, much visible energy has to be absorbed and the resulting source is of low efficacy. Unwanted infra-red radiation and insufficient ultra-violet content are other unavoidable drawbacks. As an example, a well-designed viewing cabinet producing 1500 lux with a 7400 K distribution over less than half a square metre of working surface, required the dissipation of about 2 kW in the lamps, and forced cooling as well.

A photographic application has been described by Vieth and Heiland to anticipate the expected standardization of D_{55} for testing daylight colour film.[613] Formerly the spectral distribution standard was based on Taylor and Kerr data. The authors gave formulations of four-component Schott glass filters for both old and new definitions of daylight. A similar project by Carman began with the selection of clear sky distributions from published data.[330, 335] These suggested a mean CCT of 6000 K as a standard distribution; D_{60} was chosen, extrapolated to 920 nm by Moon's curve for $m=1\cdot5$,[264] corrected for the transmission of lenses in the cameras, and finally approximated by a computed combination of a 2854 K incandescent lamp with three glass filters.[614]

Other attempts have been made to simulate D_{65} and other daylight distributions by incandescent lamps with filters. Though the visible spectrum can be closely imitated, the ultra-violet region is invariably deficient, even with tungsten halogen lamps with their longer effective spectrum and increased ultra-violet flux. These have been used as standards of radiant flux;[337] and in another application a 1 kW tungsten–halogen lamp in a ceramic aluminized reflector produced, at a distance of 400 mm, an irradiance (though not a spectral match) equal to extraterrestrial sunlight, S.[427]

DISCHARGE LAMPS: FLUORESCENT TUBES

About 1940, medium pressure discharge lamps were in active development and the low pressure fluorescent tube was in its early stages. Both types have since contributed to the artificial daylight requirement, the former type represented by the xenon arc which appeared about 1947. Up to the present time the fluorescent tube has been far more successful, not least by raising illuminance values which with incandescent or gas mantle sources had been at levels now regarded as intolerably low, whether for domestic or commercial or industrial purposes. It improved lamp efficacy by a factor of about three from the beginning, and by much more in comparison with the incandescent lamp with daylight filter; it had a wide range of possible colour appearance; and it was a cool light source of long life which could be applied to lighting large areas. Against these advantages it had, initially, a lack of emission at the red end of the spectrum owing to the lack of suitable phosphors, and this defect was worse than the excess of red emission in the incandescent lamp.

The deficiency in red emission was cured in due course, but the poor correlation between colour appearance and colour rendering which was now evident for the first time in the history of artificial light sources, and the initial enthusiasm which welcomed more and more lumens per watt regardless of colour rendering, raised difficulties which are not completely solved even now. There is also the presence of strong mercury line emission in all low pressure mercury discharge lamps. Because of this no fluorescent tube can yet produce a spectrum close to that of natural daylight, and it seems unlikely that it will ever be able to do so though theoretically the filtering process described later in this chapter could produce the desired effect. The effect of the Fraunhofer absorptions in daylight can be ignored in respect of colour rendering, except perhaps for the strong G lines near 431 nm. The same cannot be said of the strong line emissions in fluorescent tubes. They have no serious effects in many applications, but in cases of critical colour matching and appraisal this is not so certain. Some investigation of this problem is desirable, for not much is known about it. A contribution to the subject was made by Bodmann and Voit.[615]

In this country the first use of the fluorescent tube for colour matching purposes was in cold-cathode installations made to fit specified areas.[616] During the Second World War the familiar 5 ft 80 W hot-cathode tube appeared in a 'daylight' colour, approximately 4500 K. Colder colours followed, with one near 6500 K produced in 1948 ('Northlight'). This became accepted as a substitute for north sky-light in colour matching, though its colour rendering properties would now be considered only moderately good.

For reasons which are not very clear, the colour appearance standard of 6500 K or thereabouts became fairly widely accepted in this country for colour matching work. In the U.S.A., following much wartime experience there, opinion had hardened on the need for a still colder light. An example of the kind of survey made is shown in Nickerson's report on textile colour matching.[617] The chosen CCT was about 7500 K or, for calculation purposes, 7400 K by the Abbot–Gibson distribution of 15 per cent sky plus 85 per cent sunlight which became widely used in colorimetry. Fluorescent tubes of this bluish colour could not be made with efficient light output compared with those of warmer colours, and in the U.S.A. the commercial standard tube of highest CCT became the 'Daylight' tube at 6500 K. At the same time the commonest tube in general use was the 'White' tube at 3500 K. With improved phosphors the 'Cool White' tube at 4300 to 4500 K improved in efficacy and became much more popular in the U.S.A. In the U.K. the 'Warm White' (3000 K) and 'White' (3500 K) tubes still largely prevail. The similar shift in preferred colours of light for colour matching, cooler in the U.S.A. than in the U.K., may have a similar cause though it is difficult to separate visual preferences from the commercial importance of lamp efficacy. The choice of colour matching illumination has been discussed by McLaren.[618] For colour temperatures of light sources, see Appendix 1.

None of the fluorescent tubes mentioned so far would now be considered to have good colour rendering properties in comparison with daylight, and test samples would show large colour distortions in most cases. The magnitude of these distortions is surprising, and has been recorded in a colour plate by Helson, Judd and Wilson.[619] This showed the colour appearance of ten Munsell samples in daylight (as reproduced by a Macbeth lamp at 6500 K), and under fluorescent tubes of the 6500, 4500 and 3500 K colours. It also demonstrated the very great range of variations among observers, which has been described as a form of metamerism (Chapter 17): compare Warburton's results.[581]

FLUORESCENT TUBE AND LAMP COMBINATIONS

While the fluorescent tube was still suffering from restricted spectral range, it was necessary to combine other light sources with it to produce acceptable colour appearance and colour rendering in imitation of daylight. Several of these combined light sources appeared as commercial articles in the fifties and were usually assembled in more or less portable cabinets for the examination of small samples. One of the first (Siemens) was described by Harrison.[620] The components were gas-filled tungsten lamps and fluorescent tubes made with a calcium halophosphate phosphor of blue fluorescence. By altering the wattage of the incandescent lamps the proportions of the two types of light flux could be altered to span the range of CCT between 8000 and 5500 K. See Fig. 85, where the incandescent component is taken to be the full radiator at 2750 K. A section of the RD locus is included in the diagram. By the use of a magnesium tungstate phosphor a less saturated blue component can be provided, and its combinations with the incandescent source give chromaticities on a line lying more closely

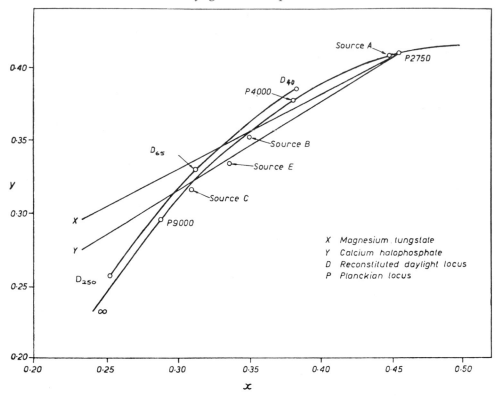

FIG. 85. Location of chromaticities produced by blending light from
a tungsten filament lamp (2750 K) and blue fluorescent tubes.

along the full radiator locus, as in the Figure.[621] Mixtures of this kind
produce some of the best artificial daylight sources in which the
unwelcome prominence of mercury lines is appreciably reduced by the
addition of the continuous tungsten spectrum. Blending the output of
the two sources is difficult, and the proportions need periodical
adjustment to compensate for the different rates at which the two
types of lamps decline in light output. The proportion of total wattage
expended in the incandescent lamps is at least half and often more.

In the U.S.A. a similar arrangement was used for a long time and
received official sanction. Most of the spectrum was covered by the
'Examolite' fluorescent tube which by itself had a CCT of 8300 to
8500 K. With a small incandescent lamp addition, dissipating about
three-quarters of the power used by the fluorescent tube, the overall
CCT became 7300 to 7500 K. The spectral power distribution is
included in Fig. 86. The Examolite unit had many uses.[598] In the

U.S.A. the best known has been the grading of raw cotton. The American specification for this included a description of a former method of assessment of lamp spectral quality, the conformity index, used as a kind of colour rendering index.[622, 350]

Some of the smaller installations for colour matching or appraisal were capable of more than one colour of illumination, arranged by switching a variety of lamps. One optional addition which came to be regarded as essential was a source of ultra-violet radiation. It was provided by a medium pressure mercury discharge tube in a Wood's glass bulb, or by a fluorescent tube with ultra-violet emitting phosphors (maximum usually near 350 nm). These tubes could have Wood's glass envelopes to remove visible light, or other arrangements of the filter glass were used. One cabinet (GEC) had fluorescent tubes of light blue emission, ultra-violet tubes, and incandescent lamps, with a diffusing screen to assist in blending the light. Varied colour temperatures could be attained, including 6500 K.

One other colour viewing cabinet (Philips) was the first example to use improvements in fluorescent tubes which made the incandescent lamp addition unnecessary. The components were fluorescent tubes of colour matching and blue types, the former having a phosphor mixture with some emission in the 660 nm region, and therefore not deficient in red light. The resulting light from the unit was at about 10 000 K.

When different light sources were combined in large units the problems of control increased. It became clear that dyehouses and printing works and other colour matching sites must have a single light source to supply artificial daylight. This seemed a distant prospect in the early fifties. The red emission of the fluorescent tube needed a substantial increase. The ultra-violet emission was inadequate to provide for the appearance changes in white fabrics and paper caused by the incorporation of fluorescent brighteners[623] (Chapter 16). In the more restricted field of colorimetry the CIE had certainly made recommendations on the desirable characteristics of artificial daylight, but no action of note followed for several years.

IMPROVEMENTS IN THE FLUORESCENT TUBE

The first improvements were directed towards the poor red emission of the early tubes, and the less obvious excess of blue emission, largely from the 435·8 nm mercury line. These faults had caused much dislike of fluorescent lighting for general purposes. Competition among lamp makers largely centred on lumen output, which is unavoidably decreased by extension of the spectrum towards the red end since the utilization of mercury resonance radiation cannot be appreciably increased. Diversion of some ultra-violet quanta to excitation of phosphors producing longer wavelength emission therefore resulted in a lower energy conversion efficiency because of the smaller quantum of red light arising from the large ultra-violet quantum (at 253·7 or 184·9 nm), and in addition the red light is of lower luminous efficiency. Less lumens were therefore produced per watt input.

It is not necessary to produce a continuous spectrum in the long wavelengths owing to the fortunate circumstance that the spectral colours appear of a nearly constant hue from about 640 nm onwards. The spectral luminous efficiency decreases steadily from 17·5 per cent of the maximum at 640 nm, to 6 per cent at 660 nm, to 0·4 per cent at 700 nm. Most red-coloured materials have wide reflectance bands in this spectral region and it is of little consequence where the red light is located, except for the efficiency changes. The best known are the manganese-activated magnesium fluorogermanate[624] and magnesium arsenate phosphors[625] with narrow band emission, both with a maximum close to 660 nm. They have been used extensively in fluorescent lamps of improved colour rendering properties. Later europium-activated yttrium oxide became a practicable phosphor for this purpose. Its narrow band, orange-red emission has a maximum near 610 nm. A further advantageous step, made possible by the yellow colour of the arsenate phosphor powder, was to coat this thinly on the tube wall before the main phosphor coating.[626] The outer filter coating removed part of the blue light from inside, particularly the 435·8 mercury line. The absorbed part was partly converted into red emission, while the filter coating was also normally fluorescent under ultra-violet radiation penetrating the inner phosphor layer, including the 365 to 366 nm mercury lines; other non-fluorescent filter coatings of slightly yellow colour have been used to remove the excess blue and ultra-violet radiation.

Attempts have been made at a more favourable compromise between extending the spectral range and so improving the colour rendering properties, and lowering the luminous efficiency. This requires new phosphor mixtures, particularly those containing rare earths in matrix or activator. A different approach is to use phosphors with narrow band spectra in approximations to the theoretically advantageous line spectra.[627, 628]

D$_{65}$

The action which led to standardization of daylight distributions in order to improve on existing standards has been described in Chapter 13. Somewhat surprisingly, but conveniently to established U.K. practice in the field of colour matching, 6500 K was selected from the whole range as a new Standard Illuminant, named D$_{65}$, to supplement Standard Illuminants A, B and C and perhaps assist in the demise of B and C. Though the primary intention was to provide standards for colorimetry, the only acceptable applications of existing sources at present are in colour matching and appraisal rather than for measurement.

An illuminant as specified is not a light source but only a design for its performance. To produce the sources is another problem. A single source was desired to embody the CIE recommendation on D$_{65}$, and a fluorescent tube seemed to be the simplest possibility. For purposes of a colorimetry standard it is unlikely that the spectrum of such a light source, overlain by strong mercury emission lines, will prove acceptable even if fluorescent tubes could be controlled at accurately constant output. The position is otherwise for colour assessment without measurement. Fluorescent tubes are easily capable of illuminating large working areas up to an illuminance of 3000 lux, which is as high as most critical colour assessments require. The problem for the lamp maker was to provide a phosphor mixture giving a colour appearance close to D$_{65}$ and with a spectral power distribution not too far removed from the nominal values, including the ultra-violet range.

Fluorescent tubes with the necessary ultra-violet emission were made in early 1963 and became commercially available in August 1965.[629] Other tubes of the same type have been produced in the U.S.A.[630] and Japan,[631] and probably many others will follow. Visible light emission is less than in tubes of conventional design because of the ultra-violet and red content of the spectrum. The specification applicable in the U.K. is BS 950, Part 1 (1967), which controls the spectral output of D_{65} sources by the bands used in the NPL colour rendering system,[582] with increased tolerances to allow for the presence of mercury lines and inevitable manufacturing variations in fluorescent tubes.[632] Of course these are not the only possible type of D_{65} sources, but the first to become available.

Fig. 86 compares the original 'artificial daylight' tube with the D_{65} specification. Though the spectral differences are large and unavoidable, the tube has proved acceptable in many kinds of colour matching and appraisal. Cases where the spectrum is not quite suitable may be discovered by trial and error in the future, and it may be helpful to alter the blend of phosphors somewhat.

The ultra-violet emission of some fluorescent and discharge lamps has been detailed by Thorington and Parascandola, with a separation into 'middle ultra-violet' (290 to 320 nm) and 'near ultra-violet' (320 to 380 nm). The amounts of ultra-violet radiation in μW per lumen were very variable and different from that in the D_{65} distribution, but these lamps were not intended to be daylight substitutes.[633]

D_{50}

The emphasis on D_{65} in the CIE recommendations, and its widespread acceptance as a colour matching standard, do not resolve all the problems which arise in colour work. In the last few years the graphic arts industry has realized the need to standardize a light source for its operations, and if possible one for international acceptance. The main operation in this field of colour is the production of colour prints from similar reflection copy, or from photographic colour transparencies. A large proportion of this work is in the colour magazine or newspaper business, mainly in advertisements. Wrappers and labels for packaged goods form another large outlet for the bulk-produced coloured print where colour standards must be preserved.

The usual process of printing from colour separation negatives in yellow, magenta and cyan (and possibly black as well) allows metamerism to occur in the printing inks, or between inks and dyes in the transparencies. A constant quality of lighting is desirable for the production stages, but the products may be viewed by the consumer by daylight or fluorescent or incandescent lighting. Metamerism seen under any one of these is likely to be reduced if matching during

manufacture is done under a light of intermediate colour, with proportionately more red in its spectral composition than there is in D_{65}.
This is the main argument for the use of D_{50} which is now preferred in
this industry, with a good deal of added weight from experience in the
use of lights of varied colours.

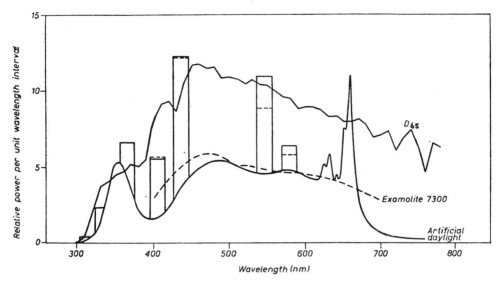

FIG. 86. Spectral power distribution of D_{65} and simulations by a
fluorescent tube (Artificial Daylight, Thorn Lighting Ltd) and by a
fluorescent tube plus a tungsten filament lamp (Examolite). Power
emitted in mercury lines is given by the areas of the blocks 20 nm
wide above the phosphor curve.

The D_{50} spectral distribution has been specified in the U.K. in
BS 950, Part 2 (1967), for colour assessment in the graphic arts
industry. Tolerances are placed on the wide spectral bands for the
same reasons as those for Part 1 of the specification. Some difficulty
has arisen because the standard is for an illumination, not a light
source as in Part 1. For a fluorescent tube to meet the specification it
must have a CCT somewhat higher than 5000 K to correct for the
selective absorption in paints on lamp fittings and reflectors, or in
plastic diffusers which are often used. Suitable fluorescent tubes were
offered by the Macbeth Corporation[598] and other examples soon
followed in the U.S.A., Japan and Holland. An account of several

new tubes in this chromaticity region was given by Hanada, Sugiyama and Kobuya.[634]

If a transparency is received for copying as a reflection print, it is now agreed that it should also be illuminated by a D_{50} source. At an earlier stage it was proposed to illuminate the transparency at 3800 K and the printed copy at (Abbot–Gibson) 7400 K.[635] It is true that most transparencies made in daylight are so balanced in colour that they look correct when projected by a lamp at about 3000 ± 200 K. There are therefore inconsistencies in the selection of illuminants, but there is some compensation by the chromatic adaptation of the eye (Chapter 16). In forthcoming British and international specifications for D_{50} illumination for viewing colour transparencies and prints the band specification is likely to be replaced by a chromaticity area and the nominal D_{50} spectral distribution, with tolerances based on colour rendering indices for the light.

The ultra-violet content of D_{50} is proportionately less than that of D_{65}. Many users find the ultra-violet component essential in D_{65}, but in D_{50} the specification makes it optional.[636] The omission of ultra-violet emitting phosphors would simplify tube manufacture, but this can hardly be a final solution since many papers contain fluorescent brighteners and are therefore slightly altered in appearance in daylight by its ultra-violet content.

THE XENON ARC

The arc in xenon in a fused silica envelope, usually operating on direct current, has attractions as a substitute for daylight. The CCT is usually about 5500 to 6000 K, but the spectral power distribution may resemble D_{50} (Fig. 87) or daylight at 5500 K.[637] The advantages of this spectrum and the consequent white appearance and good colour rendering properties of the source are offset to a serious degree by some unavoidable limitations. These are the moderate efficacy, up to 50 lm/W, caused partly by thermal losses in the unenclosed lamp and partly by high emission in the ultra-violet and infra-red regions; the cost of the control gear which has to provide a high voltage starting pulse and then high d.c. power at low voltage; the rather high cost of the lamp, and some risk in operation due to gas pressure varying from a few atmospheres upwards according to lamp type.

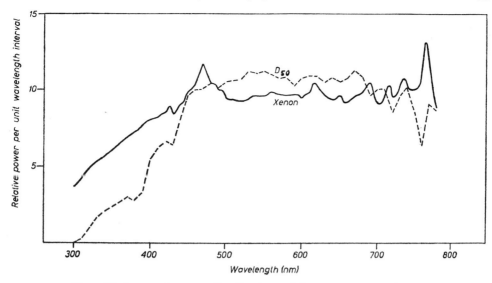

FIG. 87. Spectral power distribution of the xenon arc and D_{50}.

Lamps of moderate loading have chromaticities slightly on the purple (or pink) side of the full radiator locus. Correction is required towards the greenish daylight chromaticity, with removal of excess ultra-violet and infra-red, and numerous attempts have been made to obtain an approximation to D_{65} or other daylight distributions by means of filter combinations. In some cases the agreement is close and better than for filtered incandescent sources or fluorescent tubes. The excess of ultra-violet radiation from the xenon arc is a distinct advantage over other sources where there is too little ultra-violet, or where it has to be added by a phosphor component.

One proposal for filtering xenon lamp emission to produce a D_{65} source within the tolerances of BS 950, Part 1, was made by Thomas.[638] He used Chance HA3 glass to reduce infra-red radiation and Chance SC2 (didymium) glass for the main spectral correction. A cabinet with this lamp was used for the colour vision tests discussed in Chapter 17. Two 150 W lamps were needed to produce 1000 lux on the working surface.

At least one example of a general purpose viewing cabinet with a xenon lamp has been offered, the Siemens–Schuckert type CLX 100. This appears to have no special filtering of the emission from the 100 W lamp, except by glass to remove short wavelength ultra-violet.

Costs are relatively unimportant in the construction of solar simulators, where large areas are required to be irradiated by the equivalent of extraterrestrial sunlight for testing materials and equipment to be used outside the atmosphere. Xenon arcs are used though their spectral range is shorter in the ultra-violet range than sunlight.[639-641] They have often been used without filters to correct the spectral emission since a large reduction in irradiance is entailed if a fair match to the solar spectrum is made. In one example a simulator with a 2 m diameter beam produced twice the extraterrestrial irradiation though with a loss of 50 per cent by filtering.[642] Beauchene and Dennis reported on a spectrally corrected unit using 140 kW of xenon lamps to irradiate a cylindrical space 2 m by 2 m diameter at the equivalent of S.[643] To simulate conditions at one-tenth of the earth's distance from the sun, small areas have been irradiated at more than 100 S by a lens system and a 20 kW xenon lamp.[713] Relatively enormous amounts of power are now available in the xenon ultra-violet spectrum. A 130 kW air-cooled lamp, 3·8 m long, was reported to deliver 450 W of UV-C ($<$280 nm), 750 W of UV-B (280 to 315 nm) and 3 kW of UV-A (315 to 400 nm).[644] Larger amounts are furnished by high pressure mercury arcs where the intention is to initiate photochemical reactions.[645]

The xenon arc appears to have a promising future in the artificial daylight field. A well-established application is described in the next section. In respect of colorimetry standards, while the xenon lamp is more likely to be adapted to this use than is any fluorescent tube, the filtered light of a tungsten–halogen lamp is probably more suitable than either in the absence of any ultra-violet requirement in the standard.

PHOTOGRAPHIC LIGHT SOURCES

The electronic flash tube is the most successful development of the gas arc as an artificial daylight source. Normally the filling is xenon, though other inert gases are sometimes used. The discharge produced by a charged capacitor varies widely in the amount of energy dissipated in the many available designs of tube. Efficiency of light production may reach about 50 lm/W. One photographic studio type produces 1600 joules per flash and a light output of 75 000 lm s during a period of about 10 ms. This is one of the longest times used with this device. Most have a much shorter flash. The main use is in photography with film balanced to an approximate daylight spectrum. Other applications in signalling, in stroboscopic lighting, and in activation of laser output make use of the valuable qualities of high emission, repeatability and simple output pattern of the electronic flash tube. Goncz and Newell made tubes of varied spectral emission according to the gas composition and electrical loading, with CCT ratings up to 40 000 K.[646] Other tubes are designed to give a pulsed output at each half cycle of the supply.

Though having an entirely different mechanism from the electronic flash tube, the expendable photoflash bulb fulfils similar photographic purposes and spectral requirements, and may conveniently be considered here. Most types are small and produce light outputs at the lower end of the range covered by the flash tube, from 150 joules upwards. The flash is initiated by electrical heating of a primer containing zirconium powder on a small tungsten filament or, in a later development, by mechanical impact from an external spring which avoids the need for electrical firing. In either case the primer burns and ignites the shredded zirconium foil in the bulb atmosphere of oxygen under pressure. Formerly aluminium foil was used in oxygen at atmospheric pressure. The complete combustion takes about 40 ms in the common AG1B type used in amateur photography, with a peak of 5×10^5 lm and an output of about 8000 lm s. A mainly continuous spectrum is emitted from the incandescent metal and oxide, and its approximate CCT is 4000 K. For daylight colour film the bulbs are coated with a blue lacquer and the resulting spectrum and CCT depend on the dyes used. Typical spectra of commercial bulbs are shown in Fig. 88, and cover the usual spread of CCT values. The assessment of colour rendering properties of the bulbs may be based on the integration of output over wide bands (blue, green and red), with weighting according to a standard set of sensitivity values for the three component emulsions of colour film. The three values, expressed logarithmically, are adjusted to make the smallest equal to zero. The differences of the other two from zero give a rough indication of how closely the light flash approaches the optimum spectral distribution.

OTHER DISCHARGE TUBE LIGHT SOURCES

The extension of the properties of the medium pressure mercury arc by the addition of other metals and iodine to the discharge has not yet been fully exploited owing to problems concerned with control of lamp filling, electrical characteristics and lamp efficacy. Line spectra of apparently unlimited variation are possible, and it is reasonable to expect that lamp fillings will be devised to give closer approximations

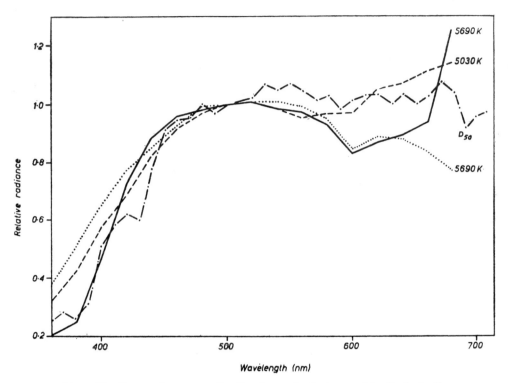

FIG. 88. Spectral power distribution of photographic flash bulbs normalized at 500 nm, compared with D_{50}.

to daylight output than is the case at present. Dobrusskin described lamps with reasonably uniform spectra, suitable colour temperatures and high colour rendering indices (R_a=85 to 95). These contained such combinations of metals as Tl+Dy or Tl+Dy+Ho+Tm.[647] The possibility of optimizing colour rendering in these lamps by three-line spectra has been demonstrated by Koedam and Opstelten,[648] but this technique is more readily applied to fluorescent tubes.[628]

An extra possibility is the use of a phosphor layer inside the lamp envelope as in earlier 'colour corrected' mercury arcs with magnesium fluorogermanate. For metal halide lamps a suitable phosphor is europium-activated yttrium vanadate having a narrow emission peak near 620 nm. Some types of lamps have replaced fluorescent tubes in stores where colour rendering is important, but the development of small lamps suitable for domestic use is not yet achieved.

A lamp with a continuous spectrum was described by Higashi, Mori and Nagano.[649] This contained mercury, argon, stannous bromide and iodide. Superimposed on the continuous spectrum were comparatively weak emission lines of mercury and tin. Chromaticity was close to the RD locus, with a CCT of 5300±200 K for the 400 W lamp. Other examples of continuous emission spectra in halide discharges containing tin or other metals have been published and may be developed into useful light sources.[650-653] Imitations of daylight are less important here than high efficacy and good colour rendering. A review by Eckhardt summarizes the development of halide discharges since 1960.[654]

Another type of discharge tube is that in which an arc is formed in an envelope of translucent, microcrystalline alumina operating at about 1500 K.[655] The first successful filling for this tube was sodium with mercury and xenon. Other fillings, possibly with halides, could well produce continuous spectra at high efficiency.

A novel attempt to construct without optical filters a laboratory source of radiation matching the solar spectrum between 250 and 2500 nm was made by Neuder. Vortex-stabilized plasmas in rare gases at several atmospheres pressure gave a better match than the compact xenon arc at a similar power rating. The best result was produced by a mixture of neon and xenon (8:1) in a 1·5 mm diameter tube at a current of 100 A.[656] Another application of a plasma discharge to provide a radiation standard is that of Günther and Radtke.[657] A silica tube containing xenon at 400 torr and carrying current pulses of 3·9 kA/cm² (1·3 kJ per 1·3 ms pulse) gave a spectrum from 260 to 600 nm closely resembling that of a full radiator at 12 000 K at wavelengths longer than 320 nm, with some falling off of the radiance at shorter wavelengths.

This chapter has shown that the possibilities of light sources are not yet exhausted. All lamps applied to the objective of artificial daylight have some disadvantage or other, leaving a strong incentive to further research.

ASSESSMENT OF ARTIFICIAL DAYLIGHT SOURCES

With so many possibilities of lamps to replace daylight for different purposes, it is necessary that standards of comparison should be agreed with respect to D_{65} etc. The test source and standard illuminant will usually form a 'metameric pair' with identical chromaticity (or nearly so) but different spectral power distributions. Of the methods proposed for assessing the closeness of fit, functions of the spectral radiance differences, alone or weighted by the spectral tristimulus values have been used.[347, 658] Several other authors have based indices of metamerism on calculations of colour differences between pairs of metameric samples which would appear identical under D_{65} but differ under the test lamp. The samples could be theoretical pairs of achromatic greys[411] or real pairs of coloured samples of known spectral luminance factors.[659, 660] In some cases colour rendering indices are the criteria,[411, 659] but this excludes consideration of the ultra-violet emission of the lamp, unless the test sample is a fluorescent material, preferably white.[661]

A limited but more direct approach by Grum *et al.* found examples of tungsten, tungsten halogen and xenon lamps with filters having colour rendering indices over 93, CCT within 90 K of the standard D_{65} and chromaticities within the limits of the British standard specification.[632] Apart from the plain tungsten lamp, they gave good colorimetric reproduction of a fluorescent white and a coloured sample.[630]

The most thorough investigations are due to Wyszecki[658] and Berger and Strocka,[660] who used the data for a collection of thirty-three lamps, mostly of commercial origin and from several countries, comprising filtered tungsten, filtered xenon, and fluorescent lamps. The authors emphasized the need for adequate ultra-violet output from substitutes for D_{65}.

Correct assessment of the ultra-violet output has been attempted by several methods: by using one white fluorescent sample to give a special colour rendering index,[661] and similarly for coloured fluorescent samples;[662] by colour difference from a standard white sample;[660] by matching one of a series of metameric fluorescent white samples against a standard non-fluorescent white;[663] or by the calculation of quantum absorption in three fluorescent samples having peak absorptions at about 340, 370 and 390 nm.[664] This last method uses colour difference formulae based on the Adams–Nickerson colour space (Appendix 1), and provides tolerances for acceptance of lamps. Direct comparisons of the spectral power distribution of lamps in the ultra-violet region is also a possible method.[347]

A striking agreement with the D_{65} spectrum was achieved by Ikeda *et al.* by modifying the emission spectrum of a 1 kW short arc xenon lamp. The 25 mm long spectrum was focused on a photographically produced filter having transverse lines grouped in width and spacing according to the absorption required in each narrow band of wavelength. Each band sampled between 390 and 760 nm was 0·25 mm wide. The optical system produced 500 lx on a circle of 55 mm diameter for colour matching or other testing. D_{55} and D_{75} were also simulated.[665]

There can be no exact correlation between the various indices of colour rendering or metamerism and the practical performance of the sources, having regard to the infinite variety of coloured and white samples. In colour matching, particularly of dyed fabrics, metamerism is an ever-present hazard, but though the materials may have their chromaticities calculated for D_{65} illumination, they are unlikely to be viewed critically under this exact spectral distribution and a real lamp matching D_{65} precisely is therefore not essential. For colorimetric purposes the requirement is different and more urgent, but no generally acceptable alternative to Source C is available, though the filtered xenon imitation of D_{65} looks promising. In the less exacting procedures of colour rendering and assessment some of the sources listed by Wyszecki appear to offer adequate service.

VIEWING CONDITIONS

To discuss the spectral power distributions of daylight and all its possible substitutes is essential for colour technologies. It omits one important difference between the light sources. This is that daylight is mostly used as a large-area diffuse source of fairly uniform brightness, whether out of doors or in a room with windows, while all artificial sources are of limited size, producing very different viewing conditions and subconscious mental attitudes to the lighting. Except in some large-area false-ceiling installations with fluorescent tubes, the feeling of daylight is rarely conveyed, and its normal illuminance levels never achieved (except in solar simulators). Even under critical colour matching conditions the illuminance is only a few per cent of the maximum value for outdoor sunshine.

The colour technologies have agreed methods for their purposes, including details of illuminance and directions of illumination and viewing. Where comparison with natural daylight is involved, differences in these factors may lead to dissatisfaction with artificial sources, attributed perhaps to poor spectral conformity with daylight, whereas the cause may lie in the inadequate size, shape, location and power output of the source.

In the non-technical uses of artificial daylight in office, school and factory it is well established that the lowest acceptable level of illumination depends directly on the colour temperature of the light.[666] If a low value of illuminance is just acceptable under given conditions with incandescent lighting, this same level will become increasingly unpleasant to the observer as the colour temperature is raised. Natural daylight normally has a higher CCT than other light sources, needs to be used at the highest illuminance values, and these are normally available. Artificial daylight installations at 5000, 6000 or 7000 K become increasingly costly in producing acceptable illuminance levels even though they are still far below the values for natural daylight out of doors. These factors of cost and 'pleasantness' no doubt account largely for the much 'warmer' colour temperatures (that is, lower CCT) in most commercial lighting. At present in this country there is a distinct tendency to prefer an intermediate colour temperature for lighting of good colour rendering quality. The preferred CCT is near 4000 K, which is not often encountered in natural daylight. This type of lighting is being increasingly used in shops, in museums and art galleries, and in meeting places which are more social than technical. It may be fulfilling a subconscious desire for lighting of a daylight type which is also of a lower colour temperature than usual. Some examples of its use are described in Chapter 19.

The idea of 'clarity' was introduced in experiments where observers were asked to consider personal satisfaction with the appearance of a scene and to discount differences of colour or brightness. In the first study they compared identical small scale models of rooms under two different types of fluorescent lighting, and adjusted the illuminance to give equal 'clarity'.[667] In the second study they made similar judgments on a pair of rooms of normal dimensions.[668] It appeared that, compared with illumination by standard high efficacy lamps of inferior colour rendering properties, equal clarity could be obtained by a lower illuminance from lamps of similar colour and good colour rendering. The satisfactory reduction of illuminance was up to 40 per cent in the first experiments and about 20 to 25 per cent in the second. The main effect was that better colour rendering allowed a lower illuminance to be used without loss of 'clarity' or 'pleasantness'.

Similar experiments with similar results to those of Bellchambers *et al.* were reported by Boyce and Lynes.[669]

19

Artificial Daylight:
Trying to Improve on Nature

A mere copier of nature can never produce anything great.
JOSHUA REYNOLDS, *Discourse to R.A. students*, 1770.

The performance of artificial daylight sources described in earlier chapters is adequate on the whole, but not impressive. The one clear advantage of these sources is that they are reproducible and nearly constant in output of light. Variations in natural daylight may well have indispensable psychological or physiological effects on man, but they are inconvenient in technology. To this extent an artificial light source may improve on the natural product.

In this chapter we consider a few cases where it is advantageous to replace daylight by sources of different spectral distribution. Most artificial sources are unable to compete with daylight in the illuminance or irradiance produced over more than quite small areas. On the other hand, possibilities of altering their spectra are endless.

PLANT CULTURE

It is likely that in future more attention will be given to the use of different spectral regions for accelerating or retarding the processes of plant life. There is no pattern of behaviour common to all plants, and investigation of individual cases is necessary. Light sources of high output and sharply limited spectral distribution are scarce, so that optimum use of the spectrum is not very probable except in small scale experiments. The importance of the longer visible wavelengths has been described in Chapter 17. One attempt in this direction was made by the Sylvania Corporation in the 'Grolux' fluorescent tube. Its basis was the assumption that most plants, being green, reflect much light of this colour and are less dependent on it than on other parts of the spectrum. A lamp for promoting indoor cultivation was designed to take account of this selective use of daylight. It contained phosphors emitting mainly in the blue and red regions of the spectrum, leaving a wide and deep minimum in the green and yellow parts of the tube emission (though mercury lines appeared as usual). The amount of ultra-violet power available for conversion to visible light could thus be concentrated on the ends of the visible spectrum, producing more light in these regions than is usual when the tube emission is largely in the green and yellow wavelengths. The advantage was appreciable from the lamp maker's point of view, though perhaps less striking in its effects on the plant's complicated radiation balance, where infra-red exchanges are important.[670] One unexpected advantage was that in indoor cultivation the peculiar spectral distribution of the tube heightened the colourfulness of displays of plants and flowers, and the same applied to coloured fish in aquaria. A similar lamp with a spectrum based on the 'photosynthetic saturation curve' had a wider emission band in the red region with a maximum near 625 nm instead of 660 nm in the Grolux lamp.[671]

In a botanical laboratory at Liège, one of the 'Phytotrons', modifications of the spectrum of fluorescent tubes were examined for their effects on plants. The results of outdoor cultivation were improved indoors under illumination by special tubes ('Phytor') made by ACEC of Charleroi.[672] The noticeable characteristic of the spectrum was the reduction in the blue region, and it was far from typical of a daylight distribution.

ULTRA-VIOLET FOR ANIMALS

Animal farming, like indoor plant cultivation, recognizes the value of light but without any subtle manipulations of the spectrum. The use of the ultra-violet component, increasing its proportion above that received in natural daylight, has long been considered beneficial. In its application to man there has been a belief in the health-giving properties of ultra-violet radiation, an interest in the amount received under different circumstances, and a search for artificial generators for controlled ultra-violet dosage. This 'ray hygiene' was discussed in the 1931 CIE report.[239] The subject has become more prominent since about 1930 when the mercury discharge lamp was approaching successful development.[673] A review was given by Dantsig, Lazarev and

Sokolov describing how ultra-violet irradiation has become a common practice in hospitals, schools and factories of the U.S.S.R., where it is also considered to be beneficial in the rearing of cattle and poultry. Xenon lamps are used for irradiation.[674] In quantitative terms, mid-day summer sunlight at sea level supplies about 0·18 W/m² of erythemal radiation, or in the customary units 1·8 E-vitons/cm² or 1·8 Finsen. These units of erythemal dose rate apply to irradiation at 297 nm. At other wavelengths the values are weighted according to a sensitivity curve or 'action spectrum' with a sharp peak at 297 nm, in a similar way to the weighting of radiant power by the $V(\lambda)$ function with a peak at 555 nm to determine light flux. Other ultra-violet effects including photochemical and germicidal action have different wavelength maxima. The dose for minimum perceptible erythema (MPE) is 2500 Finsen s or 250 J/m² on the average, though different subjects show very large differences in response. The values given imply a 23-minute exposure for MPE. If required this can be given more rapidly by medium pressure mercury discharge lamps in silica envelopes, or at the natural outdoor rate by convenient arrangements of fluorescent tubes with phosphors emitting in the erythemal region, and envelopes of glass with better ultra-violet transmission than the usual soda glass.[675] Control of dose is much more accurate than in natural outdoor conditions, being based on test exposures to determine MPE, whereas in skylight alone or with sunlight the irradiance at the end of the spectrum is extremely variable though the unaided eye can see no differences.

The advantages of rapid, controlled ultra-violet dosage are balanced by the somewhat clinical nature of the experience unless it is accompanied by exposure to adequate visible light and heat, which compose a necessary part of the agreeable circumstances of outdoor sun-bathing, and may be as much responsible for the feeling of good health as the absorption of ultra-violet radiation and its effects on the skin. Ruff described an indoor solarium with this objective.[676] If the sun-lamp is merely an ultra-violet source, it is somewhat inadequate when the object is to make the process pleasant. The lamp may emit at shorter wavelengths than are required for tanning or erythema, and besides accelerated erythema this will cause conjunctivitis if the eyes are not thoroughly protected; in fact at the shorter wavelengths one-tenth of the MPE dose may cause this trouble. In the usual erythemal band, known as UV-B, conjunctivitis still occurs but at a lower rate, and in natural sunlight its incidence is much less in normal subjects. Improvement on nature is seen to be not very simple in this case.

Medical use of ultra-violet is of course greatly assisted by sun-lamps, and the disadvantages are easily controlled. We next consider some other medical applications of light where the ultra-violet part of the spectrum is of no apparent importance.

THE MRC INVESTIGATION[677]

With a view to determining the best type of fluorescent tube or other light source in hospitals, skilled observers made judgments of the relative colour rendering properties of a number of light sources, comparing them always in pairs illuminating the same scene. The tasks included clinical diagnosis, examination of pathological and bacteriological specimens, and laboratory tests involving colour reactions. Only two of the light sources were within the normal range of daylight chromaticities. The blend of incandescent filament light with that from fluorescent tubes with blue emission (Chapter 18) received the highest score, and next in order of preference were fluorescent tubes of CCT between 3800 and 4200 K. The rating became higher with nearer approximation to full radiator distribution. On the evidence the area selected as most suitable, and later embodied in a specification for lamps to be used in hospitals, was defined by the chromaticity and spectral distribution in wide bands of a full radiator at 4000 K, with tolerances in the nominal values assessed by the NPL colour rendering index (Chapter 17). Fluorescent tubes have since been made with the object of supplying this type of lamp. Since its colour is in a region where daylight rarely occurs, the experiments may be regarded as irrelevant to the artificial daylight question. As suggested before, they may show the need for a general purpose illuminant of good colour rendering properties which appears to be superior to daylight though similar to it in general spectral distribution.

THE DETECTION OF HYPOXAEMIA

Of all the tests in the MRC experiments, probably the most critical is that for hypoxaemia, which appears as cyanosis of the lips and other parts in subjects suffering from lack of oxygen in the arterial blood. It is important to diagnose it in anaesthetized patients, when it can be measured by objective tests but is usually and more rapidly detected by skilled observation. These judgments are by memory of the typical appearance of the condition, and evidently much depends on the lighting under which the experience has been gained. If a variety of light sources including daylight is normally used, there is much to be said for a constant reproducible lighting, such as the specification mentioned above would ensure.

Some tests made by Kelman and Nunn[678] with a small selection of fluorescent tubes in the moderate to good colour rendering class showed, however, that the question is not simply of a progression through a series of light sources from poor to good colour rendering of the effect. The lighting was capable of making the subjects look either more or less cyanosed than their real condition warranted, that is, false positive or false negative diagnosis respectively. The highest false positive score from the observers was due to an artificial daylight tube, giving the only illumination near to a daylight spectral distribution in this investigation. The highest false negative score arose in the use of one of the most successful tubes in the MRC study, with a CCT near 4000 K. One tube graded fairly low in the MRC tests, of a CCT near 3700 K, gave the least distorted results by an equal and small percentage of both positive and negative errors. A light source with a special spectral distribution might be very valuable in this medical application. As its spectrum would probably not approximate to a daylight distribution, it might not be suitable for general lighting though an improvement on daylight in this particular case.

TESTING BY LIGHT

Finally we consider the exposure to the destructive effects of daylight of materials which have been coloured. The main visible effect is fading, with a wide range of sensitivity depending on the chemical structure of the dye, its spectral absorptance, the spectrum of the light applied to it, the temperature and humidity. Another significant factor is the substrate for the dye. Besides woven textiles of all kinds this includes paper, plastics, leather, metal containers, and foodstuffs. Some include resistant pigments instead of dyestuffs, as in paints, and here physical weathering effects become as important as colour change. The dyed fabric has received most attention in the past. Tests for fastness, now with scales available in standardized samples, have long been an essential part of the dyer's craft. The subject has been surveyed by McLaren.[679, 680]

Exposure to natural daylight or sunlight was naturally the first method of testing, a method becoming impracticable as dyes improved in fastness and the time required for recognizable or reproducible colour change became too long, to say nothing of the variation in the light source. Integration of received energy by pyrheliometer was possible, but spectral variations remained undetected.

Artificial sources have been used in fading tests for a long time, improving on daylight by their high and constant power output but often distorting the daylight spectrum so much that correlation was difficult. In dyed textiles the fugitive dyes often respond to longer wavelengths in the visible spectrum while fast dyes require ultra-violet radiation to affect them. The widely-used testing instrument, the Fade-Ometer, employed a carbon arc, with short ultra-violet radiation removed by glass windows. A large preponderance of ultra-violet and blue emission remained.[594] Other sources have replaced the arc for testing purposes, but the ultra-violet content has always been important and the subject of comparative measurements.[320] Mercury arcs have been used with some success, especially the tungsten-ballasted mercury lamp with germanate phosphor on the outer envelope,[681] though the specialized spectrum may not always allow correlation with testing under other sources. When the xenon arc came into use its high power possibilities and acceptable imitation of daylight removed many earlier difficulties in testing. The arc is now used in the Fade-Ometer, where its spectrum can be controlled to some extent by simple filters. Natural daylight is no longer essential except as a control when new materials need to be tested. The difference in spectra between extraterrestrial sunlight and the xenon arc or other substitute source may cause errors in the testing of materials to be exposed to sunlight outside the atmosphere.[643, 682] (See Chapter 18.)

CONSERVATION

When pigments are mixed with oils and the resulting paints applied to canvas, with varnishes on top, the pictures are particularly sensitive to light by reason of the organic constituents. In this context the sensitivity is judged in relation to an expected life which may be very long. These considerations apply to many kinds of irreplaceable museum objects, and if conservation is the primary duty of those who have to display them, the abundance of lamps for illuminating them at high levels adds to the problem. Illuminance and time of exposure may be fixed at values thought to be reasonable, but if the lighting is by fluorescent tubes, the danger is largely in the ultra-violet emission. This may exceed the amount in urban daylight after penetrating museum windows. It is now regarded as good practice to reduce or remove this part of the lamp spectrum by absorbent coatings on the glass envelope, or by plastic sleeves on the tubes, or by the prior design of the tube spectrum. Ultra-violet emission may be reduced by the double coating technique (Chapter 18), or by some non-fluorescent internal coat-ing.[675] Thomson has discussed illuminance values and the degree of protection possible for some pigments and other materials. In some instances the removal of ultra-violet radiation did not reduce the fading.[683] Improving on daylight is not completely successful so far in this application. Significant advances have been made in the investigation of colour changes in paintings, and in museum design for low illumination.[714]

These small triumphs in replacing and improving on daylight, for purposes either constructive or destructive, give some encouragement for the future. Lamp techniques are still advancing and new types will be found, their spectra controlled and shaped to the requirements of general lighting. Sooner or later artificial daylight will be achieved, undistorted and reproducible, more like the real thing than any present attempt. Artificial daylight varying at will to imitate natural changes may be another prospect. Meanwhile the search for new filters and new optical techniques must proceed as well as the search for new types of lamp, in order to emphasize the valuable features of daylight and remove its unwanted characteristics in any given applica-tion. With the known principles of light production and manipulation for guidance, there should be no lack of progress if the demand for it is sufficiently strong.

UNITS AND DEFINITIONS

SI units are used in most cases. Some older units are so familiar, or so much used, that they have been retained, for example, the torr, the calorie/cm² or langley, the rayleigh, the electron-volt.

$$
\begin{aligned}
\text{1 standard atmosphere} &= 1{\cdot}013 \times 10^6 \ \text{dyn/cm}^2 \\
&= 1{\cdot}013 \times 10^5 \ \text{Pa} \\
\text{1 torr} \qquad &= 1{\cdot}333 \times 10^2 \ \text{Pa} \\
\text{1 langley (L)/min} \quad &= 1 \ \text{mean gram calorie/cm}^2 \ \text{min} \\
&= 698{\cdot}3 \ \text{W/m}^2 \\
S(E_o, E_{eo}) \qquad &= 1{\cdot}35 \pm 0{\cdot}02 \ \text{kW/m}^2 \\
&= 1{\cdot}94 \pm 0{\cdot}03 \ \text{L/min} \\
\text{1 eV} \qquad &= 1{\cdot}602 \times 10^{-19} \ \text{J} \\
\text{1 rayleigh (R)} \quad &= 10^6 \ \text{photons/(cm}^2 \ \text{column) s.}
\end{aligned}
$$

Luminance in rayleighs per μm for monochromatic light at λ μm may be found from the relation

$$
\text{R}/\mu\text{m} \qquad\qquad = 1{\cdot}986 \ \pi \ \lambda \ L_\lambda \times 10^9
$$

where L_λ is the luminance in W/m² μm sr.[684]

RADIATION

A full or Planckian radiator is one which completely absorbs all incident radiation, and has the maximum spectral concentration of radiant exitance for any wavelength, at a given temperature.

Radiant exitance is the quotient of the radiant flux leaving a surface element by the area of the element.

Irradiance is the corresponding term for flux incident on a surface: see Chapter 6.

Laws relating to the full radiator.
Stefan–Boltzmann: Radiant exitance

$$
M_e = \sigma T^4
$$

where

$$\sigma = 2\pi^5 k^4 / 15 h^3 c^2$$
$$= 5\cdot67 \times 10^{-8}\ \text{W/m}^2\,\text{K}^4.$$

Wien, displacement on a linear scale of wavelength

$$\lambda_{\max} T = 2\cdot898 \times 10^{-3}\ \text{m K}.$$

Wien, radiation

$$M_{e,\lambda}(\lambda,\ T) = \frac{c_1}{\lambda^5 \exp{(c_2/\lambda T)}} \text{W/m}^3$$

where $M_{e,\lambda}(\lambda,\ T)$ is the spectral concentration of unpolarized radiance exitance, per unit wavelength (1 m).

$$c_1 = 2\pi h c^2$$
$$= 3\cdot741 \times 10^{-16}\ \text{W m}^2$$
$$c_2 = hc/k$$
$$= 1\cdot4388 \times 10^{-2}\ \text{m K}.$$

Planck, radiation

$$M_{e,\lambda}(\lambda,\ T) = \frac{c_1}{\lambda^5[\exp{(c_2/\lambda T)}-1]}\ \text{W/m}^3.$$

Standards based on the former value of $c_2 = 1\cdot438 \times 10^{-2}$ m K are discussed in earlier chapters.

LIGHT

Luminous intensity I: A full radiator at the solidification temperature of platinum (2045 K on the 1968 IPT scale) has an intensity of 60 candelas (cd) per cm².

Luminous flux F: A point source of 1 cd emits a flux of 4π lumens (lm), or 1 lm/steradian (sr). Within some years the lumen, defined with relation to the watt at a specified wavelength of radiation, is likely to replace the candela as the primary standard of light in SI units.[715]

Illuminance:　　1 lux (lx) $= 1$ lm/m²
　　　　　　　　　　　 $= 0\cdot0929$ lm/ft².
Luminance:　　1 stilb (sb) $= 10^4$ cd/m²
　　　　　　　　　　　 $= \pi$ lambert (L)
　　　　　　　　　　　 $= 2919$ ft-lambert.

These conversions are commonly used though it is more correct to specify the luminance of a perfectly reflecting diffuser as 10^{-4} lambert, or 1 apostilb (asb) when it receives an illuminance of 1 lux. If the diffuser has a reflectance ρ, the luminance becomes $10^{-4}\,\rho$L or ρ asb.

Conversion of radiant power to visible light: Luminous flux (in lm) is derived from the spectral power distribution by the relation

$$F = K_m \int_{\lambda_1}^{\lambda_2} S_\lambda\, V(\lambda)\, d\lambda$$

where S_λ is the radiant power in watts in the wavelength interval $d\lambda$ at wavelength λ, $V(\lambda)$ is the photopic spectral luminous efficiency of radiation at wavelength λ, λ_1 and λ_2 are the agreed wavelength limits of the visible spectrum, and K_m is the spectral luminous efficacy at the maximum of the $V(\lambda)$ curve, namely 680 lm/W at 555 nm.

In the scotopic (dark adapted) range of vision the efficacy K'_m is 1725 lm/W at the 507 nm maximum. The visible spectrum limits were set at 400 to 760 nm for the $V(\lambda)$ function defined in 1924, extended to 380 to 780 nm in 1931, and to 360 to 830 nm in 1971.[605]

Luminous efficiency of radiation is the ratio of the radiant flux weighted by $V(\lambda)$, to the flux itself, or

$$\frac{\int S_\lambda\, V(\lambda)\, d\lambda}{\int S_\lambda\, d\lambda}.$$

Luminous efficacy is the quotient of luminous flux by radiant flux, or in the case of a light source, of luminous flux to power input, in lm/W.

Luminance factor is the ratio of the luminance of a surface to that of a perfect diffuser under the same illumination. It refers to mono-chromatic light if prefixed 'spectral'. The corresponding term for a subjective estimate of the amount of light diffusely reflected from a surface is *lightness*. A scale of lightness (called *Value*) from black (0) to white (10) is provided by the Munsell colour atlas.[691]

Brightness is sometimes used to mean lightness, but is now defined as a dyer's term including lightness and saturation.

Reflectance is the ratio of reflected flux to incident flux.

For a comprehensive account of the measurement of light, see Walsh,[685] and for standards in photometry, see Jones and Preston.[686] Other useful articles deal with principles of radiometry,[687] and units and computations.[688]

In ultra-violet therapy, irradiance is called *dose-rate* (in W/m²). Radiant exposure, or surface density of received radiant energy, is called *dose* (in J/m²). Total energy received is called *integral dose* (in J). The Finsen is a unit of dose-rate equal to 0·1 W/m² at 297 nm, and to 1 E(rythemal)-viton/cm² or 0·1 er/m².[674]

The ultra-violet spectrum is conventionally divided into regions named

$$\begin{array}{ll} \text{UV-A} & \text{100 to 280 nm} \\ \text{UV-B} & \text{280 to 315 nm} \\ \text{UV-C} & \text{315 to 400 nm.} \end{array}$$

A similar division of the infra-red spectrum is

$$\begin{array}{ll} \text{IR-A} & \text{780 nm to 1·4 } \mu\text{m} \\ \text{IR-B} & \text{1·4 } \mu\text{m to 3 } \mu\text{m} \\ \text{IR-C} & > \text{3 } \mu\text{m.} \end{array}$$

COLOUR

The *CIE 1931 standard observer* and colorimetric coordinate system are fully described by Wright,[689] and in the CIE publication.[605] The necessary data include spectral tristimulus values ($\bar{x}$, $\bar{y}$, $\bar{z}$), and chromaticity coordinates (x, y, z) for wavelengths at intervals throughout the visible spectrum. Chromaticities are normally expressed and plotted by x and y only since $x+y+z=1$. The locus of the spectral chromaticity coordinates forms an area within which the chromaticities of all physically realizable colours are located. A straight line joining chromaticity points is the locus of mixtures of the two stimuli.

The *CIE 1960 uniform chromaticity scale* (UCS) was adopted from MacAdam.[690] The chromaticity coordinates (u, v, w) which it provides are derived from the tristimulus values X, Y, Z or the chromaticity coordinates x, y, z of the 1931 system by the equations

$$u = 4X/(X+15Y+3Z) = 4x/(-2x+12y+3)$$
$$v = 6Y/(X+15Y+3Z) = 6y/(-2x+12y+3)$$
$$u+v+w = 1.$$

In UCS diagrams, of which many have been proposed, it is intended that equal steps of perceptual colour difference at the same luminance should be represented by equal distance on the chart. No plane diagram fulfils this condition except approximately over limited areas.

These systems apply to colorimetry at a 2° angle of view. For larger fields of view, nominally 10°, similar relations hold but different tables of spectral tristimulus values are required.

A modification of this UCS was proposed at the 1975 CIE meeting, with the aim of improving uniformity in respect of visual colour differences. The new coordinate $u'=u$ in the above equations, but $v'=1\cdot5v$. This alters the values of CCT and other quantities based on the geometry of the CIE 1960 UCS.

AN*Lab* is a nonlinear transformation of the 1931 CIE colour space in which the tristimulus values X, Y and Z which are related by fifth order polynomials in V (the Munsell Value)[691] to V_x, V_y and V_z referred to a standard illuminant (originally Source C). The lightness coordinate L and the chromatic values a and b are respectively functions of V_y, (V_x-V_y) and (V_y-V_z). If a and b are zero the colour is neutral white; positive values of a indicate red colours, and negative green; similarly positive b is yellow, negative blue.[692] This system is likely to be replaced by a CIE-sponsored modification (CIE $L*a*b*$) with simpler calculations using the cube roots of X/X_0, Y/Y_0 and Z/Z_0 for determining $L*$, $a*$ and $b*$ where X_0, Y_0 and Z_0 are the tristimulus values of the standard white.[693] This formula derives from an earlier colorimetric system devised by Glasser *et al.*[694] As in most similar systems a colour difference is given by

$$[(\varDelta L)^2+(\varDelta a)^2+(\varDelta b)^2]^{\frac{1}{2}}.$$

Numerical coefficients are included in AN*Lab* and CIE $L*a*b*$ to give suitable units of colour difference and a lightness value of 100 for a perfect reflecting diffuser. A similar colour space was proposed in 1975, CIE $L*u*v*$, derived from the revised UCS with $v'=1\cdot5v$.

MacAdam's geodesic chromaticity diagram is a nonlinear transformation of the 1931 CIE x,y coordinates intended to be uniform for small visually equal colour differences. Straight lines on this diagram show, not loci of mixtures of different colours of light, but the shortest distance by perceptible steps between the ends of the lines. It was based on visual colour matching data.[695]

For the calculation of chromaticity from spectral power distribution see ref. 605. For the use of matrix algebra in the calculations, see Allen,[696] and Halstead *et al.*[697]

Dominant wavelength is that of a monochromatic light stimulus which matches the colour stimulus considered when combined with the correct proportion of the specified achromatic light stimulus. It is found by joining the chromaticity points for the colour stimulus and the achromatic stimulus (or 'white point') and producing this line away from the white point till it cuts the spectrum locus at the dominant wavelength. The complementary wavelength, found by producing the line in the opposite direction, is used for pink or purple colours.

Purity is often confused with *saturation*, its subjective correlate which has been defined as a judgment either of the degree of difference of a visual colour sensation from an achromatic one, or of the proportion of pure chromatic colour contained in the sensation. The usual measure of purity is excitation purity p_e, the ratio of the distance between the chromaticity points of a colour stimulus (x, y) and the white point (x_w, y_w), to the distance between the latter and the chromaticity point of the dominant wavelength (x_d, y_d); that is,

$$p_e = \frac{x - x_w}{x_d - x_w} \qquad \text{or} \qquad \frac{y - y_w}{y_d - y_w}$$

which may be expressed as a percentage. The achromatic stimulus has zero purity, all spectral wavelengths 100 per cent.

Colorimetric purity is the ratio of luminances derived from excitation purity:

$$p_c = p_e \frac{y_d}{y}.$$

For use with the UCS the term metric purity has been proposed, p_m. It is defined like p_e with u and v terms replacing x and y terms.

These definitions are valid for the two-dimensional x, y or u, v diagrams where luminance is constant and light sources are concerned, but more sophisticated definitions are needed in three-dimensional colour space and for surface colours.

Chromaticity can be specified in terms of dominant wavelength, purity, and the achromatic point.

The *Planckian or full radiator locus* shows the chromaticities of full radiators ('black bodies') of temperatures up to infinity, where the chromaticity is approximately $x = 0.240$, $y = 0.234$.

Colour temperature is the temperature of a full radiator which emits radiation matching in chromaticity (or nearly so) that of the source considered.[698] The term is convenient, though improperly used, for classifying sources like fluorescent tubes whose spectra are very dissimilar from those of full radiators.

Distribution temperature (DT) is the temperature of a full radiator whose spectral power distribution curve has ordinates proportional (or nearly so) to those of the curve for the source considered. This applies in the visible region only. The two radiations have the same chromaticity. To facilitate calculation, Robertson has suggested altering the definition so that the full radiator ordinates are 'closest to being proportional' to those of the source considered. He also proposed the use of 'daylight distribution temperatures' (DDT) using the RD series instead of full radiator distributions as reference standards[699] (see below for some values).

Correlated colour temperature (CCT) of a source is the colour temperature of the point on the Planckian locus which is nearest to the chromaticity point for the source considered, on an agreed uniform chromaticity scale. Isotemperature lines are thus normal to the Planckian locus, but not so for non-uniform chromaticity scales such as the CIE 1931 x, y, z scale. Diagrams of lines to allow CCT to be determined from an x, y specification were published by Judd[700] for his own UCS;[701] these were superseded when the CIE 1960 UCS was adopted, and the appropriate charts published by Kelly.[332] There is no agreement on the permissible distance between Planckian locus and chromaticity point, but there are fluorescent tubes and discharge lamps which are so far away that a calculated or graphically selected CCT is meaningless because of the great difference in appearance, apart from spectral distribution, between radiation from a lamp and what the full radiator (though non-existent) would emit. The objection on grounds of spectral disparity, given under 'colour temperature', also applies.

Mired scale: Colour temperatures or correlated colour temperatures are expressed in mireds (micro reciprocal degrees) by dividing 10^6 by the temperature (K) in question.[241] A number of alternative polynomials are available giving the light source chromaticity u, v in terms of M, its mired value.[702] Tabulated values of M or T_c in terms of u, v would be more useful. With degrees now called kelvins, the mired needs renaming.

The *Equi-energy spectrum* has a constant concentration of energy throughout the visible spectrum, evaluated on a wavelength basis. This is taken as achromatic for colorimetric purposes, though any other illuminant may be specified as achromatic for particular experimental conditions. The equi-energy spectrum is sometimes named Source E: it is not a Standard Illuminant.

Standard Illuminants: The chromaticities are according to the CIE 1931 standard observer and the temperatures for $c_2 = 1·438 \times 10^{-2}$ m K in order to compare Robertson's values for DT and DDT. Revised values for the Standard Illuminants are given in Chapter 18. The fourth significant figure in values of x and y given in this book is mainly of instrumental or theoretical interest. When the eye is directly involved the third significant figure may usually be taken as the limit of sensitivity.

	x	y	CCT	DT	DDT
A	0·4476	0·4074	2854	2854	
B	0·3484	0·3516	4871	4648	4924
C	0·3101	0·3162	6770	6211	6445
D_{65}	0·3127	0·3290	6500	6201	6500
E	0·3333	0·3333	5464		

THE RD DISTRIBUTIONS

Characteristic vector analysis (Chapter 13): Simmonds[407] gives references describing the mathematical procedure. Briefly, it consists of computing a variance–covariance matrix $\mathbf{S}$ from the original $n \times r$ matrix. The characteristic vectors are obtained as the latent roots of the equation

$$| \mathbf{S} - \mathbf{LI} | = 0$$

where $\mathbf{I}$ is an $r \times r$ linear matrix and $\mathbf{L}$ a diagonal $r \times r$ matrix of the latent roots.

Daylight chromaticities: For the revised value of $c_2 = 1·4388 \times 10^{-2}$ m K, equation (1) in Chapter 13 is unaltered.

Equations (4) and (5) of Chapter 13 become

$$x_D = -4·6070 \frac{10^9}{T_c^3} + 2·9678 \frac{10^6}{T_c^2} + 0·09911 \frac{10^3}{T_c} + 0·244063$$

for 4000 K $\leqslant T_c \leqslant$ 7000 K and

$$x_D = -2\cdot0064\,\frac{10^9}{T_c{}^3}+1\cdot9018\,\frac{10^6}{T_c{}^2}+0\cdot24748\,\frac{10^3}{T_c}+0\cdot237040$$

for $7000\ \mathrm{K} \leqslant T_c \leqslant 25000\ \mathrm{K}$.

CHEMICAL ACTINOMETERS

The fuchsin actinometer uses a solution of the dye previously de-colorized by sulphite. Exposure to ultra-violet radiation restores the colour in proportion to the energy absorbed, which may be estimated by colour standards (Büttner, Chapter 8).

The methylene blue actinometer uses a solution of methylene blue in aqueous acetone. Here the colour is discharged according to the ultra-violet energy absorbed at wavelengths less than about 310 nm. Standard colour samples provide a scale of comparison.[703] (Götz, Chapter 8.)

The potassium ferrioxalate actinometer has a wider spectral range with high quantum efficiency at wavelengths below 500 nm. The acid aqueous solution of the salt produces ferrous ions under irradiation, and these are estimated by the colour reaction with phenanthrolin followed by measurement of optical density of the solution.[704] (Searle, Hirt *et al.*, Chapter 10.)

A more convenient type of actinometer uses a dilute aqueous solution of benzoic acid at pH 12. It responds to radiation at less than 300 nm by a blue fluorescence proportional to the dose, which is measured by a fluorometer.[705]

ULTRA-VIOLET METERS

A simple 'Finsen meter' to measure dose-rate was described by Ruff.[706] This uses a magnesium tungstate fluorescent screen to convert ultra-violet radiation to visible light for measurement by a selenium cell, with suitable filters and a compensating cell to eliminate the effect of UV-A on the measuring cell.

Robertson used a similar screen with a single vacuum photocell.[568]

For testing organic building materials Harris designed an instrument with a zinc cadmium sulphide screen and a selenium cell, to be calibrated by the ferrioxalate actinometer.[707]

A simpler method reported by Davis *et al.* is the use of a thin film of polyphenylene oxide which darkens on exposure to radiation of wavelengths less than 400 nm. Increase of optical density is related to the darkening produced by exposure to a standard fluorescent 'black lamp' of known output. This has been used in a world-wide programme at 24 stations where the global ultra-violet radiation has been continuously monitored.[708]

CHRONOLOGICAL INDEX OF SOME CONTRIBUTORS TO THE STUDY OF DAYLIGHT

Anaximenes	*c.* 544 B.C.	William Harkness	1837–1903
Anaxagoras	500– 428	Alfred Cornu	1841–1902
Aristotle	384– 322	Hermann C. Vogel	1841–1907
Aristarchus	310– 230	John W. Strutt, Lord	
Posidonius	130– 50	Rayleigh	1842–1919
Vitruvius Pollio	*c.* 46 B.C.	William de W. Abney	1843–1920
Ovid	43 B.C.–A.D. 18	Walter Hartley	1846–1913
Pliny the Elder	23– 79	Henry A. Rowland	1848–1901
Alhazen (Ibn-al-Haitham)	965–1039?	A. Henri Becquerel	1852–1908
Leonardo da Vinci	1451–1519	Edward L. Nichols	1854–1937
Galileo Galilei	1564–1642	Johannes Wilsing	1856–1943
Marin C. de la Chambre	1594–1669?	Knut J. Ångström	1857–1910
Johannes Marcus Marci	1595–1667	William H. Pickering	1858–1938
René Descartes	1596–1650	Max Planck	1858–1947
Francesco M. Grimaldi	1618–1663	Carl Dorno	1865–1942
Isaac Newton	1642–1727	Charles Fabry	1867–1945
Edmond Halley	1656–1742	Frederick E. Fowle	1869–1940
Pierre Bouguer	1698–1758	Albert A. König	1871–1946
Johann H. Lambert	1728–1777	Charles G. Abbot	1872–1974
F. William Herschel	1738–1822	Henri A. Buisson	1873–1944
William H. Wollaston	1766–1828	Theodore Lyman	1874–1954
Thomas Young	1773–1829	Karl W. F. Linke	1878–1944
Johann W. Ritter	1776–1836	Albert H. Taylor	1879–1961
Josef Fraunhofer	1787–1826	Arthur S. Eddington	1882–1944
John F. W. Herschel	1792–1871	Herbert E. Ives	1882–1953
John W. Draper	1811–1882	Loyal B. Aldrich	1884–1965
Anders J. Ångström	1814–1874	Irwin G. Priest	1886–1932
Léon Foucault	1819–1868	Matthew Luckiesh	1888–1967
George G. Stokes	1819–1903	Sydney Chapman	1888–1970
A. Edmond Becquerel	1820–1891	Gordon M. B. Dobson	1889–1976
John Tyndall	1820–1893	F. W. Paul Götz	1891–1954
Gustav R. Kirchhoff	1824–1887	John W. T. Walsh	1891–1962
Jules Janssen	1824–1907	Marcel G. Minnaert	1893–1970
William Huggins	1824–1910	Fritz H. W. Albrecht	1896–1965
Johann J. Balmer	1825–1898	Deane B. Judd	1900–1972
André P. P. Crova	1833–1907	Zdenek Sekera	1905–1973
Samuel P. Langley	1834–1906	Rudolf Schulze	1906–1974
Charles A. Young	1834–1908	Andrew J. Drummond	1917–1972
J. Norman Lockyer	1836–1920		

References

1. ARISTOTLE. *Meteorologia*, part 3. Trans. H. D. P. LEE (Heinemann, 1962).
2. GALILEI, G. *Il Saggiatore*, 1624. From *A Short History of Scientific Ideas*, C. SINGER, p. 246 (Clarendon Press, Oxford, 1959).
3. NEWTON, I. *Opticks*, Book 1, part 2 (London, 1704).
4. NEWTON, I. *Phil. Trans. R. Soc., Lond.*, 1672, **6**, no. 80, 3075–3087.
5. EDDINGTON, A. S. *The Nature of the Physical World*, p. 94 (Cambridge University Press, 1929).
6. SARTON, G. *Isis*, 1947, **37**, 69–71.
7. BOYER, C. B. *The Rainbow* (T. Yoseloff, London, 1959).
8. DA VINCI, L. *The Notebooks*, vol. 2, p. 299. Trans. E. MCCURDY (Jonathan Cape, 1938).
9. DESCARTES, R. *Les Météores*, Discours 8, p. 255 (Leyden, 1637).
10. MARCI, J. M. M. *Thaumantias*, p. 108 (Prague, 1648).
11. DE LA CHAMBRE, M. C. *Nouvelles Observations et Coniectures sur l'Iris*, pp. 266, 276 (Paris, 1650).
12. GRIMALDI, F. M. *Physico-mathesis de Lumine, Coloribus et Iride*, p. 257 (Bologna, 1665).
13. NEWTON, I. *Opticks*, Book 1, part 2, fig. 16.
14. WOLLASTON, W. H. *Phil. Trans. R. Soc., Lond.*, 1802, **92**, 365–380.
15. FRAUNHOFER, J. *Denkschr. Königl. Akad. Wiss. München*, 1814–1815, **5**, 193–226.
16. MINNAERT, M., MULDERS, G. F. W., and HOUTGAST, J. *Photometric Atlas of the Solar Spectrum*, $\lambda 3612$–$\lambda 8771$ (Amsterdam, 1940).
17. MOORE-SITTERLY, C. E. *The Solar Spectrum*, pp. 89–101. Ed. C. DE JAGER (Reidel, Dordrecht, 1965); *Mém. Soc. R. Liège*, 1976, **9**, 59–76.
18. MITCHELL, W. E., and MOHLER, O. C. *Astrophys. J.*, 1969, suppl. ser. 18, no. 165, 379–428.
19. LOCKYER, J. N. *Recent and Coming Eclipses*, 2nd edn, chap. 18. pp. 205–210 (Macmillan, 1900).
20. FERENCZ, C., and TARCSAI, G. *Acta Tech. Acad. Sci. Hung.*, 1972, **72**, 171–181.
21. HERSCHEL, J. F. W. *Phil. Trans. R. Soc., Lond.*, 1840, **130**, 1–59, articles 54–56.
22. DRAPER, J. W. *Phil. Mag.* 1881, **11** [5], 157–169.
23. TOUSEY, R., WATANABE, K., and PURCELL, J. D. *Phys. Rev.* 1951, **83**, 792–797.
24. KAZACHEVSKAYA, T. V., IVANOV-KHOLODNYI, G. S., MEDVEDEV, M. S., RAZUMOVA, T. K., and CHUDAIKIN, A. V. *Planet. Space Sci.*, 1964, **12**, 167–174.
25. ROBINSON, M. S., and RUMPEL, W. F. *Rev. Sci. Instrum.*, 1963, **34**, 794–797.
26. KARANDIKAR, R. V. *J. Opt. Soc. Am.*, 1955, **45**, 483–488.
27. DRUMMOND, A. J. *Arch. Meteorol. Geophys. Bioklimatol.*, 1958, **B9**, 149–163.
28. DUNKELMAN, L., and SCOLNIK, R. *J. Opt. Soc. Am.*, 1959, **49**, 356–367.
29. JUAN, J., PLAZA, L., and CRUZ, C. *Opt. Pura y Aplicada*, 1970, 3, 145–158.

30. DOVAN, P., and KROCHMANN, J. *Lichttechnik*, 1974, **26**, 245.
31. ALLEN, C. W. *Astrophysical Quantities*, 3rd edn (Athlone Press, 1973).
32. HERSCHEL, F. W. *Phil Trans. R. Soc., Lond.*, 1795, **85**, 46–72; 1801, **91**, 265–318.
33. HERSCHEL, J. F. W. *Outlines of Astronomy*, 10th edn, chap. 6 (Longmans, Green, 1869).
34. THOURET, W. E., STRAUSS, H. S., CORTORILLO, S. F., and KEE, H. *Illum. Eng.*, 1965, **60**, 339–346.
35. LIENHARD, O. E. *Illum. Eng.*, 1965, **60**, 348–352.
36. WILDT, R. *Astrophys. J.* ,1939, **90**, 611–620; 1941, **93**, 47–51.
37. UNSÖLD, A. *The New Cosmos*, pp. 162–178. Trans. by W. H. McCREA (Longmans, 1969).
38. HALLEY, E. *Phil. Trans. R. Soc., Lond.*, 1714–1716, **29**, 245–262.
39. PICKERING, W. H. *Proc. Am. Acad. Arts Sci.*, 1879–1880, **15**, 236–250.
40. ROSETTI, F. *Phil. Mag.*, 1879, **8** [5], 324–332, 438–449, 537–550.
41. GOODY, R. M. *Atmospheric Radiation*, part 1, pp. 417–426 (Clarendon Press, Oxford, 1964).
42. ALLEN, C. W. *Quart. J. R. Meteorol. Soc.*, 1958, **84**, 307–318.
43. HULBURT, E. O. *J. Opt. Soc. Am.*, 1947, **37**, 405–415.
44. GOLDBERG, L. *Ann. Rev. Astron. Astrophys.*, 1967, **5**, 278–324.
45. REBER, G. *Astrophys. J.*, 1944, **100**, 279–287.
46. HEY, J. S. *The Evolution of Radio-Astronomy* (Elek Science, 1973).
47. LINDSAY, J. C., NEUPERT, W. M., and STONE, R. G. *Space Science*, 1st edn, chap. 16 (*The Sun*). Ed. W. N. HESS (Blackie, 1965).
48. GIBSON, E. G. *Rev. Geophys. Space Phys.*, 1972, **10**, 395–461.
49. WILLSTROP, R. V. *Mem. R. Astron. Soc.*, 1965, **69**, 83–143.
50. LABS, D., and NECKEL, H. *Z. Astrophys.*, 1968, **69**, 1–73.
51. RATCLIFFE, J. A. *Sun, Earth and Radio* (Weidenfeld and Nicolson, 1970).
52. BEARD, D. B. *Rep. Prog. Phys.*, 1967, **30**, 404–444.
53. FABRY, C., and BUISSON, H. *J. Phys. Théor. Appl.*, 1913, **3** [5], 196–206.
54. CORNU, A. *J. Phys. Théor. Appl.*, 1881, **10**, 5–17.
55. FABRY, C., and BUISSON, H. *Astrophys. J.*, 1921, **54**, 297–322.
56. RICE, M. L., and THURONYI, G. *Meteor. Abstr.*, 1959, **10**, 759–794, 920–957.
57. CHAPMAN, S. *Mem. R. Met. Soc.*, 1930, **3**, no. 26, 103–125.
58. HOUZEAU, A. *C. R. Acad. Sci., Paris*, 1858, **46**, 89–91.
59. JOHNSON, G. R. A., and WARMAN, J. M. *Discuss. Faraday Soc.*, no. 37 (*Chemical Reactions in the Atmosphere*), 1964, 87–95.
60. HUNT, B. G. *J. Geophys. Res.*, 1966, **71**, 1385–1398.
61. DOBSON, G. M. B. *Appl. Opt.*, 1968, **7**, 387–405.
62. DÜTSCH, H. U. *Annals IQSY*, chap. 15, pp. 234–245 (MIT Press, 1968).
63. WARDLE, D. I., WALSHAW, C. D., and WORMELL, T. W. *Nature*, 1963, **199**, 1177–1178.
64. LONDON, J. *Beitr. Phys. Atmos.*, 1963, **36**, 254–263.
65. STICKSEL, P. R. *Monthly Weather Rev.*, 1970, **98**, 787
66. BERKNER, L. V., and MARSHALL, L. C. *Discuss. Faraday Soc.*, 1964, no. 37, 122–141.

67. RATNER, M. I. *J. Atmos. Sci.*, 1972, **29**, 803–808.

68. SCHIDLOWSKI, M., and YUNGE, C. *Phys. Bl.*, 1973, **29**, 203–212.

69. CADLE, R. D. *J. Colloid Interface Sci.*, 1972, **39**, 25–31.

70. KERR, J. A., CALVERT, J. G., and DEMERJIAN, K. L. *Chemistry in Britain*, 1972, **8**, 252–257.

71. LEVY, H. *Planet. Space Sci.*, 1972, **20**, 919–935.

72. DERWENT, R. G., and STEWART, H. N. M. *Nature*, 1972, **241**, 342–343.

73. CORNU, A. *C. R. Acad. Sci., Paris*, 1890, **111**, 941–947.

74. ABBOT, C. G., FOWLE, F. E., and ALDRICH, L. B. *Ann. Smithson. Astrophys. Obs.*, 1913, **3**, 12.

75. MIETHE, A., and LEHMANN, E. *Sitzungsber. Königl. Preuss. Akad. Wiss., Berlin*, 1909, **8**, 268–277.

76. HOELPER, O. *Gerlands Beitr. Geophys.*, 1929–1930, **24**, 26–29.

77. WIGAND, A. *Verh. Deutsch. Phys. Ges.*, 1913, **15**, 1090–1099.

78. BUISSON, H. *Gerlands Beitr. Geophys.*, 1929–1930, **24**, 30.

79. REGENER, E., and REGENER, V. H. *Physik. Z.*, 1934, **35**, 788–793.

80. CORNU, A. *C. R. Acad. Sci., Paris*, 1879, **88**, 1101–1108.

81. GÖTZ, F. W. P. *Gerlands Beitr. Geophys.*, 1929–1930, **24**, 31–33.

82. GÖTZ, F. W. P. *Strahlentherapie*, 1931, **40**, 690–695.

83. GÖTZ, F. W. P., and CASPARIS, P. *Z. Angew. Photogr.*, 1942, **4**, 65–67.

84. KONDRATIEV, K. Y. *Radiation in the Atmosphere* (Academic Press, 1969).

85. BENER, P. *Approximate values of intensity of natural ultraviolet radiation for different amounts of atmospheric ozone*, June 1972 (U.S. Army contract DAJA 37-68-C-1017).

86. NAZAROVA, L. G. *Radiotekh. Elektron.*, 1969, **14**, 1731–1743.

87. STAUFFER, F., and STRONG, J. *Appl. Opt.*, 1962, **1**, 129–130.

88. THEKAEKARA, M. P., *et al. The Solar Constant, etc.* (*Report X-322-68-304*) (Goddard Space Flight Center, 1968).

89. THEKAEKARA, M. P., KRUGER, R., and DUNCAN, C. H. *Appl. Opt.*, 1969, **8**, 1713–1732.

90. SINTON, W. M. *J. Opt. Soc. Am.*, 1955, **45**, 975–979.

91. CURCIO, J. A., *et al. NRL Report 4669*, 1955.

92. CURCIO, J. A., *et al. Appl. Opt.*, 1964, **3**, 1401–1409.

93. CURCIO, J. A., *et al. NRL Report 6352*, 1965.

94. FARMER, C. B. and TODD, S. J. *Appl. Opt.*, 1964, **3**, 453–458.

95. BRADFORD, W. R., FARMER, C. B., and TODD, S. J. *Appl. Opt.*, 1964, **3**, 459–465.

96. DELBOUILLE, L., and MIGEOTTE, M. *J. Opt. Soc. Am.*, 1960, **50**, 1305–1307.

97. DELBOUILLE, L., and ROLAND, G. *Photometric Atlas of the Solar Spectrum from 7498 to 12016 Å* (Liège, 1963).

98. SWENSON, J. W., BENEDICT, W. S., DELBOUILLE, L., and ROLAND, G. *The Solar Spectrum from λ7498 to λ12016* (Liège, 1970).

99. HOPKINSON, R. G., PETHERBRIDGE, P., and LONGMORE, J. *Daylighting*, chap. 2 (Heinemann, 1966).

100. GOODY, R. M. *Atmospheric Radiation*, part 1, chap. 5 (Clarendon Press, Oxford, 1964).

101. ALLEN, C. W. *Astrophysics Quantities*, 3rd edn, pp. 126–130 (Athlone Press, 1973).
102. BELL, E. E., EISNER, L., YOUNG, J., and OETJEN, R. A. *J. Opt. Soc. Am.*, 1960, **50**, 1313–1320.
103. JASTROW, R., and RASOOL, S. I. *Introduction to Space Science*, 2nd edn, chap. 18 (*Planetary Atmospheres*). Ed. W. N. HESS and G. D. MEAD (Blackie, 1968).
104. RASCHKE, E., and PASTERNAK, M. *Proc. 10th Meet. Committee on Space Research, London*, 1967, 1033–1043.
105. MÖLLER, F. *Appl. Opt.* ,1964, **3**, 157–166.
106. HUNTEN, D. M. *Appl. Opt.*, 1964, **3**, 167–174.
107. BARRY, R. G., and CHORLEY, R. J. *Atmosphere, Weather and Climate*, 2nd edn (Methuen, 1971).
108. STRUTT, J. W. (Lord Rayleigh). *Phil. Mag.*, 1871, **41** [4], 107–120, 274–279.
109. BOUGUER, P. *Traité d'Optique sur la Gradation de la Lumière*, Book 3, sect. 2, p. 248 (Paris, 1760).
110. BEMPORAD, A. *Meteorol. Z.*, 1907, **24**, 306–313.
111. KASTEN, F. *Technical Report AD610554* (Cold Regions Res. Eng. Lab. Hanover, N.H., 1964).
112. CROVA, A. *C. R. Acad. Sci., Paris*, 1891, **112**, 1176–1179.
113. CHANDRASEKHAR, S. *Radiative Transfer* (Clarendon Press, Oxford, 1950).
114. CHANDRASEKHAR, S., and ELBERT, D. D. *Phil. Trans. Am. Phil. Soc.*, 1954, **44**, 643–729.
115. COULSON, K. L., DAVE, J. V., and SEKERA, Z. *Tables related to Radiation emerging from a Planetary Atmosphere with Rayleigh Scattering* (Univ. California Press, 1960).
116. SCHÜEPP, W. *Arch. Meteorol. Geophys. Bioklimatol*, 1949, **B1**, 257–346.
117. LINKE, F. *Handb. Geophys.*, 2nd edn, **8**, chap. 6, p. 252. Ed. B. GUTENBERG (Borntraeger, 1942).
118. PENNDORF, R. *J. Opt. Soc. Am.*, 1957, **47**, 176–182.
119. FOWLE, F. E. *Astrophys. J.*, 1914, **40**, 435–442.
120. DUCLAUX, J., and GINDRE, R. *Bull. Obs. Lyon*, 1929, **11**, 5–13, 69–84, 183–204; 1930, **12**, 255–271.
121. DUCLAUX, J. *Bull. Obs. Lyon*, 1931, **13**, 239–254.
122. TOUSEY, R., and HULBURT, E. O. *J. Opt. Soc. Am.*, 1947, **37**, 78–92.
123. KONDRATIEV, K. Y., NICOLSKY, G. A., BADINOV, I. Y., and ANDREEV, S. D. *Appl. Opt.*, 1967, **6**, 197–207.
124. COULSON, K. L. *J. Quantum Spectrosc. Radiat. Transfer*, 1971, **11**, 739–755.
125. FOGLE, B. *J. Geophys. Res.*, 1972, **77**, 720–725.
126. MIE, G. *Ann. Phys., Lpz.*, 1908, **25** [4], 377–445.
127. ÅNGSTRÖM, A. *Geogr. Annlr*, 1929, **11**, 156–166.
128. ÅNGSTRÖM, A. *Geogr. Annlr*, 1930, **12**, 130–159.
129. BULLRICH, K., DE BARY, E., and MÖLLER, F. *Geofis. Pura Appl.*, 1952, **23**, 69–110.
130. GÖTZ, F. W. P. *Gerlands Beitr. Geophys.*, 1931, **31**, 119–154.
131. LINKE, F. *Handb. Geophys.*, 2nd edn, **8**, chap. 6, p. 266. Ed. B. GUTENBERG (Borntraeger, 1942).

132. DOGNIAUX, R., and DOYEN, P. *Publ. Inst. R. Météorol. Belg.*, 1968, Sér. A, no. 65, 1–43.

133. ELTERMAN, L. *Environ. Res. Papers*, no. 46 (Air Force Cambridge Res. Labs., 1964).

134. FOITZIK, L. *Gerlands Beitr. Geophys.*, 1965, **74**, 199–206.

135. FOITZIK, L., HEBERMEHL, G., and SPÄNKUCH, D. *Optik*, 1965–1966, **23**, 268–278.

136. SHAH, G. M. *Tellus*, 1970, **22**, 82–93.

137. FISCHER, K. *Beitr. Phys. Atmos.*, 1970, **43**, 244–254.

138. HEGER, K. *Beitr. Phys. Atmos.*, 1971, **44**, 201–212.

139. DE BARY, E. *Contrib. Atmos. Phys.*, 1973, **46**, 213–221.

140. GOLDBERG, B., and KLEIN, W. H. *Solar Energy*, 1971, **13**, 311–321.

141. TYNDALL, J. *Fragments of Science*, 6th edn, **1**, chaps 4 and 5. (Longmans, Green, 1899).

142. GEHRELS, T. *J. Opt. Soc. Am.*, 1962, **52**, 1164–1173.

143. FOITZIK, L., and LENZ, K. *Monatsber. Dtsch. Akad. Wiss., Berlin*, 1960, **2**, 682–684.

144. FOITZIK, L., and LENZ, K. *Optik und Spectroskopie aller Wellenlängen* (Jena Meeting 1960), pp. 615–622, ed. P. GÖRLICH (Akademie Verlag, Berlin, 1962).

145. BENER, P. *Solar intensity and intensity and polarization of sky radiation for 347·0, 488·0 and 533·5 nm . . . at 2818m a.s.l.*, August 1970 (U.S. Army contract DAJA 37-68-C-1017).

146. NOWAK, W. *Optik*, 1970, **32**, 22–49.

147. VAN DE HULST, H. C. *Light Scattering by Small Particles*, chaps 19 and 20 (Chapman and Hall, 1957).

148. SELIGER, H. H., and MCELROY, W. D. *Light: Physical and Biological Action*, chap. 5 (Academic Press, 1965).

149. POUILLET, C. S. M. *C. R. Acad. Sci., Paris*, 1838, **7**, 24–65.

150. HERSCHEL, J. F. W. *Results of astronomical observations made in 1834–8 at the Cape of Good Hope etc.*, Appendix C, pp. 444–446 (Smith, Elder & Co., London, 1847).

151. MORIKÖFER, W. *Das Physik.-Meteorol. Observatorium Davos. Schweiz. Meteorol. Zentralanstalt* (Zürich, 1964).

152. ALDRICH, L. B., and HOOVER, W. H. *Ann. Smithson. Astrophys. Obs.*, 1954, **7**, 99–105, 138–143.

153. *Ann. Int. Geophys. Yr.*, 1958, **5**, part 6, 367–466.

154. ABBOT, C. G., ALDRICH, L. B., and FOWLE, F. E. *Ann. Smithson. Astrophys. Obs.*, 1932, **5**, 16–18, 28–29, 36–37, 120–121, 247–259.

155. *Recommendations for the integrated irradiance and the spectral distribution of simulated solar radiation for testing purposes* (*CIE Publ. 20*) (Paris, 1972).

156. JOHNSON, F. S. *J. Meteorol.*, 1954, **11**, 431–439.

157. NICOLET, M. *Ann. d'Astrophys.*, 1951, **14**, 249–265.

158. JOHNSON, F. S., PURCELL, J. D., TOUSEY, R., and WILSON, N. *Rocket Exploration of the Upper Atmosphere*, pp. 279–288 (Pergamon Press, 1954).

159. THEKAEKARA, M. P. *Solar Energy*, 1965, **9**, 7–20.

160. THEKAEKARA, M. P. *NASA Report SP 74*, 1965, 1–43.

161. STAIR, R., and ELLIS, H. T. *J. Appl. Meteorol.*, 1968, **7**, 635–644.

162. DRUMMOND, A. J., HICKEY, J. R., SCHOLES, W. J., and LAUE, E. G. *Nature*, 1968, **218**, 259–261.

163. LAUE, E. G., and DRUMMOND, A. J. *Science*, 1968, **161**, 888–891.

164. MURCRAY, D. G., KYLE, T. G., KOSTERS, J. J., and GAST, P. R. *Tellus*, 1969, **21**, 620–624.

165. KONDRATIEV, K. Y., and NIKOLSKY, G. A. *Quart. J. R. Meteorol. Soc.*, 1970, **96**, 509–522.

166. THEKAEKARA, M. P., and DRUMMOND, A. J. *Nature (Phys. Sci.)*, 1971, **229**, 6–9.

167. THEKAEKARA, M. P. *Solar Energy*, 1973, **14**, 109–127.

168. WEBB, J. J., DUNCAN, C. H., McINTOSH, R., and LESTER, D. *Appl. Opt.*, 1970, **9**, 345–349.

169. ARVESEN, J. C., GRIFFIN, R. N., and PEARSON, B. D. *Appl. Opt.*, 1969, **8**, 2215–2232.

170. NECKEL, H., and LABS, D. *Int. Astron. Union Symp. Sept. 1972*, pp. 149–152 (D. Reidel, 1973).

171. *Solar Electromagnetic Radiation*, May 1971 (*NASA Report SP-8005*).

172. LABS, D., and NECKEL, H. *Z. Astrophys.*, 1967, **65**, 133–155.

173. LABS, D., and NECKEL, H. *Solar Phys.*, 1970, **15**, 79–87.

174. LABS, D., and NECKEL, H. *Solar Phys.*, 1971, **19**, 3–15.

175. SMITH, E. V. P., and GOTTLIEB, D. M. *Space Sci. Rev.*, 1974, **16**, 771–802.

176. MAKAROVA, E. A. and KHARITONOV, A. V. *Sov. Astron.*, 1969, **12**, 599–609.

177. ABBOT, C. G. *Smithson. Misc. Coll.*, 1955, **128**, no. 4 (Publ. 4213), 1–20.

178. ALBRECHT, R., MAITZEN, H. M., and RAKOS, K. D. *Astron. Astrophys.*, 1969, **3**, 236–242.

179. BALASUBRAHMANYAN, V. K., and VENKATESAN, D. *Solar Phys.*, 1971, **19**, 257–263.

180. BRAY, R. J., and LOUGHHEAD, R. E. *Sunspots* p. 237 (Chapman and Hall, 1964).

181. ÅNGSTRÖM, A. *Astrophys. J.*, 1922, **55**, 24–29.

182. ABBOT, C. G. *Smithson. Contrib. Astrophys.*, 1958, **3**, 13–21.

183. BOSSOLASCO, M., CICCIONI, G., DAGNINO, I., ELENA, A., and FLOCCHINI, G. *Pure Appl. Geophys.*, (*a*) 1964/I, **57**, 221–224; (*b*) 1965/III, **62**, 207–214.

184. ÅNGSTRÖM, A. (K). *Appl. Opt.*, 1974, **13**, 474–486, 1477–1480.

185. VITINSKY, Y. I. *Izv. Gl. Astron. Obs. Pulkove*, 1971, no. 186, 20–25.

186. COHEN, T., and LINTZ, P. *Nature*, 1974, **250**, 389–399.

187. DILKE, F., and GOUGH, D. *New Scientist*, 1972, **56**, 562–564.

188. DRUMMOND, A. J., THEKAEKARA, M. P., eds, *The Extraterrestrial Solar Spectrum*. (Inst. Environ. Sci., Mount Prospect, Ill., 1973).

189. THEKAEKARA, M. P. *NASA Report X-912-75-155*, June 1975.

190. COLLINGBOURNE, R. H. *Meteorol. Mag.*, 1969, **98**, 223–230.

191. MacDOWELL, J., and TRIBBLE, D. T. *R. Soc. Int. Geophys. Yr. Exped. Halley Bay*, 1962, **3**, 111–160.

192. COULSON, K. L. *Solar and Terrestrial Radiation* (Academic Press, 1975).

193. SHEKLEIN, A. V. *Appl. Solar Energy*, 1966, **2**, no. 1, 34–38.

194. EJDER, E. *J. Opt. Soc. Am.*, 1969, **59**, 223–224.

195. NICHOLS, E. L. *Phys. Rev.*, 1905, **21**, 147–165.
196. VOGEL, H. C. *Monatsber. Königl. Preuss. Akad. Wiss. Berlin*, 1880, 801–811.
197. VOGEL, H. C. *Monatsber. Königl. Preuss. Akad. Wiss. Berlin*, 1877, 104–142.
198. CROVA, A. *C. R. Acad. Sci., Paris*, 1881, **93**, 512–513.
199. CROVA, A., and LAGARDE. *C. R. Acad. Sci., Paris*, 1881, **93**, 959–961.
200. PITCHER, F. B. *Am. J. Sci.*, 1888, **36** [3], 332–336.
201. NICHOLS, E. L., and FRANKLIN, W. S. *Am. J. Sci.*, 1889, **38** [3], 100–114.
202. KÖTTGEN, E. *Wied. Ann. Phys. Chem.*, 1894, **53**, 793–811.
203. KÖNIG, A. *Wied. Ann. Phys. Chem.*, 1894, **53**, 785–792.
204. LANGLEY, S. P. *Phil. Mag.*, 1883, **15** [5], 153–183.
205. ABBOT, C. G., FOWLE, F. E., and ALDRICH, L. B. *Ann. Smithson. Astrophys. Obs.*, 1913, **3**, 18.
206. ABBOT, C. G., FOWLE, F. E., and ALDRICH, L. B. *Smithson. Misc. Coll.*, 1923, **74**, no. 7, 1–30 (*Publ. 2714*).
207. HARDY, A. C. *MIT Handbook of Colorimetry*, 1936, p. 22, Table 6.
208. ABNEY, W. de. W. *Br. Assoc. Rep.*, 1883, no. 53. 422–425.
209. ABNEY, W. de W. *Proc. R. Soc., Lond.*, 1893, **54**, 2–4.
210. ZETTWUCH, G. *Phil. Mag.*, 1902, **4** [6], 199–202.
211. MILLOCHAU, G. *J. Phys. Théor. Appl.*, 1909, **8**, 347–360.
212. IVES, H. E. *Trans. Illum. Eng. Soc., N. Y.*, 1910, 189–208.
213. NICHOLS, E. L. *Phys. Rev.*, 1908, **26**, 497–511.
214. NICHOLS, E. L. *Phys. Rev.*, 1909, **28**, 122–131.
215. IVES, F. E. *J. Franklin Inst.*, 1907, **164**, 421–423.
216. MÜLLER, G., and KRON, E. *Publ. Astrophys. Obs., Potsdam*, 1912, **22**, no. 64, 1–92.
217. WILSING, J. *Publ. Astrophys. Obs., Potsdam*, 1913, **22**, no. 66, 1–80.
218. WILSING, J. *Publ. Astrophys. Obs., Potsdam*, 1917, **23**, no. 72, 1–101.
219. PLASKETT, H. H. *Publ. Dom. Astrophys. Obs., Victoria*, 1923, **2**, no. 12, 213–260.
220. ÅNGSTRÖM, K. (J). *Phys. Rev.*, 1903, **17**, 302–314.
221. JONES, L. A. *Trans. Illum. Eng. Soc., N. Y.*, 1914, **9**, 716–733.
222. JONES, L. A. *Trans. Illum. Eng. Soc., N. Y.*, 1914, **9**, 687–709.
223. WAYNE, R. P. *Photochemistry*, chap. 8 (Butterworth, 1970).
224. ABBOT, C. G., FOWLE, F. E., and ALDRICH, L. B. *Ann. Smithson. Astrophys. Obs.*, 1913, **3**, 35.
225. ROBERTS, W. O. *Smithson. Contrib. Astrophys.*, 1956, **1**, 99–101.
226. PARANJPE, M. M. *Quart. J. R. Meteorol. Soc.*, 1938, **64**, 459–474.
227. HOELPER, O. *Z. Geophys.*, 1927, **3**, 184–195.
228. DORNO, C., MEISSNER, K. W., and VAHLE, W. *Meteorol. Z.*, 1924, **41**, 234–239, 269–277.
229. DORNO, C., and LINDHOLM, F. *Meteorol. Z.*, 1929, **46**, 281–292.
230. KALITIN, N. N. *Gerlands Beitr. Geophys.*, 1927, **18**, 383–397.
231. AURÉN, T. E. *Ark. Mat. Astron. Fys.*, 1933, **24a**, no. 4, 1–55.
232. KALITIN, N. N. *Strahlentherapie*, 1931, **39**, 717–728.
233. PETTIT, E. *Astrophys. J.*, 1932, **75**, 185–221.
234. PETTIT, E. *Astrophys. J.*, 1940, **91**, 159–185.

235. SCHLÖMER, W. *Veröff. Meteorol. Inst.*, *Berlin*, 1938, **3**, Heft 2, 1–30.
236. BÜTTNER, K. *Gerlands Beitr. Geophys.*, 1939, **55**, 13–51.
237. ORTH, R. *Gerlands Beitr. Geophys.*, 1939, **55**, 52–102.
238. BABCOCK, H. D., MOORE, C. E., and COFFEEN, M. F. *Astrophys. J.*, 1948, **107**, 287–302.
239. *Compte Rendu CIE* (Cambridge, 1931), pp. 615–646 (*Report on ultra-violet illumination*).
240. PRIEST, I. G. *J. Opt. Soc. Am.*, 1923, **7**, 1175–1209.
241. PRIEST, I. G. *J. Opt. Soc. Am.*, 1933, **23**, 41–45.
242. CUNLIFFE, P. W. *J. Text. Inst.*, 1929, **20**, T34–46.
243. DORNO, C. *Strahlentherapie*, 1929, **31**, 330–348.
244. GREIDER, C. E., and DOWNES, A. C. *Trans. Illum. Eng. Soc.*, *N.Y.*, 1930, **25**, 378–396.
245. GREIDER, C. E., and DOWNES, A. C. *Trans. Illum. Eng. Soc.*, *N.Y.*, 1931, **26**, 561–571.
246. ORNSTEIN, L. S., EYMERS, J. G., VERMEULEN, D., and POSTMA, G. W. *Verh. Konink. Nederl. Akad. Wet. Naturk.*, Sectie 1, 1936, **16**, 1–79.
247. HERZING, F. *Gerlands Beitr. Geophys.*, 1937, **49**, 71–96.
248. HESS, P. *Gerlands Beitr. Geophys.*, 1939, **55**, 204–220.
249. REINER, H. *Gerlands Beitr. Geophys.*, 1939, **55**, 234–248.
250. BULLRICH, K. *Meteorol. Z.*, 1942, **59**, 256–262.
251. REESINCK, J. M. M. *Physica*, 1944, **11**, 61–77.
252. REESINCK, J. M. M. *Physica*, 1946, **12**, 296–300.
253. HERRMANN, R. *Optik*, 1947, **2**, 384–395.
254. NAGEL, M. *Optik*, 1950, **6**, 349–375; **7**, 1–12.
255. TAYLOR, A. H. *Trans. Illum. Eng. Soc.*, *N.Y.*, 1930, **25**, 154–160.
256. TAYLOR, A. H., and KERR, G. P. *J. Opt. Soc. Am.*, 1941, **31**, 3–8.
257. TAYLOR, A. H. *Illum. Eng.*, 1949, **44**, 201–203.
258. KIMBALL, H. H. *Monthly Weather Rev.*, 1925, **53**, 112–115.
259. BERLAGE, H. P. *Meteorol. Z.*, 1928, **45**, 174–180.
260. ALBRECHT, F. *Meteorol. Z.*, 1935, **52**, 454–460.
261. ELVEGARD, E. and SJÖSTEDT, G. *Illum. Eng.*, 1941, **36**, 767–785.
262. KASTROW, W. *Meteorol. Z.*, 1928, **45**, 377–381.
263. GÖTZ, F. W. P., and SCHÖNMANN, E. *Helv. Phys. Acta*, 1948, **21**, 151–168.
264. MOON, P. *J. Franklin Inst.*, 1940, **230**, 583–617.
265. GIBSON, K. S. *J. Opt. Soc. Am.*, 1940, **30**, 88.
266. NICKERSON, D. *Illum. Eng.*, 1958, **53**, 77.
267. MOON, P., and SPENCER, D. E. *J. Opt. Soc. Am.*, 1945, **35**, 399–427.
268. MOON, P., and SPENCER, D. E. *J. Appl. Phys.*, 1946, **17**, 506–514.
269. MOON, P., and SPENCER, D. E. *J. Franklin Inst.*, 1947, **244**, 441–464.
270. BLEVIN, W. R., and BROWN, W. J. *J. Sci. Instrum.*, 1964, **42**, 19–23.
271. STERNE, T. E., and DIETER, N. *Smithson. Contrib. Astrophys.*, 1958, **3**, 9–12.
272. JONES, B. Z. *Lighthouse of the Skies* (Smithsonian Inst., 1965).
273. DRUMMOND, A. J., and ÅNGSTRÖM, A. K. *Solar Energy*, 1967, **11**, 133–141.

274. STAIR, R. *J. Res. N.B.S.*, 1951, **46**, 353–357.
275. STAIR, R. *J. Res. N.B.S.*, 1952, **49**, 227–234.
276. STAIR, R., and JOHNSTON, R. *J. Res. N.B.S.*, 1953, **51**, 81–84.
277. STAIR, R., JOHNSTON, R. G. and BAGG, T. C. *J. Res. N.B.S.*, 1954, **53**, 113–119.
278. STAIR, R. *J. Opt. Soc. Am.*, 1953, **43**, 971–974.
279. STAIR, R., and JOHNSTON, R. G. *J. Res. N.B.S.*, 1956, **57**, 205–211.
280. SCHULZE, R. *Compte Rendu CIE* (Stockholm, 1951), **2**, paper Nn.
281. HISDAL, V. *Arch. Meteorol. Geophys. Bioklimatol.*, 1961, **B10**, 59–68.
282. HULL, J. N. *Trans. Illum. Eng. Soc., London*, 1954, **19**, 21–28.
283. COOPER, E. R., and PROBINE, M. C. *Trans. Illum. Eng. Soc., London*, 1954, **19**, 8–20.
284. PEYTURAUX, R. *Ann. d'Astrophys.*, 1952, **15**, 302–351.
285. HISDAL, B. *Arch. Math. Naturvidensk.*, 1959, **54**, no. 3, 35–44.
286. MACADAM, D. L. *J. Opt. Soc. Am.*, 1958, **48**, 832–840.
287. LENZ, K. *Monatsber. Dtsch. Akad. Wiss. Berlin*, 1960, **2**, 720–721.
288. HARDING, H. G. W. *J. Sci. Instrum.*, 1952, **29**, 145–148.
289. HARDING, H. G. W., and LAMBERT, G. E. V. *Nature*, 1951, **167**, 436–437.
290. ALBRECHT, F. *Arch. Meteorol. Geophys. Bioklimatol.*, 1951, **B3**, 220–243.
291. MIDDLETON, W. E. K. *J. Opt. Soc. Am.*, 1954, **44**, 793–798.
292. VIGROUX, E. *Ann. Phys., Paris*, 1953, **8**, 709–762.
293. *Compte Rendu CIE* (Zürich, 1955) p. 10 (*E-1.3.1. Report*).
294. LENOBLE, J. *C. R. Acad. Sci., Paris*, 1957, **244**, 647–650.
295. LENOBLE, J. *Ann. Géophys.*, 1954, **10**, 117–147.
296. LENOBLE, J. *Rev. Opt.*, 1957, **36**, 343–356.
297. NICOLET, M. *Arch. Meteorol. Geophys. Bioklimatol.*, 1951, **B3**, 209–219.
298. HINZPETER, H. *Z. Meteorol.*, 1955, **9**, 308–315.
299. DEIRMENDJIAN, D., and SEKERA, Z. *Tellus*, 1954, **6**, 382–398.
300. HINZPETER, H. *Z. Meteorol.*, 1956, **10**, 100–110; 1957, **11**, 1–10.
301. LILJEQUIST, G. H. *Norw. Brit. Swed. Antarct. Exped.*, *1949–1952*, **2**, part 1 (Norsk Polarinstitutt, Oslo, 1956).
302. DAVE, J. V., and SEKERA, Z. *J. Meteorol.*, 1959, **16**, 211–212.
303. DAVE, J. V. *J. Appl. Meteorol.*, 1975, **14**, 388–395.
304. BENER, P. *Arch. Meteorol. Geophys. Bioklimatol.*, 1963, **B12**, 442–457.
305. BENER, P. *Investigation on the spectral intensity of ultraviolet sky and sun + sky radiation . . . at 1590m above sea level*, December 1960 (AFCRL-626, Contract AF 61(052)-54).
306. BENER, P. *Comparison of measured and theoretical values of the spectral intensity of ultraviolet sky radiation*, September 1962 (Contract AF 61(052)-618).
307. BENER, P. *The diurnal and annual variations of the spectral intensity of ultraviolet sky and global radiation . . . at Davos, 1590m a.s.l.*, January 1963 (Contract AF 61(052)-618).
308. BENER, P. *The biologic effects of ultraviolet radiation.* Ed. F. URBACH, pp. 351–358 (Pergamon Press, 1969).
309. BENER, P. *Investigation on the influence of clouds on ultraviolet sky radiation*, March 1964 (Contract AF 61(052)-618).

310. BENER, P. *A new spectrometer for measuring ultraviolet sky brightness and direct solar radiation*, April 1967 (Contract AF 61(052)-618).

311. BENER, P. *Measured and theoretical values of the spectral intensity of ultraviolet . . . at 316, 1580, and 2818m a.s.l.*, July, 1970 (Contract F 61052-67-C-0029).

312. DAVE, J. V., and FURUKAWA, P. M. *Meteorol. Monographs*, 1966, **7**, no. 29, (Am. Meteorol. Soc., Boston, Mass.).

313. BENER, P., FRÖHLICH, C., and VALKO, P. *Congress on the Sun in the Service of Mankind*, paper E31 (Paris, 1973).

314. BENER, P. *Investigation on the spectral intensity of ultraviolet . . . at 1590m a.s.l.*, May 1962 (Contract AF 61(052)-54).

315. ANDREYCHIN, R. *Dtsch. Akad. Wiss. Berlin, 2nd Int. UV Colloq.*, 1963, pp. 5–12.

316. KECHLIBAROV, T., and ANDREYCHIN, R. *Dtsch. Akad. Wiss. Berlin, 2nd Int. UV Colloq.*, 1963, pp. 23–28.

317. LENOBLE, J. *Ann. Géophys.*, 1954, **10**, 187–225.

318. LENOBLE, J. *Rev. Opt.*, 1959, **38**, 282–289.

319. LENOBLE, J., ARCEDUC, J. P. C., and COQUELLE, J. M. *Rev. Opt.*, 1963, **42**, 291–297.

320. HIRT, R. C., SCHMITT, R. G., SEARLE, N. Z., and SULLIVAN, A. P. *J. Opt. Soc. Am.*, 1960, **50**, 706–713.

321. SEARLE, N. Z., and HIRT, R. C. *J. Opt. Soc. Am.*, 1965, **55**, 1413–1421.

322. GREENE, A. E. S., SAWADA, T., and SHETTLE, E. P. *Photochem. Photobiol.*, 1974, **19**, 201–259; GREEN, A. E. S., MO, T., and MILLER, J. H. *Photochem. Photobiol.*, 1974, **20**, 473–482; MO, T., and GREEN, A. E. S. *Photochem. Photobiol.*, 1974, **20**, 483–496.

323. TOUSEY, R. *J. Opt. Soc. Am.*, 1961, **51**, 384–395.

324. MASSEY, H. S. W. *Space Physics* (Cambridge University Press, 1964).

325. TOUSEY, R. *The Middle Ultraviolet*, chap. 1, pp. 1–39. Ed. A. E. S. GREEN (Wiley, 1966).

326. DETWILER, C. R., GARRETT, D. L. PURCELL, J. D., and TOUSEY, R. *Ann. Géophys.*, 1961, **17**, 263–271.

327. HINTEREGGER, H. E. *The Extraterrestrial Solar Spectrum*, chap. 2 (Inst. Environ. Sci., Mount Prospect, Ill., 1973).

328. HEATH, D. F. *The Extraterrestrial Solar Spectrum*, chap. 3 (Inst. Environ. Sci., Mount Prospect, Ill., 1973).

329. BREWER, A. W., and WILSON, A. W. *Quart. J. R. Meteorol. Soc.*, 1965, **91**, 452–461.

330. HENDERSON, S. T., and HODGKISS, D. *Br. J. Appl. Phys.*, 1963, **14**, 125–131; 1964, **15**, 947–952.

331. BODMANN, H. W., and JANTZEN, R. *Lichttechnik*, 1964, **16**, 20–27.

332. KELLY, K. L. *J. Opt. Soc. Am.*, 1963, **53**, 999–1002.

333. CONDIT, H. R., and GRUM, F. *J. Opt. Soc. Am.*, 1964, **54**, 937–944.

334. NAYATANI, Y., and WYSZECKI, G. *J. Opt. Soc. Am.*, 1963, **53**, 626–629.

335. JUDD, D. B., MACADAM, D. L., and WYSZECKI, G. *J. Opt. Soc. Am.*, 1964, **54**, 1031–1040.

336. WINCH, G. T., BOSHOFF, M. C., KOK, C. J., and DU TOIT, A. G. *J. Opt. Soc. Am.*, 1966, **56**, 456–464.

337. STAIR, R., SCHNEIDER, W. E., and JACKSON, J. K. *Appl. Opt.*, 1963, **2**, 1151–1154.
338. BOSHOFF, M. C., and KOK, C. J. *Plastics, Paint and Rubber*, March–April 1969, 76–77.
339. KOK, C. J. *J. Phys. D: Appl. Phys.*, 1972, **5**, 1513–1520.
340. KOK, C. J. *J. Phys. D: Appl. Phys.* 1972, **5**, L85–88 and p. 2300.
341. DAS, S. R., and SASTRI, V. D. P. *J. Opt. Soc. Am.*, 1965, **55**, 319–323.
342. SASTRI, V. D. P., and DAS, S. R. *J. Opt. Soc. Am.*, 1966, **56**, 829–830.
343. SASTRI, V. D. P., and DAS, S. R. *J. Opt. Soc. Am.*, 1968, **58**, 391–398.
344. SASTRI, V. D. P., and MANAMOHANAN, S. B. *J. Phys. D: Appl. Phys.*, 1971, **4**, 381–386.
345. SASTRI, V. D. P. *J. Phys. D: Appl. Phys.*, 1976, **9**, L1–3.
346. NAYATANI, Y., HITANI, M., and MINATO, H. *Bull. Electrotech. Lab.*, *Japan*, 1967, **31**, 1127–1135.
347. NIMEROFF, I., and YUROW, J. A. *J. Opt. Soc. Am.*, 1965, **55**, 185–190.
348. HITANI, M., KURIOKA, Y., and MINATO, H. *Bull. Electrotech. Lab.*, 1973, **37**, 733–743.
349. ANDO, I., IKEMORI, T., and SEKINE, S. *CIE* (Washington, 1967), preprint and private communication.
350. *A.S.T.M. Specification D 1729–60T*, 1960.
351. KAWAKAMI, G., ANDO, I., and IKEMORI, T. *J. Illum. Eng. Inst., Japan*, 1965, **49**, 521–526.
352. TARRANT, A. W. S. *Die Farbe, Tagungsband Luzern*, 1965, 689–696.
353. TARRANT, A. W. S. *Thesis, University of Surrey: Some work on the spectral power distribution of daylight* (Nov. 1967).
354. TARRANT, A. W. S. *Trans. Illum. Eng. Soc., London*, 1968, **33**, 75–82.
355. TARRANT, A. W. S. *Compte Rendu CIE* (Barcelona, 1971), **21A**, 219–222.
356. TARRANT, A. W. S., and BROCK, J. R. *Compte Rendu CIE* (London, 1975), 384–392.
357. KNESTRICK, G. L., and CURCIO, J. A. *NRL Report 6615*, 1967.
358. KNESTRICK, G. L., and CURCIO, J. A. *Appl. Opt.*, 1970, **9**, 1574–1576.
359. HISDAL, V. *Årbok* 1967, pp. 7–27 (Norsk Polarinstitutt, Oslo, 1969).
360. GAST, P. R. Solar electromagnetic radiation. *Handbook of Geophysics and Space Environments*. Ed. S. L. VALLEY (McGraw Hill, 1965).
361. BUDDE, W. *Appl. Opt.*, 1964, **3**, 939–941.
362. HISDAL, V. Private communication, to be published.
363. KONDRATIEV, K. Y., ANDREEV, S. D., BADINOV, I. Y., GRISHECHKIN, V. S., and POPOVA, L. V. *Appl. Opt.*, 1965, **4**, 1069–1076.
364. TOOLIN, R. B. Atmospheric Optics. *Handbook of Geophysics and Space Environments*. Ed. S. L. VALLEY (McGraw Hill, 1965).
365. THEKAEKARA, M. P. *Optical Spectra*, March 1972, **6**, 32–35.
366. THEKAEKARA, M. P. *Appl. Opt.*, 1974, **13**, 518–522.
367. BRANDHORST, H., HICKEY, J., CURTIS, H., and RALPH, E. *NASA Report TM X-71771* (Lewis Research Center, 1975).
368. SITNIK, G. F. *Sov. Astron.*, 1965, **9**, 44–49.

369. PIERCE, A. K. *Astrophys. J.*, 1954, **119**, 312–317.
370. SITNIK, G. F. *Sov. Astron.*, 1965, **9**, 768–774.
371. PEYTURAUX, R. *C. R. Acad. Sci., Paris*, 1961, **252**, 668–669.
372. PEYTURAUX, R. *Ann. d'Astrophys.*, 1968, **31**, 227–235.
373. LABS, H. *Z. Astrophys.*, 1957, **44**, 37–55.
374. LABS, D., and NECKEL, H. *Z. Astrophys.*, 1962, **55**, 269–289.
375. HOLWEGER, H. *Z. Astrophys.*, 1967, **65**, 365–417.
376. GINGERICH, O., and DE JAGER, C. *Solar Phys.*, 1968, **3**, 5–25.
377. HOUTGAST, J. *Proc. K. Ned. Acad. Wet.*, 1965, **68B**, 306–308.
378. HOUTGAST, J. *Solar Phys.*, 1968, **3**, 47–54; 1970, **15**, 273–287.
379. HOLWEGER, H. *Astron. Astrophys.*, 1970, **4**, 11–17.
380. GINGERICH, O., NOYES, R. W., KALKOFEN, W., and CUNY, Y. *Solar Phys.*, 1971, **18**, 347–365.
381. LABS, D., and NECKEL, H. *Solar Phys.*, 1972, **22**, 64–69.
382. ELTERMAN, L. *Appl. Opt.*, 1975, **14**, 1262–1263.
383. GUTTMANN, A. *Appl. Opt.*, 1968, **7**, 2377–2381.
384. PACKER, D. M., and LOCK, C. *J. Opt. Soc. Am.*, 1951, **41**, 473–478.
385. CHAMBERLIN, G. J., LAWRENCE, A., and BELBIN, A. A. *Light and Lighting*, 1963, **56**, 70–72.
386. COLLINS, J. F. *Br. J. Appl. Phys.*, 1965, **16**, 527–532.
387. ANDO, I., IKEMORI, T., and KAWAKAMI, G. *J. Illum. Eng. Soc., Japan*, 1962, **46**, 218–224.
388. OKADA, Y., ANDOW, I., and SEKINE, S. *JCIE Symposium Color Sci.*, paper 73.13 (Tokyo, 1973).
389. LOPUKHIN, E. A. *Appl. Solar Energy*, 1965, **1**, no. 1, 36–40.
390. ELNESR, M. K., and HEGAZY, N. A. *Pure Appl. Geophys.*, 1964, **59**, 267–271.
391. GATES, D. M. *Science*, 1966, **151**, no. 3710, 523–529.
392. DOGNIAUX, R. *Ciel et Terre*, 1966, **81**, 1–16.
393. NICOLET, M., and DOGNIAUX, R. *Mem. Inst. R. Météorol. Belg.*, 1951, 47.
394. DOGNIAUX, R. *Variations qualitatives et quantitatives des composants du rayonnement solaire etc.* (*Publ. 62*). (Inst. R. Météorol. Belg, 1970).
395. SCHULZE, R. *Meteorol. Rundsch.*, 1967, **20**, 149–151.
396. SCHULZE, R. *Meteorol. Rundsch.*, 1970, **23**, 56–58.
397. *Compte Rendu CIE* (Washington, 1967), p. 233.
398. KROCHMANN, J., MÜLLER, K., and RETZOW, U. *Lichttechnik*, 1970, **22**, 551–554.
399. *Standardization of luminance distribution on clear skies* (*CIE publ. 22*) (Paris, 1973).
400. *Compte Rendu CIE* (Zürich, 1955), **2**, 3-2-A/35.
401. KITTLER, R. *Proc. CIE Intersessional Conf., Newcastle on Tyne, 1965* (Bouwcentrum International, Rotterdam, 1967).
402. JUAN, J., and CRUZ, C. *Opt. Pura y Aplicada*, 1973, **6**, 151–164.
403. KROCHMANN, J., and SEIDL, M. *Light. Res. Technol.*, 1974, **6**, 165–171.
404. SIVKOV, S. I. *Computation of solar radiation characteristics, 1968* (Israel Program for Scientific Translation, 1971).
405. DRUMMOND, A. J., and ÅNGSTRÖM, A. K. *Appl. Opt.*, 1971, **10**, 2024–2030.

406. *Compte Rendu CIE* (Vienna, 1963), p. 109 (*E-1.3.1 Report*).
407. SIMMONDS, J. L. *J. Opt. Soc. Am.*, 1963, **53**, 968–974.
408. CONDIT, H. R. *Appl. Opt.*, 1972, **11**, 74–86.
409. *Compte Rendu CIE* (Washington, 1967), pp. 95–97 (*E-1.3.1 Report*).
410. NIMEROFF, I. private communication.
411. NAYATANI, Y., and TAKAHAMA, K. *J. Opt. Soc. Am.*, 1972, **62**, 140–143.
412. WILSON, N. L. TOUSEY, R. PURCELL, J. D., JOHNSON, F. S., and MOORE, C. E. *Astrophys. J.*, 1954, **119**, 590–612.
413. BENER, P. *The Biologic Effects of Ultraviolet Radiation*, Ed. F. URBACH pp. 409–415 (Pergamon Press, 1969).
414. JACQUINOT, P. *Appl. Opt.*, 1969, **8**, 497–499.
415. GEBBIE, H. A. *Appl. Opt.*, 1969, **8**, 501–504.
416. SCHINDLER, R. A. *Appl. Opt.*, 1970, **9**, 301–306.
417. MACADAM, D. L. *J. Opt. Soc. Am.*, 1963, **53**, 397–398.
418. GRAINGER, J. F., and RING, J. *Nature*, 1962, **193**, 762.
419. GRAINGER, J. F., and RING, J. *Space Res.*, 1963, **3**, 989–996.
420. BRINKMAN, R. T. *Astrophys. J.*, 1968, **154**, 1087–1093.
421. HARRISON, A. W. *Can. J. Phys.*, 1974, **52**, 2030–2036.
422. DE VOS, J. C. *Physica*, 1954, **20**, 690–714.
423. LARRABEE, R. D. *J. Opt. Soc. Am.*, 1959, **49**, 619–625.
424. PRESTON, J. S. *Br. J. Appl. Phys.*, 1963, **14**, 43–45.
425. SCHURER, K. *Thesis, University of Utrecht: The tungsten strip lamp and the anode of the carbon arc as radiometric standards* (Rotterdam, May 1969).
426. JONES, O. C. *J. Phys. D: Appl. Phys.*, 1970, **3**, 1967–1976.
427. SCHNEIDER, W. E. *Appl. Opt.*, 1970, **9**, 1410–1418.
428. SUTTER, E. *Optik*, 1973, **38**, 73–79.
429. VAN DEN HOEK, W. J., and BERNS, E. G. *Light. Res. Technol.*, 1975, **7**, 143–146.
430. WRIGHT, W. D. *The Measurement of Colour*, 4th edn, (Adam Hilger, 1969).
431. HALLEY, E. *Phil. Trans. R. Soc., Lond.*, 1714–1716, **29**, 245–262.
432. SHARP, W. E., LLOYD, J. W. F., and SILVERMAN, S. M. *Appl. Opt.*, 1966, **5**, 787–792.
433. SHARP, W. E., SILVERMAN, S. M., and LLOYD, J. W. F. *Appl. Opt.*, 1971, **10**, 1207–1210.
434. DEEHR, C. S., and REES, M. H. *Planet. Space Sci.*, 1964, **12**, 875–888.
435. DANDEKAR, B. S. *Appl. Opt.*, 1968, **7**, 705–710.
436. MOORE, J. G., and LAMB, D. *Ann. Géophys.*, 1967, **23**, 339–344.
437. VELASQUEZ, D. A. *Appl. Opt.*, 1971, **10**, 1211–1214.
438. LLOYD, J. W. F., and SILVERMAN, S. M. *Appl. Opt.*, 1971, **10**, 1215–1219.
439. DANDEKAR, B. S., and TURTLE, J. P. *Appl. Opt.*, 1971, **10**, 1220–1224.
440. SHAW, G. E. *Appl. Opt.*, 1975, **14**, 388–394.
441. HALL, W. N. *Appl. Opt.*, 1971, **10**, 1225–1231.
442. SASTRI, V. D. P. *Planet. Space Sci.*, 1968, **16**, 647–651.
443. JAGANNATHAN, P., CHACKO, O., and VENKITESHWARAN, S. P. *Indian J. Meteorol. Geophys.*, 1957, 893.

444. LANGLEY, S. P., and VERY, F. W. *Mem. Nat. Acad.*, 1889, **4**, part 2, 107.
445. ADEL, A. *Astrophys. J.*, 1946, **103**, 19–24.
446. HAPKE, B. *Physics and Astronomy of the Moon*, 2nd edn, chap. 5, Ed. Z. KOPAL (Academic Press, 1971).
447. PICKERING, W. H. *Pop. Astron.*, 1919, **27**, 579–583; 1921, **29**, 404–423; 1922, **30**, 257–262; 1924, **32**, 69–78, 302–312, 393–404.
448 ROOSEN, R. G. *Earth Extraterr. Sci.*, 1970, **1**, 151–154.
449. MINNAERT, M. *Light and Colour in the Open Air* (Bell, 1959).
450. O'CONNELL, D. J. K. *Endeavour*, 1961, **20**, 131–137.
451. TAYLOR, J. H., and MATTHIAS, B. T. *Nature*, 1969, **222**, 157.
452. WHYMPER, E. *Travels amongst the Great Andes of the Equator*, chap. 18, p. 324 (John Murray, 1892).
453. ANON. *Nature*, 1884, **30**, 155–156.
    BISHOP, S. E. *Nature*, 1884, **30**, 537; 1885, **31**, 288–289.
    NEUMAYER. *Meteorol. Z.*, 1884, **1**, 1–4, 49–65, 156–163, 181–198, 277–282, 311–319.
454. BARNETT, E. W. *Solar Energy*, 1971, **13**, 323–337.
455. WEICKMANN, H. K., and PUESCHEL, R. F. *Beitr. Phys. Atmos.*, 1973, **46**, 112–118.
456. KARYAGINA, Z. V. *Sov. Astron.*, 1961, **4**, 828–832.
457. PARIISKII, N. N., and GINDILIS, L. M. *Sov. Astron.*, 1960, **3**, 992–1003.
458. GINDILIS, L. M. *Sov. Astron.*, 1962, **6**, 67–76.
459. JACOBSEN, T. S. *J. R. Astron. Soc., Canada*, 1952, **46**, 93–102.
460. WILSON, R. *Mon. Not. R. Astron. Soc.*, 1951, **111**, 478–489.
461. BOYD, R. L. F. *Space Research by Rocket and Satellite* (Arrow Books, 1960).
462. HÖHN, D. H., and BÜCHTEMANN, W. *Appl. Opt.*, 1973, **12**, 52–61.
463. SCHIMPF, R., and ASCHENBRENNER, C. *Z. Angew. Photogr.*, 1940, **2**, 41–45.
464. WENZEL, A. *Optik*, 1953, **10**, 330–338.
465. MACADAM, D. L. *Proc. I.R.E.*, 1954, **42**, 166–174.
466. MASAKI, H. *Science of Light*, 1959, **8**, 67–86.
467. KONDRATIEV, K. Y. *Naturwiss.*, 1971, **58**, 529–541.
468. STAMM, G. L., and LANGEL, R. A. *J. Opt. Soc. Am.*, 1961, **51**, 1090–1094.
469. NEWTON, I. *Opticks*, Book 1, part 2, p. 139.
470. LENOBLE, J. *C. R. Acad. Sci., Paris*, 1954, **239**, 1831–1833; 1955, **241**, 1407–1409; 1956, **243**, 668–672, 1781–1783.
471. LE GRAND, Y., LENOBLE, J., and SAINT-GUILY, B. *Ann. Géophys.*, 1954, **10**, 59–63.
472. CHALONGE, D., and SERVIGNE, M. *Ann. d'Astrophys.*, 1952, **15**, 151–153.
473. LENOBLE, J. *Ann. d'Astrophys.*, 1956, **12**, 16–31.
474. LENOBLE, J. *C. R. Acad. Sci., Paris*, 1956, **242**, 662–664.
475. SHULEIKIN, V. V. *Bull. Acad. Sci. USSR, Geophys. Ser. no. 10*, 1962, 875–881.
476. SULLIVAN, S. A. *J. Opt. Soc. Am.*, 1963, **53**, 962–968.
477. DUNTLEY, S. Q. *J. Opt. Soc. Am.*, 1963, **53**, 214–233.
478. JERLOV, N. G. *Optical Oceanography* (Elsevier, 1968).
479. KAMPA, E. M. *J. Marine Biol. Assoc. UK.*, 1970, **50**, 397–420.
480. TYLER, J. E. *Nature*, 1964, **202**, 1262–1264.
481. TYLER, J. E. *J. Opt. Soc. Am.*, 1965, **55**, 800–805.

482. TYLER, J. E., and SMITH, R. C. *J. Opt. Soc. Am.*, 1966, **56**, 1390–1396.

483. TYLER, J. E. *Nature*, 1965, **208**, 549–550.

484. SMITH, R. C., and TYLER, J. E. *J. Opt. Soc. Am.*, 1967, **57**, 589–595.

485. KULLENBERG, G. *Deep Sea Res.*, 1968, **15**, 423–432.

486. RASCHKE, E. *Beitr. Phys. Atmos.*, 1972, **45**, 1–19.

487. McCLUNEY, W. R. *Appl. Opt.*, 1974, **13**, 2422–2429.

488. KINNEY, J. A. S., LURIA, S. M., and WEITZMAN, D. O. *J. Opt. Soc. Am.*, 1967, **57**, 802–809.

489. GELLER, M., ALTMAN, D. E., and BARSTOW, G. J. *Appl. Opt.*, 1972, **11**, 1439–1441.

490. LARSON, D. A., RIXTON, F. H., and UNGLERT, M. C. *Illum. Eng.*, 1970, **65**, 644–648.

491. LYTHGOE, J. N., and NORTHMORE, D. P. M. *Colour 73* (*AIC Congress, York, 1973*), pp. 77–98 (Adam Hilger, 1973).

492. MIDDLETON, W. E. K. *J. Opt. Soc. Am.*, 1960, **50**, 97–100.

493. VOLZ, F. *Umschau*, 1956, **19**, 593–596.

494. HAKIM MOHAMMED SAID, ed., *Ibn al-Haitham* (*Alhazen*), *Proc. 1000th Anniversary Celebrations, 1969* (Hamdard Academy, Pakistan, 1971).

495. DA VINCI, L. *The Notebooks*, vol. 1, pp. 418–420.

496. NEWTON, I. *Opticks*, Book 2, part 3, p. 60.

497. BOUGUER, P. *Traité d'Optique sur la Gradation de la Lumière*, Book 3, pp. 240–241. Trans. W. E. K. MIDDLETON (University of Toronto Press, 1961).

498. HERSCHEL, J. F. W. *Meteorology*, para. 233 (Edinburgh, 1861).

499. KOCINSKI, J. *Acta Phys. Pol.*, 1969, **36**, 633–658.

500. STRUTT, J. W. (Lord Rayleigh). *Phil. Mag.* 1871, **41** [4], 447–454.

501. NICHOLS, E. L. *Proc. Am. Phys. Soc.*, 1908, **26**, 497–511.

502. NICHOLS, E. L. *Phil. Mag.*, 1879, **8** [5], 425–433.

503. BOCK, A. *Ann. d. Phys. u. Chem.*, 1899, **68**, 674–687.

504. CROVA, A. *Ann. Chim. Phys.*, 1890, **20** [6], 480–504.

505. BROWN, T. G., and DE BEER, G. *The First Ascent of Mont Blanc*, p. 40, 44 (Oxford, 1957).

506. FORBES, J. D. *Travels through the Alps*, chap. 12 (A. & C. Black, 1900).

507. HENDLEY, C. D., and HECHT, S. *J. Opt. Soc. Am.*, 1949, **39**, 870–873.

508. LINKE, F. *Handb. Geophys.*, 2nd edn, **8**, chap. 7, pp. 389–396. Ed. B. GUTENBERG (Borntraeger, 1942).

509. RÖSCH, S. *Appl. Opt.*, 1968, **7**, 233–239.

510. BILLET, F. *Ann. Sci. Ecole Norm. Supér.*, 1868, **5**, 67.

511. GREENLER, R. G. *Science*, 1971, **173**, 1231–1232.

512. TOUSEY, R. *J. Opt. Soc. Am.*, 1953, **43**, 113–118.

513. LE GRAND, Y., and LENOBLE, J. *C. R. Acad. Sci.*, Paris, 1954, **238**, 2435–2437.

514. GADSDEN, M. *J. Atmos. Terr. Phys.*, 1957, **10**, 176–180.

515. SIEDENTOPF, H., and HOLL, H. *Reichsber. Phys.*, 1944, **1**, 32.

516. ROZENBERG, G. V. *Twilight*, pp. 23–46 (Plenum Press, New York, 1966).

517. ASHBURN, E. V. *J. Opt. Soc. Am.*, 1953, **43**, 805–806.

518. KONDRATIEV, K. Y., VOLYNOV, B. V., GAITSEV, A. P., SMOKTY, O. I., and KHRUNOV, E. V. *Appl. Opt.*, 1971, **10**, 2521–2533.

519. ELTERMAN, L. *Environ. Res. Papers*, no. 285 (Air Force Cambridge Res. Labs, 1968).

520. KROCHMANN, J., ÖZVER, Z., and STOCKMAR, A. *Lichttechnik*, 1974, **26**, 318–319, 347–349.

521. AYRTON, W. E. *J. Inst. Electr. Eng.*, 1892, **21**, 404.

522. PRIEST, I. G. *J. Opt. Soc. Am.*, 1921, **5**, 205–209.

523. PRIEST, I. G. *J. Opt. Soc. Am.*, 1926, **13**, 306.

524. HURVICH, L. M., and JAMESON, D. *J. Opt. Soc. Am.*, 1951, **41**, 521–527, 528–536, 787–801.

525. SPROSON, W. N. *BBC Quarterly*, 1953, **8**, 176–192.

526. HONJYO, K., and NONAKA, M. *J. Opt. Soc. Am.*, 1970, **60**, 1690–1694.

527. VALBERG, A. *Vision Res.*, 1971, **11**, 157–160.

528. HUNT, R. W. G., and WINTER, L. M. *J. Photogr. Sci.*, 1975, **23**, 112–115.

529. HOWETT, G. L. *J. Opt. Soc. Am.*, 1970, **60**, 951–958.

530. BRITISH STANDARDS INSTITUTION. *Colours of Light Signals*, BS 1376:1974.

531. *Colors of Light Signals. (CIE Publ. 2.2)* (Paris, 1975).

532. MACADAM, D. L. *J. Opt. Soc. Am.*, 1942, **32**, 247–274.

533. BARTLESON, C. J. *J. Soc. Motion Pict. Telev. Eng.*, 1968, **77**, 1–12.

534. TERSTIEGE, H. *Colour* 73 (*AIC Congress, York*, 1973), pp. 340–341 (Adam Hilger, 1973).

535. GRUM, F., and SALTZMANN, M. *Compte Rendu CIE* (London, 1975), pp. 91–98.

536. DONALDSON, R. *Br. J. Appl. Phys.*, 1954, **5**, 210–214.

537. EITLE, D., and GANZ, E. *Textilveredlung*, 1968, **3**, 389–392.

538. SIMON, F. T. *J. Color Appearance*, 1972, **1**, no. 4, 5–11.

539. ALLEN, E. *Appl. Opt.*, 1973, **12**, 289–293.

540. BERGER, A. *Description of samples used and their colorimetric measurement*, to be published in *Die Farbe*.

541. MORI, L. *Acta Chromatica*, 1969, **2**, 25–41.

542. THIELERT, R., and SCHLIEMANN, G. *J. Opt. Soc. Am.*, 1973, **63**, 1607–1612.

543. CIBA-GEIGY AG. *Ciba-Geigy Review*, 1973, no. 1, 12–13.

544. COATES, E., KING, M. G., and RIGG, B. *Colour* 73 (*AIC Congress, York*, 1973), pp. 447–449 (Adam Hilger, 1973).

545. JUDD, D. B. *J. Opt. Soc. Am.*, 1941, **31**, 462–463.

546. GANZ, E. *Appl. Opt.*, 1976, **15**, 2039–2058.

547. GANZ, E. *J. Color Appearance*, 1972, **1**, no. 5, 33–41.

548. MCCONNELL, D. J. *Proc. AIC Congress* (Stockholm, 1969), pp. 329–334; *Colour* 73 (*AIC Congress, York*, 1973), pp. 401–403 (Adam Hilger, 1973).

549. JUDD, D. B. *J. Opt. Soc. Am.*, 1968, **58**, 1638–1649.

550. GRUM, F., WITZEL, R. F., and STENSBY, P. *J. Opt. Soc. Am.*, 1974, **64**, 210–215.

551. WILTSHIRE, T. J., and SAVAGE, R. D. *Compte Rendu CIE* (Barcelona, 1971), **21A**, 119 and paper P.71.35.

552. STENIUS, Å. S. *J. Opt. Soc. Am.*, 1975, **65**, 213–216.

553. JUDD, D. B. *J. Opt. Soc. Am.*, 1960, **50**, 254–268.

554. WEIR, J. *Light Res. Technol.*, 1975, **7**, 209–225.

555. VAN DER VEEN, R., and MEIJER, G. *Light and Plant Growth* (Philips Tech. Library, 1962).
556. CANHAM, A. E. *Artificial Light in Horticulture* (Centrex, Eindhoven, 1966).
557. BICKFORD, E. D., and DUNN, S. *Lighting for Plant Growth* (Kent State Univ. Press, 1972).
558. OTT, J. N. *Illum. Eng.*, 1965, **60**, 254–261.
559. OTT, J. N. *Optometric Weekly*, 5.9.1968.
560. THORINGTON, L., PARASCANDOLA, L., and CUNNINGHAM, L. *J. Illum. Eng. Soc.*, 1971, **1**, 33–41.
561. BIRREN, F. *Colour 73* (*AIC Congress, York*, 1973), pp. 179–189 (Adam Hilger, 1973).
562. BENOIT, J. *Lux*, 1972, no. 69, 341–345.
563. ASSENMACHER, I., and BOISSIN, J. *Lux*, 1972, no. 69, 346–350.
564. REINBERG, A. *Lux*, 1972, no. 69, 351–355.
565. LOGAN, H. L. *Illum. Eng.*, 1967, **62**, 159–167.
566. HOLLWICH, F., DIECKHUES, B., and MEINERS, C. O. *Lichttechnik*, 1975, **27**, 388–394.
567. MAGNUS, I. A. *Penguin Science Survey* ,1968, *Biology*, pp. 162–183.
568. ROBERTSON, D. F. *The Biologic Effects of Ultraviolet Radiation*, ed. F. URBACH, pp. 433–436 (Pergamon Press, 1969).
569. DANZ, E. *Architecture and the Sun* (Thames and Hudson, 1967).
570. JACKSON, G. *Opt. Acta*, 1969, **16**, 1–16.
571. HOPKINSON, R. G., and KAY, J. D. *The Lighting of Buildings*, 2nd edn (Faber and Faber, 1972).
572. BELL, J. A. M. *Light. Res. Technol.*, 1973, **5**, 173–185.
573. BELLCHAMBERS, H. E. *Lamps and Lighting*, 2nd edn, chap. 23. Ed. S. T. HENDERSON and A. M. MARSDEN (Arnold, 1972).
574. *Daytime Lighting in Buildings* (*IES Tech. Report no. 4*, 2nd edn) (IES, London, 1972)
575. WALSH, J. W. T. *The Science of Daylight* (Macdonald, 1961).
576. NE'EMAN, E. *Light. Res. Technol.*, 1974, **6**, 159–164.
577. NE'EMAN, E., and HOPKINSON, R. G. *Compte Rendu CIE* (London, 1975), pp. 431–444.
578. BRINKWORTH, B. J. *Solar Energy for Man* (Compton, 1972).
579. LAUNER, H. F. *Nature*, 1968, **218**, 160–161.
580. VITRUVIUS. *De Architectura*, Book 6, chap. 4. Trans. F. GRANGER (Heinemann, 1932).
581. WARBURTON, F. L. *Proc Phys. Soc.*, 1954, **67**, 477–484.
582. CRAWFORD, B. H. *Br. J. Appl. Phys.*, 1963, **14**, 319–328.
583. *Method of measuring and specifying colour rendering of light sources* (*CIE Publ. 13.2*) (Paris, 1974).
584. ARISTOTLE. *Meterologia*, part 3. Trans. H .D. P. LEE (Heinemann, 1962).
585. OVID. *Ars Amatoria*, **1**, 250.
586. TROTTER, A. P. *J. Inst. Electr. Eng.*, 1892, **21**, 360–384.

587. MILES, E. E., and PEACH, D. C. *Trans. Illum. Eng. Soc., London*, 1956, **21**, 135–147.

588. MOORE, D. M. *Trans. Illum. Eng. Soc., N. Y.*, 1910, **5**, 209–241.

589. *Shirley Inst. Bull.*, 1932, 221–224.

590. IVES, H. E. *Trans. Illum. Eng. Soc., N. Y.*, 1912, **7**, 62–72.

591. IVES, H. E., *J. Franklin Inst.*, 1914, **177**, 471–499.

592. RUBENS, H. *Ann. Phys., Lpz.*, 1905, **18** (4), 725–738.

593. TROLAND, L. T. *J. Opt. Soc. Am.*, 1922, **6**, 527–596.

594. CUNLIFFE, P. W., and LANIGAN, H. *Shirley Inst. Mem.*, 1928, **7**, no. 7, 101–110.

595. PRIEST, I. G. *J. Opt. Soc. Am.*, 1927, **15**, 131–136.

596. HARRINGTON, R. E. *Proc. 13th Ann. Tech. Meet. Inst. Environ. Sci.*, 1967, **1**, 241–243.

597. LANGMUIR, I., and ORANGE, J. A. *Trans. Am. Inst. Electr. Eng.*, 1913, **32**, 1935–1946.

598. MACBETH, N., and REESE, W. B. *Illum. Eng.*, 1964, **59**, 461–471.

599. NICKERSON, D. *Reference Data for Light Sources etc.* [used in computations for colour rendering studies], Table 3, curve 61 (Colour Research Laboratory, MQRD, A.R.S., U.S. Department of Agriculture, 1962 with later additions).

600. CUNLIFFE, P. W., and LANIGAN, H. *Shirley Inst. Mem.*, 1928, **7**, no. 8, 111–113.

601. ORD, P. R. *Illum. Eng.*, July 1923, 167–177.

602. *Compte Rendu CIE* (Cambridge, 1931), p. 22.

603. PRESTON, J. S. *Light. Res. Technol.*, 1969, **1**, 248–250.

604. MORREN, L. *Lux*, 1970, no. 59, 459–461.

605. *Colorimetry (CIE Publ. 15)* (Paris, 1971).

606. *Compte Rendu CIE* (Vienna, 1963), p. 96, 109 (*E-1.3.1. Report*).

607. HARRISON, W. *J. Sci. Instrum.*, 1959, **36**, 234–236.

608. HISDAL, B. *Optica Acta*, 1956, **3**, 139–144; 1961, **8**, 199–212.

609. HISDAL, B. *J. Opt. Soc. Am.*, 1958, **48**, 608–613.

610. GAGE, H. P. *J. Opt. Soc. Am.*, 1933, **23**, 46–54.

611. ESTEY, R. S. *J. Opt. Soc. Am.*, 1936, **26**, 293–297.

612. BLOTTIAU, F., PENCIOLELLI, G., and SLANSKY, S. *Rev. d'Opt.*, 1967, **46**, 83–92; 1968, **47**, 19–26.

613. VIETH, G., and HEILAND, W. *Appl. Opt.*, 1968, **7**, 1043–1046.

614. CARMAN, P. D. *Photogr. Sci. Eng.*, 1969, **13**, 376–381.

615. BODMANN, H. W., and VOIT, E. *Lichttechnik*, 1960, **12**, 359–361.

616. JENKINS, H. G., and BOWTELL, J. N. *Trans. Illum. Eng. Soc., London*, 1948, **13**, 61–87.

617. NICKERSON, D. *Illum. Eng.*, 1948, **43**, 416–467.

618. MCLAREN, K. *J. Soc. Dyers Colour.*, 1962, **78**, 261–274; 1967, **83**, 438–444.

619. HELSON, H., JUDD, D. B., and WILSON, M. *Illum. Eng.*, 1956, **51**, 329–346.

620. HARRISON, W. *J. Text. Inst.*, 1949, **40**, P1025–1030.

621. WINCH, G. T., HARRISON, W., and RUFF, H. R. *Trans. Illum. Eng. Soc., London*, 1951, **16**, 3–14.

622. ASTM. D1684–61, *Recommended Practice for Lighting Cotton Classing Rooms*, 1961.

623. ALLEN, E. *J. Opt. Soc. Am.*, 1957, **47**, 933–943.

624. WILLIAMS, F. E. *J. Opt. Soc. Am.*, 1947, **37**, 302–307.

625. TRAVNICEK, M., KRÖGER, F. A., BOTDEN, T. P. J., and ZALM, P. *Physica*, 1952, **18**, 33–42.

626. KRUITHOFF, A. A., and OUWELTJES, J. L. *Philips Tech. Rev.*, 1956–1957, **18**, 249–260.

627. THORNTON, W. A. *J. Opt .Soc. Am.*, 1971, **61**, 1155–1163.

628. VERSTEGEN, J. M. P. J. *Light. Res. Technol.*, 1974, **6**, 31–32.

629. HENDERSON, S. T. *Trans. Illum. Eng. Soc.*, *London*, 1968, **33**, 83–91.

630. GRUM, F., SAUNDERS, S., and WIGHTMAN, T. *T.A.P.P.I.*, 1970, **53**, 1264–1268.

631. SOBAYAKI, H., YAMANAKA, T., TAKAHAMA, K., and NAYATANI, Y. *J. Opt. Soc. Am.*, 1974, **64**, 743–749.

632. BRITISH STANDARDS INSTITUTION. *Artificial Daylight for the Assessment of Colour: Illuminant for Colour Matching and Colour Appraisal.* BS 950: part 1: 1967.

633. THORINGTON, L., and PARASCANDOLA, L. *J. Illum. Eng.*, 1967, **62**, 674–681.

634. HANADA, T., SUGIYAMA, H., and KOBUYA, T. *Toshiba Rev.*, March–April 1968, 19–24.

635. IES (N.Y.) COMMITTEE. *Illum. Eng.*, 1957, **52**, 493–500.

636. BRITISH STANDARDS INSTITUTION. BS 950: part 2, 1967; Amendment no. 1, 10.5, 1968.

637. GRUM, F. *Colour Group (GB) Symp.: Colour Measurement in Industry*, 1967, pp. 109–122.

638. THOMAS, D. L. *J. Sci. Instrum.*, 1969, ser. 2, **2**, 1139–1140.

639. DRUMMOND, A. J., and HICKEY, J. R. *Solar Energy*, 1967, **11** (1), 14–24.

640. LIEBMANN, R. *Appl. Opt.*, 1968, **7**, 315–323.

641. THOURET, W. E., LEYDEN, J., STRAUSS, H. S., SCHAFFER, G., and KEE, H. *J. Illum. Eng. Soc.*, 1972, **2**, 8–18.

642. BARTERA, R. E. *Proc. 13th Ann. Tech. Meet. Inst. Environ. Sci.*, 1967, **1**, 681–685.

643. BEAUCHENE, J. H., and DENNIS, P. R. *Opt. Spectra*, 1970, **4**, 48–52.

644. *Compte Rendu CIE* (Vienna, 1963), p. 188 (*E-2.1.2. Report*).

645. *Compte Rendu CIE* (Washington, 1967), p. 219 (*E-2.1.2. Report*).

646. GONCZ, J. H., and NEWELL, P. B. *J. Opt. Soc. Am.*, 1966, **56**, 87–92.

647. DOBRUSSKIN, A. *Lichttechnik*, 1971, **23**, 135–140.

648. KOEDAM, M., and OPSTELTEN, J. J. *Light. Res. Technol.*, 1971, **3**, 205–210.

649. HIGASHI, T., MORI, L., and NAGANO, S. *Compte Rendu CIE* (Washington, 1967), pp. 208–213, paper P-67.22.

650. SPEROS, D. M., CALDWELL, R. M., SMYSER, W. G., SPRINGER, R. H., and TAYLOR, R. P. *Illum. Eng.*, 1970, **65**, 641–643.

651. FISCHER, E., LORENZ, R., and REHDER, L. *Lichttechnik*, 1972, **24**, 513–516.

652. DROP, P. C., DE GROOT, J. J., JACK, A. G., and ROUWELER, G. C. J. *Light. Res. Technol.*, 1974, **6**, 212–216.

653. CHALMERS, A. G., WHARMBY, D. O., and WHITTAKER, F. L. *Light. Res. Technol.*, 1975, **7**, 11–18.

654. ECKHARDT, K. *Lichttechnik*, 1975, **27**, 407–410.

655. MILES, E. E. *Light and Lighting*, 1969, **62**, 84–89.
656. NEUDER, S. M. *Appl. Opt.*, 1970, **9**, 1014–1018.
657. GUNTHER, K., and RADTKE, P. *J. Phys. E: Sci. Instrum.*, 1975, **8**, 371–376.
658. WYSZECKI, G. *Die Farbe*, 1970, **19**, 43–76.
659. RICHTER, K. *Lichttechnik*, 1972, **24**, 370–373.
660. BERGER, A., and STROCKA, D. *Appl. Opt.*, 1973, **12**, 338–348.
661. GANZ, E., and EITLE, D. *Die Farbe*, 1970, **19**, 103–108.
662. TERSTIEGE, H., and MALLWITZ, E. *Compte Rendu CIE* (London, 1975), pp. 276–282.
    GANZ, E., *Appl. Opt.*, 1977, **16**, 806.
663. ANDERS, G., and GANZ, E. *Eine Methode zur Bestimmung des relativen, ultravioletten Strahlungsanteils von Lichtquellen* (preprint, August 1975).
664. BERGER, A., and STROCKA, D. *Appl. Opt.*, 1975, **14**, 726–733.
665. IKEDA, K., NAKAYAMA, M., and OBARA, K. *Light source simulating CIE standard illuminants D_{55}, D_{65} and D_{75} for colorimetry*. Paper presented at CIE, London, 1975 (Science Univ. of Tokyo).
666. KRUITHOFF, A. A. *Philips Tech. Rev.*, 1941, **6**, 65–73.
667. ASTON, S. M., and BELLCHAMBERS, H. E. *Light. Res. Technol.*, 1969, **1**, 259–261.
668. BELLCHAMBERS, H. E., and GODBY, A. C. *Light. Res. Technol.*, 1972, **4**, 104–106.
669. BOYCE, P. R., and LYNES, J. A. *Compte Rendu CIE* (London, 1975), pp. 290–297.
670. GATES, D. M., KEEGAN, H. J., SCHLETER, J. C., and WEIDNER, V. R. *Appl. Opt.*, 1965, **4**, 11–20.
671. CORTH, R., JIVIDEN, G. M., and DOWNS, R. J. *J. Illum. Eng. Soc.*, 1973, **2**, 139–142.
672. BOUILLENNE, R., and FOUARGE, M. *Bull. Horticole, Liège*, March 1953, **8**, no. 3.
673. LUCKIESH, M. *Artificial Sunlight* (Crosby Lockwood, 1930).
674. DANTSIG, N. M., LAZAREV, D. N., and SOKOLOV, M. V., *Appl. Opt.*, 1967, **6**, 1872–1876.
675. AZUMA, T., and SHIRAISHI, H. *Toshiba Rev.*, Summer 1965, 30–35.
676. RUFF, H. R. *Light. Res. Technol.*, 1972, **4**, 9–17.
677. *Medical Research Council. Memo. no. 43* (HMSO, 1965).
678. KELMAN, G. R., and NUNN, J. F. *Lancet*, 25.6.1966, 1400–1403.
679. MCLAREN, K. *J. Soc. Dyers Colour.*, 1956, **72**, 86–99.
680. MCLAREN, K. *Can. Inst. Textiles, 11th Seminar Book of Papers*, August 1968, pp. 70–76.
681. GILES, C. H., SHAH, C. D., and BAILLIE, D. *J. Soc. Dyers Colour.*, 1969, **85**, 410–417.
682. LILLYWHITE, M., MCINTOSH, R., DUNCAN, C., and COLONY, J. *Proc. 13th Ann. Tech. Meet. Inst. Environ. Sci.*, 1967, **1**, 557–564.
683. THOMSON, G. *Studies in Conservation*, 1961, **6**, 49–70.
684. BAKER, D. J. *Appl. Opt.*, 1974, **13**, 2160–2164.
    BAKER, D. J., and ROMICK, C. J., *Appl. Opt.*, 1976, **15**, 1966–1968.
685. WALSH, J. W. T. *Photometry* (Constable, 1965).
686. JONES, O. C., and PRESTON, J. S. *Photometric Standards and the Unit of Light* (NPL Notes on Applied Science, no. 24, 2nd edn, HMSO, 1969).
687. PRESTON, J. S. *Light. Res. Technol.*, 1969, **1**, 95–97.

688. PRESTON, J. S. *Light. Res. Technol.*, 1974, **6**, 89–94.
689. WRIGHT, W. D. *The Measurement of Colour*, 4th edn, (Adam Hilger, 1969).
690. MACADAM, D. L. *J. Opt. Soc. Am.*, 1937, **27**, 294–299.
691. NEWHALL, S. M., NICKERSON, D., and JUDD, D. B. *J. Opt. Soc. Am.*, 1943, **33**, 385–418.
692. MCLAREN, K. *J. Soc. Dyers Colour.*, 1970, **86**, 354–356; *Compte Rendu CIE* (London, 1975), pp. 180–184.
693. CIE *J. Opt. Soc. Am.*, 1974, **64**, 896.
694. GLASSER, L. J., MCKINNEY, D., REILLY, C. D., and SCHNELLE, P. D. *J. Opt. Soc. Am.*, 1958, **48**, 736–740.
695. MACADAM, D. L. *Appl. Opt.*, 1971, **10**, 1–7.
696. ALLEN, E. *Color Engineering*, July–August 1966.
697. HALSTEAD, M. B., HENDERSON, S. T., LARGE, F. E., and SPROSON, W. N. *Light. Res. Technol.*, 1973, **5**, 84–94.
698. HARDING, H. G. W. *Proc. Phys. Soc.*, 1950, **63B**, 685–698.
699. ROBERTSON, A. R. *J. Opt. Soc. Am.*, 1968, **58**, 1528–1535.
700. JUDD, D. B. *J. Opt. Soc. Am.*, 1936, **26**, 421–426.
701. JUDD, D. B. *J. Opt. Soc. Am.*, 1935, **25**, 24–35.
702. MAHR, K., MÜNCH, W., and SCHULTZ, U. *Tech. Wiss. Abh. Osram Ges.*, 1969, **10**, 283–292.
703. HILL, L. *Proc. R. Soc., Lond.*, 1927, **116A**, 268–277.
704. HATCHARD, C. G., and PARKER, C. A. *Proc. R. Soc., Lond.*, 1956, **235A**, 518–536.
705. MOROSON, H., and GREGORIADES, A. *Nature*, 1964, **204**, 676–678.
706. RUFF, H. R. *Light. Res. Technol.*, 1970, **2**, 43–46.
707. HARRIS, P. B. *J. Sci. Instrum.*, 1968, ser. 2, **1**, 1007–1010.
708. DAVIS, A., DEANE, G. H. W., GORDON, D., HOWELL, G. V., and LEDBURY, K. J. *J. Appl. Polymer Sci.*, 1976, **20**, 1165–1174.
709. NICOLET, M. *Rev. Geophys. Space Phys.*, 1975, **13**, 593–636.
710. OERTAL, G. K., and EPSTEIN, G. L., *Appl. Spectrosc. Rev.*, 1975, **10**, 139–200.
711. MAKAROVA, E. A., and KHARITONOV, A. V. *Sov. Astron.*, 1975, **19**, 585–587.
712. ROULET, R. R., MAYKUT, G. A., and GRENFELL, T. C. *Appl. Opt.*, 1974, **13**, 1652–1659.
713. BUCHELE, D. R. *Appl. Opt.*, 1973, **12**, 355–358.
714. ALLEN, W., and THOMSON, G. *Compte Rendu CIE* (London, 1975), pp. 464–472.
715. BLEVIN, W. R., and STEINER, B. *Metrologia*, 1975, **11**, 97–104.

AUTHOR INDEX

SUBJECT INDEX

TABLES

PLATES

The plates are to be found between pages 38 and 39.